Visual Group Theory

© *2009 by the Mathematical Association of America, Inc.*

Library of Congress Catalog Card Number 2009923532
ISBN 978-0-88385-757-1
Printed in the United States of America
Current Printing (last digit):
10 9 8 7 6 5 4

Visual Group Theory

Nathan C. Carter
Bentley University

Published and Distributed by
The Mathematical Association of America

Council on Publications
Paul Zorn, *Chair*

Classroom Resource Materials Editorial Board
Gerald M. Bryce, *Editor*

William C. Bauldry
Diane L. Herrmann
Loren D. Pitt
Wayne Roberts
Susan G. Staples
Philip D. Straffin
Holly S. Zullo

CLASSROOM RESOURCE MATERIALS

Classroom Resource Materials is intended to provide supplementary classroom material for students—laboratory exercises, projects, historical information, textbooks with unusual approaches for presenting mathematical ideas, career information, etc.

101 Careers in Mathematics, 2nd edition edited by Andrew Sterrett

Archimedes: What Did He Do Besides Cry Eureka?, Sherman Stein

Calculus Mysteries and Thrillers, R. Grant Woods

Conjecture and Proof, Miklós Laczkovich

Creative Mathematics, H. S. Wall

Environmental Mathematics in the Classroom, edited by B. A. Fusaro and P. C. Kenschaft

Exploratory Examples for Real Analysis, Joanne E. Snow and Kirk E. Weller

Geometry From Africa: Mathematical and Educational Explorations, Paulus Gerdes

Historical Modules for the Teaching and Learning of Mathematics (CD), edited by Victor Katz and Karen Dee Michalowicz

Identification Numbers and Check Digit Schemes, Joseph Kirtland

Interdisciplinary Lively Application Projects, edited by Chris Arney

Inverse Problems: Activities for Undergraduates, Charles W. Groetsch

Laboratory Experiences in Group Theory, Ellen Maycock Parker

Learn from the Masters, Frank Swetz, John Fauvel, Otto Bekken, Bengt Johansson, and Victor Katz

Mathematical Evolutions, edited by Abe Shenitzer and John Stillwell

Math Made Visual: Creating Images for Understanding Mathematics, Claudi Alsina and Roger B. Nelsen

Ordinary Differential Equations: A Brief Eclectic Tour, David A. Sánchez

Oval Track and Other Permutation Puzzles, John O. Kiltinen

A Primer of Abstract Mathematics, Robert B. Ash

Proofs Without Words, Roger B. Nelsen

Proofs Without Words II, Roger B. Nelsen

She Does Math!, edited by Marla Parker

Solve This: Math Activities for Students and Clubs, James S. Tanton

Student Manual for Mathematics for Business Decisions Part 1: Probability and Simulation, David Williamson, Marilou Mendel, Julie Tarr, and Deborah Yoklic

Student Manual for Mathematics for Business Decisions Part 2: Calculus and Optimization, David Williamson, Marilou Mendel, Julie Tarr, and Deborah Yoklic

Teaching Statistics Using Baseball, Jim Albert

Visual Group Theory, Nathan C. Carter

Writing Projects for Mathematics Courses: Crushed Clowns, Cars, and Coffee to Go, Annalisa Crannell, Gavin LaRose, Thomas Ratliff, Elyn Rykken

MAA Service Center
P.O. Box 91112
Washington, DC 20090-1112
1-800-331-1MAA   FAX: 1-301-206-9789

Acknowledgments

I am grateful to God for life and breath and mathematics, as well as my ability to write it, draw it, and enjoy it. I am grateful to my family, especially Lydia, who put up with my absence during times I needed to work on this manuscript.

Many thanks to Doug Hofstadter, who showed me the power of group theory visualization and supported my work on it in several ways, including much good advice. Many thanks to Charlie Hadlock, who kept me away from not-so-good advice (such as "Don't write a book before you get tenure") and who guided many of the early steps of this project. I also thank Don Albers, the MAA, and Bentley University (particularly Rick Cleary and Kate Davy), all of whom encouraged, supported, and took a chance on a first-time author.

I must also thank those whose mathematical work helped me in mine. One of my early exposures to group theory visualization techniques was Magnus and Grossman's *Groups and Their Graphs.* In writing this book I made frequent reference to the excellent texts by Michael Artin, John Fraleigh, Charles Hadlock, and Thomas Hungerford. The powerful and well-designed software *Asymptote* by John Bowman helped me make the more than 300 figures herein. Thank you!

Lastly, there are many people who read early chapters and drafts and provided much helpful feedback. These include Doug Hofstadter, Jon Zivan, an anonymous referee, my parents, and my Fall 2006 Discrete Mathematics class, especially Kathryn Ogorzalek. Thank you for helping me see my work through your eyes, and improve it.

Preface

If you are interested in learning about group theory in a relaxed, intuitive way, then this book is for you. I say learning *about* group theory because this book does not aim to cover group theory comprehensively. Herein you will find clear, illustrated exposition about the basics of the subject, which will give you a solid foundation of intuitions, images, and examples on which you can build with further study.

This book is ideal for a student beginning a first course in group theory. It can be used in place of a traditional textbook, or as a supplement to one, but its aim is quite different than that of a traditional text. Most textbooks present the theory of groups using theorems, proofs, and examples. Their exercises teach you how to make conjectures about groups and prove or refute them. This book, however, teaches you to *know* groups. You will see them, experiment with them, and understand their significance. The mental library of images and intuitions you gain from reading this book will enable you to appreciate far better the facts and proofs in a traditional text.

This book is also appropriate for recreational reading. If you want an overview of the theory of groups, or to learn key principles without going as deep as some upper-level undergraduate mathematics courses, you can read this book by itself. Only a typical high school mathematics education is assumed, but you should have a willingness to think analytically.

My work on this book stems from *Group Explorer*, a software package I wrote that creates illustrations for finite groups, and allows the user to interact and experiment with them. Many of the illustrations in this text were generated with the help of *Group Explorer*, and the investigations possible in *Group Explorer* can help you with some of this book's exercises.

You do not need *Group Explorer* to benefit from this book; very few exercises specifically direct you to *Group Explorer*. But I recommend taking full advantage of hands-on, interactive learning experiences when they're available; the more involved we are, the more we tend to learn. *Group Explorer* is free software, available for all major operating systems. You can find it online at

```
http://groupexplorer.sourceforge.net
```

Contents

Acknowledgments vii

Preface ix

Overview 1

1 What is a group? **3**
    1.1   A famous toy 3
    1.2   Considering the cube 4
    1.3   The study of symmetry 5
    1.4   Rules of a group 6
    1.5   Exercises 7

2 What do groups look like? **11**
    2.1   Mapmaking 11
    2.2   A not-so-famous toy 14
    2.3   Mapping a group 15
    2.4   Cayley diagrams 18
    2.5   A touch more abstract 19
    2.6   Exercises 21

3 Why study groups? **25**
    3.1   Groups of symmetries 25
    3.2   Groups of actions 34
    3.3   Groups everywhere 36
    3.4   Exercises 37

4 Algebra at last **41**
    4.1   Where have all the actions gone? 41
    4.2   Combine, combine, combine 44
    4.3   Multiplication tables 45
    4.4   The classic definition 48
    4.5   Exercises 52

5 Five families — 63
- 5.1 Cyclic groups — 64
- 5.2 Abelian groups — 68
- 5.3 Dihedral groups — 74
- 5.4 Symmetric and alternating groups — 78
- 5.5 Exercises — 87

6 Subgroups — 97
- 6.1 What multiplication tables say about Cayley diagrams — 97
- 6.2 Seeing subgroups — 99
- 6.3 Revealing subgroups — 101
- 6.4 Cosets — 102
- 6.5 Lagrange's theorem — 105
- 6.6 Exercises — 108

7 Products and quotients — 117
- 7.1 The direct product — 117
- 7.2 Semidirect products — 128
- 7.3 Normal subgroups and quotients — 132
- 7.4 Normalizers — 140
- 7.5 Conjugacy — 142
- 7.6 Exercises — 147

8 The power of homomorphisms — 157
- 8.1 Embeddings and quotient maps — 157
- 8.2 The Fundamental Homomorphism Theorem — 167
- 8.3 Modular arithmetic — 169
- 8.4 Direct products and relatively prime numbers — 172
- 8.5 The Fundamental Theorem of Abelian Groups — 175
- 8.6 Semidirect products revisited — 177
- 8.7 Exercises — 179

9 Sylow theory — 193
- 9.1 Group actions — 194
- 9.2 Approaching Sylow: Cauchy's Theorem — 199
- 9.3 p-groups — 205
- 9.4 Sylow Theorems — 208
- 9.5 Exercises — 217

10 Galois theory — 221
- 10.1 The big question — 221
- 10.2 More big questions — 225
- 10.3 Visualizing field extensions — 228
- 10.4 Irreducible polynomials — 231
- 10.5 Galois groups — 233
- 10.6 The heart of Galois theory — 243
- 10.7 Unsolvability — 247
- 10.8 Exercises — 252

| A | Answers to selected Exercises | 261 |

Bibliography 285

Index of Symbols Used 287

Index 289

About the Author 297

Overview

I highlight here three essential aspects of this book's nonstandard approach to group theory, and briefly discuss its organization.

First and foremost, images and visual examples are the heart of this book. There are more than 300 images, an average of more than one per page. The most used visualization tool is Cayley diagrams (defined in Chapter 2) because they represent group structure clearly and faithfully. But multiplication tables and objects with symmetry also appear regularly, and to a lesser extent cycle graphs, Hasse diagrams, action diagrams, homomorphism diagrams, and more. As you can tell by flipping through the pages, visualization is the name of the game.

Second, I focus more on finite groups than on infinite ones. This is partially because they are easier to diagram, but more so because they give a solid foundation of intuition for group theory in general. Understanding finite groups well makes the infinite a natural generalization. This approach sacrifices little, if anything, because so much remains to study in the realm of the finite. I cover the most common infinite groups, and each chapter's exercises includes some on infinite groups.

Lastly, this book approaches groups from the opposite direction of many traditional textbooks. The usual paradigm defines a group as a set with a binary operation, and later proves Cayley's Theorem, that every group is a collection of permutations (or you could say that every group acts on some set, most notably itself). The traditional definition does not appear in this book until Chapter 4; I define groups in Chapter 1 as collections of actions, and later prove that they can also be thought of as sets with binary operations. This nonstandard paradigm facilitates my introduction in Chapter 2 of Cayley diagrams, which depict groups as collections of actions.

The book's structure is linear, to be read in order; later sections usually depend on earlier ones. There are two exceptions. Chapter 5, which gives broad exposure to finite groups, is helpful but not strictly necessary for what follows. You could skip most of it (except the definition of abelian in Section 5.2) and turn back to it later as needed. The other exception is that Chapter 10 depends only slightly on Chapter 9; Cauchy's Theorem from Section 9.2 is used in Section 10.7, and the remainder of Chapter 9 may be useful in a few of the exercises in Chapter 10.

Chapter 10, on Galois theory, aims to show the power of group theory and some of its historical roots. It includes an introduction to fields, but several theorems are stated without proof. It gives enough understanding to see how group theory ties in, and points the reader elsewhere for more details on fields. The beautiful and historic result of the unsolvability of the quintic is the focus and endpoint of that final chapter.

1

What is a group?

1.1 A famous toy

In 1974, Ernö Rubik of Budapest, Hungary unleashed his fascinating invention called Rubik's Cube. It infiltrated popular culture, appearing in feature films, inspiring competitions, and captivating children and geniuses alike. Mathematics journals carried research articles analyzing the cube and its patterns. Those unable to solve the cube could learn solutions from any of dozens of books.

A new Rubik's Cube comes out of the box looking like Figure 1.1. Each face of the cube contains nine smaller faces of smaller cubes, with the colors arranged to agree. You begin playing with the cube by rotating its faces to mix up the colors. Figure 1.2 shows how two different rotations in succession begin to disorder the cube. After playing with a cube idly for a few minutes, an innocent customer finds that the colors are completely shuffled, and there is no obvious way of returning to the original, pristine state. The challenge of Rubik's Cube is to restore a scrambled cube to its original state.

Rubik's Cube has a unique flavor among puzzles and games, and much of this flavor comes from the mathematical patterns inherent in the cube itself. At first the cube may

Figure 1.1. An untouched Rubik's Cube

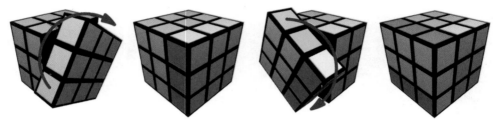

Figure 1.2. The leftmost cube shows the green face rotating 90 degrees clockwise; the next cube shows the result of that move. The third cube shows the white face rotating 90 degrees clockwise; the final cube shows the result of that move.

not seem very mathematical, given that it doesn't seem to require any skill with numbers, equations, or quantities in order to play with it, or even to solve it. But that is because the mathematics covered in this book, group theory, is not primarily about numbers; it is about patterns. Group theory studies symmetry, which can be found not only in the cube itself, but also in the patterns of its movements.

Marketing slogans like "Easy to learn, a challenge to master" adorned several puzzles invented by Ernö Rubik. This tagline certainly describes the cube, and we can get an excellent start to our group theory studies by examining why this is so. In this chapter we will examine those aspects of the cube that make it easy to learn and in Chapter 2 we will look at what makes it a challenge to master.

1.2 Considering the cube

To explore why Rubik's Cube is easy to learn, let's make several observations.

In part, the cube is easy to play with because users don't need to learn a complex list of rules. In contrast, players wanting to play chess must first learn the different rules for each piece. Furthermore, some of the available moves may become unavailable to the player as the game progresses. Rubik's Cube does not have this type of intricacy; we might even say it has only six moves—rotating any one of the six faces 90 degrees clockwise.[1] By combining these six moves, players can explore the full (enormous) gamut of cube configurations. This accessibility leads us to our first observation about the cube.

Observation 1.1. There is a predefined list of moves that never changes.

Another agreeable aspect of Rubik's Cube is that it is somewhat forgiving of mistakes. If a player rotates a face and immediately regrets the move, no great harm has been done. The player can simply rotate the face the other way to undo the mistake. Let us make this our second observation.

Observation 1.2. Every move is reversible.

Another helpful aspect of the cube is a bit less noticeable, but becomes clearer in contrast with other games: Rubik's Cube is not a game of chance. A tennis player may have every intention of hitting a great shot, but fail to do so because their body does not execute their wishes. A poker player may have an excellent strategy, but not win because of the cards they were dealt. Unpredictability and chance give a certain flavor to a game, and

---

[1] Although a player can also rotate faces 90 degrees counterclockwise, that movement is the same as three sequential 90 degree turns clockwise.

Rubik's Cube is devoid of that flavor. The turning of a face of the cube has a predictable outcome, depending on neither skill nor luck. An action whose outcome can be determined in advance, free from influences that are random or uncertain, is called deterministic.

Observation 1.3. Every move is deterministic.

But we must be fair. On their own, these observations seem to imply that the cube has no complexity. To give an accurate account, we must admit that the seemingly limitless combinations of moves make the cube challenging. Thus we balance our first three observations with one more.

Observation 1.4. Moves can be combined in any sequence.

Note the impact of Observation 1.2 on Observation 1.4. If a player were so meticulous as to write down carefully every move that he or she did, then even a long sequence of moves could be carefully reversed one at a time.

1.3 The study of symmetry

We could add many other observations to the four above. We could mention the colors of the faces, the physical construction of the cube, the number of moves it typically takes to arrive at a solution, or any other aspect of the puzzle or the experience of playing with it. But we will concentrate on the four above because they highlight those aspects of the cube that are most relevant to this book. Group theory is the study of symmetry, and we will see how Observations 1.1 through 1.4 describe the symmetry in Rubik's Cube.

All cubes are symmetrical in that they have all sides the same, all angles the same, and all edges the same. But Rubik's Cube has the added intricacy of its moving parts. The set of possible motions and configurations of the smaller cubes that make up the whole also has a great deal of symmetry. But to clearly explain this, we need to know what "symmetry" means.

This brings us to the gateway of this book, because group theory is the branch of mathematics that answers the question, "What is symmetry?" The first three chapters of this book give a careful answer to that question by introducing groups and showing how they describe symmetry. That introduction is already underway—Rubik's Cube is the first example we will see of objects and patterns that exhibit symmetry. As we analyze them, we will be learning group theory. Our explorations in future chapters will lead us to other practical examples of symmetry, including ones as diverse as molecular crystals and dancing.

But we will not limit ourselves only to seeing examples of symmetrical objects. Our investigations will use many other tools as well; the purpose of this book is to focus on tools that are *visual*. This includes common group theory tools such as the multiplication table as well as less common ones such as Cayley diagrams and cycle graphs. I will introduce each of these in chapters to come.

Beginning in Chapter 2, you will see examples of objects that have symmetry. Some of them are physical objects, some of them are actions or behaviors of physical objects, and some of them are purely imaginary situations. But what they have in common is that Observations 1.1 through 1.4 apply to all of them. Group theory studies the mathematical consequences of those observations, and therefore can help answer interesting questions about symmetrical objects. For instance, how many different configurations does Rubik's

Cube have? Although we will not do the computation of that number in this book, interested readers can refer to [21].

1.4 Rules of a group

Let's therefore return to those observations on which group theory is founded. But instead of considering them as descriptions of Rubik's Cube, let's rephrase them as rules that will define the boundaries of our study. This is how mathematical subject areas are typically introduced. A set of rules is laid down, then mathematicians proceed to study things that obey those rules. The mathematical term for such rules is *axioms,* but I will call them rules as long as our discussion remains informal.

Laying down such rules has some noteworthy advantages. First, they make the boundaries of our study clear. Things that fit the rules introduced below are part of the study of group theory, and things that don't are not. Second, by agreeing on the rules in advance as a mathematical community, we can ensure that we all have the same ideas when we discuss the subject. In other words, codifying our ideas into rules helps us be sure we are speaking the same language and minimizes communication problems. Third, we can use the rules as a basis from which to make logical deductions, and thereby unearth new facts (in our case, facts about symmetry) that we may not have anticipated from the rules alone. The computation mentioned above about Rubik's Cube depends on just such facts. You will do some of these deductions yourself in the exercises for this chapter.

So let us rephrase the *observations* about Rubik's Cube into *rules*. As mentioned above, think of these rules as the requirements that something must meet to belong in our study of symmetry.

Rule 1.5. There is a predefined list of actions that never changes.
Rule 1.6. Every action is reversible.
Rule 1.7. Every action is deterministic.
Rule 1.8. Any sequence of consecutive actions is also an action.

My rephrasing involved making only two changes. First, I changed the word *move* to the word *action* to make it sound less like a game. In the Rubik's Cube context, the former wording was appropriate, but we do not want to restrict our attention only to games. Second, when writing Rule 1.8, I rephrased Observation 1.4 to make the terminology clearer; not only is every sequence of actions possible, but we will go so far as to call every sequence an action in its own right. This does not mean that such an action has to appear on the list required by Rule 1.5, but rather that the (usually short) list in Rule 1.5 is a starting point from which we can create new actions using Rule 1.8, that is, by chaining actions together into sequences.

In the case of Rubik's Cube, although I list only six basic actions, you can form many actions by combining these six. As a simple example, combining two 90-degree rotations of the front face in sequence creates a new action, a 180-degree rotation of the front face. More complicated examples can be created from sequences of three or four or more different basic actions chained together. The standard way to say this is that Rule 1.5 gives us actions that generate all the others, and are therefore called **generators.**

We are now ready for our first definition, which marks your passage into the study of group theory. This is not the usual mathematical definition of a group, but it expresses the same idea as the usual one, as we will see later. For now, let's call this our unofficial definition.

Definition 1.9 (group, unofficially). A *group* is a system or collection of actions satisfying Rules 1.5 through 1.8.

You may now be wondering what things besides Rubik's Cube fit this definition. What other things deserve to be called groups? As we will see in Chapters 2 and 3, anything in which symmetry arises will satisfy Rules 1.5 through 1.8, including important examples from science, art, and mathematics. But for now, the following exercises will allow you to play with the rules themselves a bit, and see a few simple contexts in which they apply. They will give you a firmer grasp on the meaning of the rules before we proceed.

1.5 Exercises

The following exercises are thought experiments to help you understand the concepts just discussed. With mathematics and similar scientific endeavors, the exercises usually require some thought, and therefore take more time than the reading itself. Although you can appreciate the material with a quick reading, you can know it intimately only with lengthier consideration. Therefore, do not be discouraged if the exercises take some time; this is normal.

Also, feel no shame looking up the answers to a few exercises in the Appendix to get the idea for how to complete them. Because these are not typical mathematical exercises, they may seem unfamiliar or strange. Therefore looking up an answer or two to get the hang of them is a reasonable strategy.

1.5.1 Satisfying the rules

Exercise 1.1. Place a penny and a nickel side by side on a table or desk. Consider just one action, the action of swapping the places of the two coins. Is this a group? (Check each of the four rules to see if they are true for this situation. Explain your conclusion.)

Exercise 1.2. Consider the same situation as in Exercise 1.1, but add a dime to the right of the other two coins. The only action is still that of swapping the places of the penny and nickel. Is this a group?

Exercise 1.3. Imagine that you have five marbles in your left pocket. Consider two actions, moving a marble from your left pocket into your right pocket and moving a marble from your right pocket into your left pocket. Is this a group?

Exercise 1.4. Three walls in your bedroom hold pieces of art, one hung on each wall. You are rearranging them to see which arrangement best suits your taste. You cannot use the fourth wall, because it has a window.

(a) Count the number of ways there are to rearrange the pictures, as long as only one is hung on each wall.

(b) Consider two actions: You may swap the art on the left wall with the art on the center wall, and you may swap the art on the center wall with the art on the right wall. Can these two actions alone generate all of the configurations you counted?

(c) Does part (b) describe a group? If not, what rule or rules were broken?

(d) Now add a new action, moving all art from the right wall to the center wall, even if this causes there to be more than one piece of art there. Is this new situation a group? If not, what rule or rules were broken?

1.5.2 Consequences of the rules

Exercise 1.5. Does Rule 1.8 imply that every group must contain infinitely many actions? Explain your reasoning carefully.

Exercise 1.6. For each of Exercises 1.1 through 1.4 that actually described a group, determine exactly how many actions there are in that group. That is, do not count only the generators, but all possible actions obtainable using Rule 1.8.

Exercise 1.7. Consider again the situation from Exercise 1.1.

(a) Only one action was given. By Rule 1.8, performing this action twice in a row is also a valid action. Describe it.

(b) Will every group have an action like this? Explain your reasoning carefully.

Exercise 1.8. For each of the following requirements (a) through (e), devise a group that meets that requirement. (Groups always satisfy Rules 1.5 through 1.8; the requirements below are additional ones just for this exercise.) If some groups you've already encountered fit these requirements, you may feel free to reuse them.

(a) The order in which actions are performed impacts the outcome.

(b) The order in which actions are performed does not impact the outcome.

(c) There are exactly three actions.

(d) There are exactly four actions.

(e) There are infinitely many actions.

Exercise 1.9. Can you devise a plan for creating a group with any given number of actions? (I do not mean a plan for creating a group with a specified number of generators, but rather a specified number of actions, once Rule 1.8 has been applied.)

1.5.3 Breaking the rules

Rules 1.6 through 1.8 would not make much sense without Rule 1.5, because it introduces the notion of a list of actions. But for each of the other rules, we can ask what it would be like if that rule were not present.

Devising a context in which all but one of the four rules is true shows something significant—that the missing rule is not redundant. For instance, let's say someone suggested to you that Rule 1.8 could be dropped, because the combination of the other three rules implies that Rule 1.8 must be true. By finding a context in which the other three rules are true, but Rule 1.8 is false, you can prove that the combination of Rules 1.5, 1.6, and 1.7 does not imply the truth of Rule 1.8. Thus by doing Exercises 1.10 through 1.12, you will have shown that all the rules are necessary; that is, none are redundant.

1.5. Exercises

Exercise 1.10. Devise a situation that satisfies all of the four rules except Rule 1.6.

Exercise 1.11. Devise a situation that satisfies all of the four rules except Rule 1.7.

Exercise 1.12. Devise a situation that satisfies all of the four rules except Rule 1.8.

1.5.4 Groups of numbers

Since group theory is a mathematical subject, it should not surprise us that numbers can be formed into groups, in various ways.

Exercise 1.13. Pick any whole number and consider this set of actions: adding any whole number to the one you chose. This is an infinite set of actions; we might name them things like "add 1" and "add -17." Is it a group? If so, how small a set of generators can you find?

Exercise 1.14.

(a) What would be the answer to the previous exercise if we allowed only even numbers?

(b) What would be the answer to the previous exercise if we allowed only whole numbers from 0 to 10?

(c) What would be the answer to the previous exercise if we allowed all whole numbers, but changed the set of actions to *multiplying* by any whole number? (I do not ask you to consider those actions as generators, but as the complete set of actions.)

(d) What would be the answer to the previous exercise if we allowed only the numbers 1 and -1, and just two actions, multiplying by either 1 or -1?

2

What do groups look like?

2.1 Mapmaking

Chapter 1 introduced group theory by examining those properties of Rubik's Cube that make it attractive to beginners. Let us now investigate those aspects of the cube that make it difficult to solve.

When you have in your hands an unsolved Rubik's Cube, and you do not know a method for solving it, experimentally twisting the faces quickly begins to seem pointless. You are wandering aimlessly through the multitude of possible configurations of the cube. Somewhere in this enormous wilderness of jumbled cube configurations is the one oasis you want to find—the solved cube. But without any idea of where you are, where the oasis is, or what direction you're walking as you make moves, the situation is (almost) completely hopeless.

What would be very helpful to someone lost in the endless reaches of Rubik's wilderness is a map—a reference that could say, "You are here," and "The solution is over here," and show you the path to get there. Such a map would eliminate all your navigational problems. Anyone could solve the cube quickly with such a map (although it may not be much fun to do so).

Many books on Rubik's Cube teach solution techniques, and some academic papers use group theory to provide detailed analyses of the cube [21]. My purpose is not to duplicate their efforts here by providing you with a map of Rubik's wilderness. Instead, I want to conduct a thought experiment about what a fully detailed map of that wilderness would be like. Such a map is not just a technique for getting to the solved state, but a complete map of *every* configuration of the cube and how those configurations relate to one another. Since all this information will obviously not fit on an ordinary-sized map, let's say we'll put it in a big book. Because this is only a thought experiment, let's call our book *The Big Book*, and consider what this book will need to contain.

Imagine that you have both a jumbled Rubik's Cube and a copy of *The Big Book*. *The Big Book* should first help you determine where you are in the wilderness. To make this possible, let's organize *The Big Book* to have one cube configuration on each page, and let's order the pages in a dictionary-like way. Just as all the words starting with "a" in the dictionary are together, perhaps all cubes with a red sticker on the leftmost square of their top face would be together. Within that group of cubes, another sticker's color would organize, like the second letter of a dictionary word, and so on. The details of such an organization are not important, because this is only a thought experiment, after all. Let's agree that we could come up with such an organization if necessary, enabling the readers of *The Big Book* to look up their jumbled cubes in the book (with a good dose of patience and care).

Now let's say you have found the page in *The Big Book* showing your cube's configuration. That page should provide some navigational help, the details of which are best shown by an example. Figure 2.1 shows a page from *The Big Book*. The page shows all faces of a certain cube configuration, front and back, and tells the reader how far that cube is from being solved. A table then provides navigational help; it tells the reader what each available move would accomplish. For instance, the first row of the table tells us that twisting the front face of the cube clockwise would result in the configuration on page 36, 131, 793, 510, 312, 058, 964 of the book, which is closer to the solution than the configuration shown on the current page.

You can see how you could use such a book to solve your cube. Once you have found the page on which your cube is shown, simply make a move that puts you closer to being solved, and turn to the corresponding page. Then repeat the process from there. You always know how far you are from the solution, and you can always compare your cube against the picture on the page to be sure you haven't made a navigational error.

The Big Book has one significant problem, which you may have already noticed from Figure 2.1. Rubik's Cube has more than 4×10^{19} potential configurations, which would make a prohibitively large book. We might suggest putting more than one cube state per page, but in fact even if we could fit one cube state in a square inch, the amount of paper needed to print the book would cover the surface of the earth many times over. Storing the book electronically won't help either; even using a very efficient encoding scheme, no computer in existence at the time of this writing has sufficient storage to contain it.[1] So *The Big Book* is a thought experiment only, but the remainder of this chapter will show how valuable a thought experiment it has been.

The most important thing for *The Big Book* to teach us is that a map of Rubik's wilderness is really a map of a particular group. Let me explain. The moves in Rubik's Cube form a group because they satisfy Definition 1.9. (In fact, those moves were our first example of a group, motivating that very definition.) The rules in Definition 1.9 refer only to the cube's moves, and combinations thereof. Because *The Big Book* contains complete data on the moves in Rubik's Cube and how they combine, it is a map of *the group constituted by those combinations*.

So the mapmaking ideas introduced by our discussion about *The Big Book* do not need to be abandoned simply because the group exhibited by Rubik's Cube is too large. We can use the same ideas to map out any group, and in the next section we do just that.

---

[1] To store only the data (from which to later reconstruct text and images) would require approximately 10^{21} bytes, or one billion terabytes.

2.1. Mapmaking

Page 12, 574, 839, 257, 438, 957, 431

Cube front	Cube back

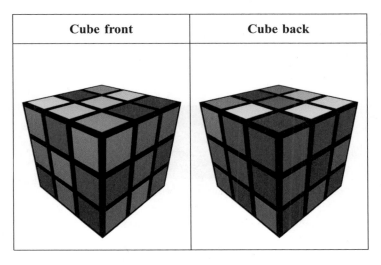

You are **15** steps from the solution.

Face	Direction	Destination page	Progress
Front	Clockwise	36, 131, 793, 510, 312, 058, 964	Closer to solved
Front	Counterclockwise	12, 374, 790, 983, 135, 543, 959	Farther from solved
Back	Clockwise	26, 852, 265, 690, 987, 257, 727	Closer to solved
Back	Counterclockwise	41, 528, 397, 002, 624, 663, 056	Farther from solved
Left	Clockwise	6, 250, 961, 334, 888, 779, 935	Closer to solved
Left	Counterclockwise	10, 986, 196, 967, 552, 472, 974	Farther from solved
Right	Clockwise	26, 342, 598, 151, 967, 155, 423	Farther from solved
Right	Counterclockwise	40, 126, 637, 877, 673, 696, 987	Closer to solved
Top	Clockwise	35, 275, 154, 257, 268, 472, 234	Closer to solved
Top	Counterclockwise	33, 478, 478, 689, 143, 786, 858	Farther from solved
Bottom	Clockwise	20, 625, 256, 145, 628, 342, 363	Farther from solved
Bottom	Counterclockwise	7, 978, 947, 168, 773, 308, 005	Closer to solved

Figure 2.1. Sample page from *The Big Book*

2.2 A not-so-famous toy

Allow me to introduce you to a puzzle much like Rubik's Cube. This puzzle has been around much longer, but has a much less impressive name. It is called *the rectangle,* and I'm sure you've heard of it before! Even though nearly everyone's mathematical education includes rectangles at some early point, I provide a useful illustration in Figure 2.2, with the corners of the rectangle conveniently numbered. The first thing to notice about this puzzle is that it is much less complicated than Rubik's Cube, and so it might cooperate better with our mapmaking aspirations.

Figure 2.2. A rectangle with its corners numbered

But first, to fully benefit from the upcoming discussion of the rectangle, you should make one for yourself. Take an ordinary sheet of paper and label it with numbers in the corners as shown in Figure 2.2. Making your own rectangle may sound unnecessary, but we will be investigating flips and twists of this object in space, which are difficult to picture accurately with the mind's eye alone. So go ahead and grab a piece of paper and make your own personal (numbered) rectangle.

Here are the rules for the rectangle puzzle. Begin with the rectangle flat on a table, on a desk, or in your lap so that you can read the four numbers, as in Figure 2.2. Like the Rubik's Cube, the rectangle puzzle starts out in its solved state—the way you have it now. You will mix up the puzzle and your job will then be to return it to this original state.

The rectangle puzzle has two legal moves. You can flip the paper over horizontally and you can flip it over vertically, as shown in Figure 2.3.[2] In Chapter 3 I'll talk about

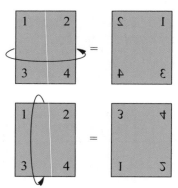

Figure 2.3. The top arrow illustrates a horizontal flip; the upper right rectangle shows the result of such a flip. The bottom arrow illustrates a vertical flip; the lower right rectangle shows the result of such a flip.

---

[2]Readers familiar with axes of rotation may consider the naming convention in Figure 2.3 backwards. Because I have not yet introduced axes of rotation, I use a simpler naming convention. "Horizontal" and "vertical" describe the motion of the player's hands, which follow the arrows in Figure 2.3.

2.3. Mapping a group

why these moves (and not others) make sense for the rectangle puzzle, but for now, let's just call them the rules of the game. As with Rubik's Cube, you may feel free to repeat and combine moves as much as you like. (In fact, you may have already noticed that the moves in the rectangle puzzle form a group.)

Take a moment now to mix up your rectangle puzzle, and then solve it. This should not take very long, but be sure you do not (accidentally) cheat! A natural mistake is to pick up the rectangle and inspect it a bit, rotating it freely and experimentally until it is solved. This move is rarely valid, and thus is not what you're supposed to be doing. Making such a move is analogous to disassembling Rubik's Cube and reassembling it solved. Remember that you are limited to the two moves in Figure 2.3 as you mix up the puzzle *and* as you solve it.

Let's take a moment to verify that the moves in the rectangle puzzle form a group, and to compare it to Rubik's Cube. Rule 1.5 requires a predefined list of moves, which we have given: horizontal and vertical flips. Rule 1.6 requires that each move be reversible. It holds true for the rectangle because in fact each move undoes itself. For example, if you have performed a horizontal flip and wish to return to the previous state, simply perform another horizontal flip. The same is true of vertical flips. Rule 1.7 requires that these moves be deterministic, free from randomness, and they are because they are completely within your control. Rule 1.8 requires that any possible combination of moves also be a valid move. This rule is satisfied because a horizontal or vertical flip puts the rectangle into a physical position from which either valid move is still possible. In contrast, if we were to add the move "rip up the rectangle and throw it away," that would bring about a dead end; no subsequent moves would be possible. Neither of our two moves has this problem; they always permit further moves and thus allow us to string them together in any sequence.

In the next section we will map out the rectangle puzzle's group. But first, notice how its physical construction contrasts with that of Rubik's Cube. The rectangle puzzle requires you to place the rectangle on a flat surface, to remember its original orientation, and to try to return to it. Rubik's Cube is not this way; you can toss it around the room, drop it in your sock drawer, and when you come back to it, its configuration will not have changed. The cube's moving parts internalize the puzzle, so that no external reference (to a table, or an original orientation) is required. Because the rectangle puzzle is a simple do-it-yourself rectangle made of paper, it has no moving parts and thus to use it as a puzzle required external landmarks (the table and the original orientation you remembered). We could design a puzzle that behaves identically to the rectangle puzzle, but which is self-contained; it would just not be as easy to make at home.

2.3 Mapping a group

Mapmaking begins with exploration. We need to know the lay of the land if we are to draw a useful map of it. Let's therefore explore the realm of the rectangle puzzle, and map it out as we go. We will need to ensure that our exploration is thorough—that we find all possible configurations of the rectangle puzzle—so we should conduct our search systematically.

Begin with the rectangle puzzle in its solved state (Figure 2.2). Our two moves (horizontal and vertical flips) are our only means for exploring. They are the group's generators, and our map will show *how* they generate the group.

Start exploring by performing a horizontal flip. Because our exploration has just begun, this of course leads us to a configuration we have not yet mapped. But it is a configuration that was originally introduced in Figure 2.3. Performing another horizontal flip returns the rectangle to its original, solved state. Therefore let's begin making our map with this information. Figure 2.4 shows a map of the terrain we have explored so far. I use a two-way arrow to mean that from either of the two configurations in the figure, a horizontal flip leads to the other configuration.

Figure 2.4. Partial map of the configurations of the rectangle puzzle, using only horizontal flips

These configurations are as far as we can go with horizontal flips alone. Keeping with the exploration metaphor, we can say that we have found two places in the rectangle realm. From each of these places, the map in Figure 2.4 tells us where a horizontal flip will take us, but there is (so far) no information about where vertical flips lead. Our map is not complete without such information, and so we must explore further.

Let's return to the rectangle in its solved state, and explore the results of vertical flips. Figure 2.3 already tells us what vertical flips do, but let's be thorough and explore those states as we add them to our map. From the solved state, a vertical flip leads to a new state we have not yet visited, and from there a vertical flip returns the rectangle to the solved state. We augment our map as shown in Figure 2.5.

Our map is still not complete, because we have not yet recorded where a vertical flip leads from the upper right configuration in Figure 2.5. To explore from that configuration, we first need to get our rectangle in that configuration. If you've been following along with your own rectangle, we left it in the solved state, and can get to the upper right configuration from Figure 2.5 by doing a horizontal flip—that's what our map tells us! After that horizontal flip, we perform a vertical flip to see where it leads. This move is the first exploration we've done whose outcome we could not predict from Figure 2.3. Perform this move yourself and see where it leads. Does it lead to a new location we must add to our map, or to a location we've already been? (You'll find the answer in Figure 2.6.)

Figure 2.6 contains four states of the rectangle, but the lower two have no "horizontal flip" arrows leading to or from them. Therefore our map is still incomplete. For instance, if your rectangle is like the lower left one in Figure 2.6, where does a horizontal flip lead you? The map does not say; we still have work to do.

Use the map to get your rectangle to look like the lower left rectangle in Figure 2.6, then perform a horizontal flip and see what configuration results. It is not a new configuration; it is the same one you discovered moments ago. Performing another horizontal flip will, of course, get you back to the lower left rectangle from Figure 2.6. The final map for the rectangle puzzle then looks like Figure 2.7. We can tell that our explorations are complete because there are no unanswered questions. From every location, it is clear from the map where every given move leads.

2.3. Mapping a group

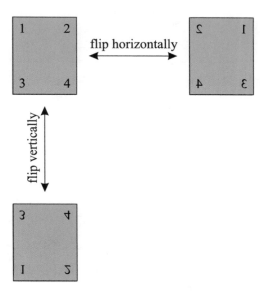

Figure 2.5. Partial map of the configurations of the rectangle puzzle, exploring from the solved state using one type of move at a time

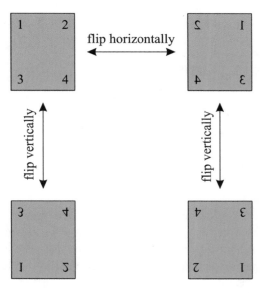

Figure 2.6. Partial map of the configurations of the rectangle puzzle, continuing to explore from Figure 2.5. The lower right state shown in this map does not appear in Figure 2.3.

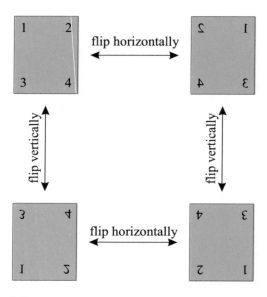

Figure 2.7. Full map of the configurations of the rectangle puzzle

You have just created your first map of a group! We have to admit that the map in Figure 2.7 is a bit unnecessary, because the rectangle puzzle is easy to solve without a map. But the map does serve to show us exactly the structure of the rectangle puzzle, and to let us see why it is easy. (For instance, from the map, you can see that alternating horizontal and vertical flips will walk you through every location in the rectangle realm.) The map we made is also an excellent first example of how to map a group, and allows us get our feet wet before we come upon more complicated groups.

2.4 Cayley diagrams

Maps like Figure 2.7 are called ***Cayley diagrams,*** after their inventor, Arthur Cayley, a nineteenth century British mathematician. We will use Cayley diagrams extensively throughout this book; they can be very potent visualization tools. It will help to begin by noting some important facts about the Cayley diagram we just made. These facts may seem obvious or uninteresting as far as they apply to the rectangle puzzle, because it is a puzzle that is so easy to solve. But we will be making Cayley diagrams for more complicated groups, and these facts will remain true.

The map in Figure 2.7 allows us to get from any place in the rectangle realm to any other without any guesswork. For instance, suppose you wanted to get from the lower right configuration to the solved configuration. From the map, you can see that there are two different (short) paths you could follow (up and then left or left and then up). To make use of the map, as you trace either of these paths on the map with your finger or your eyes, obey the instructions on that path using your rectangle. Going up from the lower right configuration, flip your rectangle vertically; then going left, flip it horizontally. Doing so successfully navigates to your desired destination; the map could help you plan and execute any such journey.

Recall also that we took pains to ensure that the map in Figure 2.7 is comprehensive. There is no location in the rectangle realm that does not appear on the map. Our construction of the map ensured this. We branched out from the starting position using each generator, and then branched out from each of those positions using each generator again. If the puzzle had been more complicated, we could have continued this process further, exploring farther and wider until we had found every location in the realm. We know our explorations are incomplete if our map fails to answer a question like "Where does a horizontal flip take me from this location?" That is, if there is a location on your map from which you have not yet explored where all moves lead, then your map is incomplete. When all such questions are answered, the map is complete.

Cayley diagrams have the two important properties just discussed: they clearly show all possible paths and they include every configuration. Just as the rectangle puzzle has a map, every other group also has a map with the same two properties. From now on, I will call such maps by their official name, Cayley diagrams. The most useful aspect of Cayley diagrams is that they give a clear picture of the structure of a group. Seeing the Cayley diagram of a group gives a much more immediate and complete idea of the group's size, complexity, and structure than a simple prose description can. This illustrative power is why we use them so frequently for learning about groups hereafter.

You can make a Cayley diagram for any group the way we made the one in Figure 2.7. Beginning at any one configuration or situation, explore carefully using each generator, one at a time. Explore thoroughly and carefully, making a map as you go and labeling the transitions between states with the moves that cause them. Continue until your map contains no unanswered questions, as described above. Although Figures 2.4 through 2.7 are laid out nicely, the first time you draw a Cayley diagram it will probably be disorganized. As you explore a realm for the first time, the Cayley diagram that evolves is messy because you do not know in advance the simplest way to lay it out. Cayley diagrams created by exploration need to be reorganized into a more symmetric or aesthetically pleasing arrangement after the exploration is complete.

The exercises at the end of this chapter encourage you to make a few Cayley diagrams using this exploratory technique. You will appreciate the previous paragraph more after some personal involvement with this type of mapmaking. Feel free to jump ahead and do the first few exercises now, and then return to this point in your reading.

2.5 A touch more abstract

It is important to get to the heart of the mapmaking concepts that we have just learned. As Definition 1.9 makes clear, what is important about a group is the interactions of its actions, not the specific situation that gave rise to those actions. Let me illustrate this fact with an example. Consider two light switches side-by-side on a wall. You are allowed two actions, flipping the first switch and flipping the second switch. This collection of (two) actions generates a group; you can check the rules yourself. The map of this group is shown in Figure 2.8.

You will notice that it has the same *structure* as the map of the configurations of the rectangle puzzle from Figure 2.7. The four rectangle configurations have become four light switch configurations, and the arrows labeled "horizontal flip" and "vertical flip"

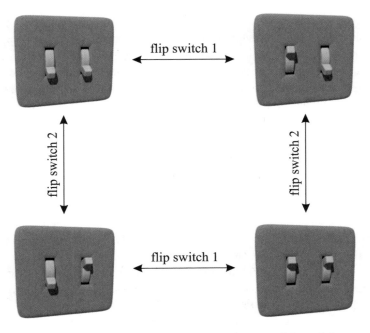

Figure 2.8. Full map of the configurations of the two-light switch group

have become arrows labeled "flip switch 1" and "flip switch 2" respectively. But the arrows connect the configurations in the same pattern as before, making a clear analogy between Figures 2.7 and 2.8.

So although these two groups are superficially different, they are structurally the same. The important lesson to learn here is that two different groups may have the same structure, and that the Cayley diagrams help us see this. Therefore, in order for us to study groups in the abstract, we wish to remove from our Cayley diagrams the details of the practical situation from which they were constructed. After all, a group is a mathematical structure, and mathematicians study groups as abstract (purely mathematical) objects. The rectangle puzzle and the light switches example just help us anchor our abstract study in something familiar.

So let's replace each rectangle in Figure 2.7 with something purely meaningless, a spot we will call a ***node.*** And we will replace the two different types of arrows (distinguished by their labels "horizontal flip" and "vertical flip") with different colored connectors that have no labels. (In fact, we can simplify even further by removing the arrowheads, since all arrows point in both directions anyway; I'll still refer to these headless connectors as "arrows.") The result is Figure 2.9, a Cayley diagram of a group, now shown pure and without any trappings of the example from which we learned it. You will note that the structure shown in Figure 2.9 is not only the heart of Figure 2.7, but also that of Figure 2.8.

This group in Figure 2.9 is called the Klein 4-group[3]. I chose it as our first group to visualize because it was simple enough for us to map quickly and easily, and yet still have some interesting structure. Figure 2.10 shows several other Cayley diagrams, to give you

---

[3]Also simply called the 4-group, and denoted V or V_4 for *vierergruppe*, "four-group" in German. It is named for the mathematician Felix Christian Klein.

2.6. Exercises

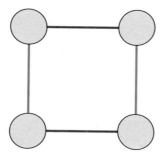

Figure 2.9. Cayley diagram of the Klein 4-group

a broader idea of what they tend to look like. As you can see, some are very simple, and others very complex. This diversity in the diagrams is indicative of a range of complexities in the underlying groups as well. You should not feel as if you must understand every part of Figure 2.10 already; it is present as an example of what is to come.

My interest in group theory visualization led me to write *Group Explorer*, a free software package that draws Cayley diagrams (and other illustrations you'll learn in later chapters). *Group Explorer* is an optional (but helpful) interactive companion to this book, and you can retrieve it from `http://groupexplorer.sourceforge.net`. It provides a list of groups, and extensive information about each one, including at least one Cayley diagram. *Group Explorer* creates Cayley diagrams using much the same algorithm we did—it uses the rules of the group to follow arrows, exploring the realm, and after it has found every location, it makes an attempt to arrange them presentably.

This chapter taught you your first technique for visualizing groups—the Cayley diagram. We will explore applications of this technique in chapters to come, and will use it extensively throughout our group theory studies. First, take some time to build your understanding of Cayley diagrams using the following exercises.

2.6 Exercises

2.6.1 Basics

Exercise 2.1. In the rectangle puzzle, what actions were the generators? What other actions are there besides the generators?

Exercise 2.2. In the light switch puzzle, what actions were the generators? What other actions are there besides the generators?

Exercise 2.3. Can an arrow in a Cayley diagram ever connect a node to itself?

2.6.2 Mapmaking

Exercise 2.4. Exercise 1.1 of Chapter 1 defined a group. Create its Cayley diagram using the technique from this chapter. (Hint: This group is simpler than even those done so far; the diagram will be small.)

Exercise 2.5. Exercise 1.4 of Chapter 1 defined a group. Create its Cayley diagram using the technique from this chapter.

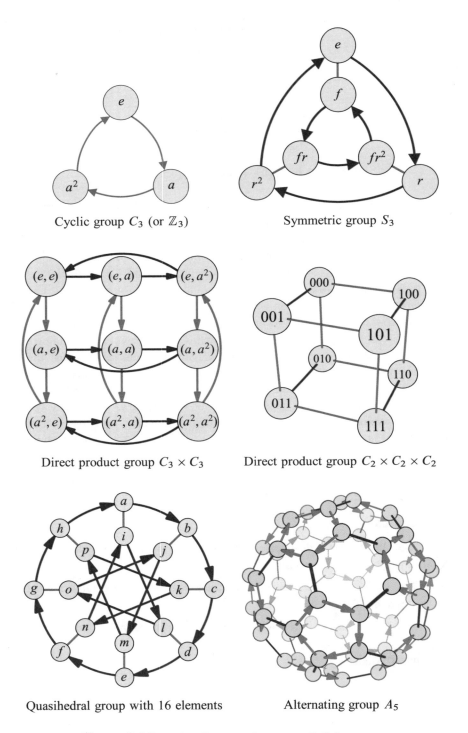

Figure 2.10. Cayley diagrams of some small, finite groups

2.6. Exercises

Exercise 2.6. Exercise 1.13 described an infinite group which can be generated with just one generator. Can you draw an infinite Cayley diagram for it? (Just draw a portion of the diagram that makes the infinite repeating pattern clear.)

How does that Cayley diagram compare to one for the group in Exercise 1.14 part (a)?

Exercise 2.7. Exercise 1.14 part (d) described a two-element group. Can you draw a Cayley diagram for it? Which arrow or arrows should you use and why?

Exercise 2.8. Section 2.2 introduced the rectangle puzzle. Imagine instead a *square* puzzle with its corners labeled the same way. Such a puzzle would allow a new move that was not possible with the rectangle puzzle; you could rotate a quarter-turn clockwise.

(a) Make the map of this group.

(b) Why is the quarter-turn move not "allowed" in the rectangle puzzle?

Exercise 2.9. Most groups can be generated many different ways, and each way gives rise to a corresponding way to connect a Cayley diagram with arrows. For example, consider the group V_4, which we met in the rectangle puzzle. Let's shorten the names of its actions to n, h, v and b, meaning (respectively) no action, horizontal flip, vertical flip, and both (a horizontal flip followed by a vertical flip).

We saw that h and v together generate V_4. But it is also true that h and b together would generate V_4, or v and b together. (You can verify these facts by exploring the rectangle realm using these generators on your own numbered rectangle.)

(a) Make a copy of Figure 2.9 and add to it a new type of arrow, representing the action b.

(b) Make a copy of your answer to part (a), with the arrows representing h removed. How does your diagram show that v and b are sufficient to generate V_4?

(c) Make a copy of your answer to part (a), with the arrows representing v removed. How does your diagram show that h and b are sufficient to generate V_4?

2.6.3 Going backwards

Exercise 2.10. If you've done all the exercises to this point, you've encountered two different Cayley diagrams that have the two-node form shown here.

Can you come up with another group whose Cayley diagram has this form?

Exercise 2.11. If you've done all the exercises to this point, you've encountered two different Cayley diagrams that have the four-node form shown here.

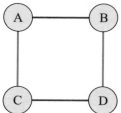

Can you come up with another group whose Cayley diagram has this form?

Exercise 2.12. We have not yet seen a group whose Cayley diagram has the three-node form called C_3, shown in the top left of Figure 2.10. Can you come up with a group whose Cayley diagram has that form?

2.6.4 Rules

Exercise 2.13. A group's generators have a special status in a Cayley diagram for the group. What is that special status?

Exercise 2.14. Chapter 1 required groups to satisfy Rule 1.5, which states, "There is a predefined list of actions that never changes." How does this rule impact the appearance of Cayley diagrams? (Or how would diagrams be different if this rule were not a requirement?)

Exercise 2.15. Chapter 1 required groups to satisfy Rule 1.6, which states, "Every action is reversible." What constraint does this place on the arrows in a Cayley diagram? Can you draw a diagram that does not fit this constraint? (That is, draw a diagram that almost deserves the name "Cayley diagram," except for that one rule violation.)

Exercise 2.16. Chapter 1 required groups to satisfy Rule 1.7, which states, "Every action is deterministic." What constraint does this place on the arrows in a Cayley diagram? Can you draw a diagram that does not fit this constraint? (That is, draw a diagram that almost deserves the name "Cayley diagram," except for that one rule violation.)

Exercise 2.17. Chapter 1 required groups to satisfy Rule 1.8, which states, "Any sequence of consecutive actions is also an action." How do we depend upon this fact when using a Cayley diagram as a map?

2.6.5 Shapes

Exercise 2.18. If we created an equilateral triangle puzzle, like the square puzzle in Exercise 2.8, what would the valid moves be? Map the group of such a puzzle.

Exercise 2.19. A regular n-gon is a polygon with n equal sides and n equal angles. You have already analyzed regular n-gons with $n = 3$ (equilateral triangle, Exercise 2.18) and $n = 4$ (square, Exercise 2.8).
(a) Based on what you know about the cases when $n = 3$ and $n = 4$, make a conjecture about how many actions will be in the group of the regular n-gon for any $n > 2$.
(b) Test your conjecture by making the map of the group for a regular pentagon ($n = 5$).
(c) Find the equilateral triangle group, the square group, and the regular pentagon group in *Group Explorer*'s library. (Hint: Use the search feature.)
   (i) Do your Cayley diagrams look like those in *Group Explorer*?
   (ii) Does your conjecture hold up against all the data you can find in *Group Explorer*?
(d) Write a paragraph giving as convincing an argument as you can in favor of your conjecture. Try to anticipate the counterarguments of a skeptical reader.

The groups you studied in this exercise are called *dihedral groups*. We will return to them in Exercise 4.9, and then formally study them in Chapter 5.

3
Why study groups?

The groups you've seen in this book so far may leave you wondering what the purpose of group theory is. After all, the things we've studied have little value other than intellectual amusement—games, puzzles, and imaginary situations like coins on a table. If group theory were just a collection of intellectual amusements, anyone could be forgiven for asking, "So what?"

The two chapters we've spent learning the basics have positioned us to learn some of group theory's more practical applications. This chapter tours a few of those applications, and provides references to where interested readers can go to explore any of them further. After this chapter, our study of group theory will get more advanced, and I will return to choosing examples based on how well they illustrate the material (rather than based on how useful they are in the real world).

We will see applications of group theory in three different areas: science, art, and mathematics. Further examples could have been included, but these are a good start. Woven throughout all the examples you will see the notion of *symmetry*. After reading this chapter, you will be well-equipped to identify and classify symmetry in the world around you. Some of this chapter's exercises will encourage you to spot new objects and situations to which group theory applies.

3.1 Groups of symmetries

We say something is symmetrical when it looks the same from more than one point of view. For instance, humans have bilateral symmetry; our left and right sides are similar, so we look (basically) the same in a mirror as we do face-to-face. Sea stars (starfish) have pentaradial symmetry; their five alike arms radiate outward, separated by equal angles. You can turn one a bit and it looks the same after the turn as before. (To be precise, turn it 72 degrees, one fifth of a full 360 degree rotation.)

How does this relate to groups? The example groups we've looked at so far deal with arrangements of similar things—the numbered corners of a rectangle, the colored squares on Rubik's Cube, the several paintings on a room's walls (Exercise 1.4)—and in Chapter 5, we will cover a foundational fact of group theory, that every group can be viewed as a collection of ways to rearrange some set of things. Groups relate to symmetry because an object's symmetries can be described using rearrangements of the object's parts. My example of turning a sea star put each arm in a new location, and is thus a (simple) rearrangement of the arms.

Mathematicians and scientists use groups to describe the symmetry in physical objects. The technique for finding a group that describes an object's symmetry has already appeared in this book; you have seen it used with the rectangle puzzle in Chapter 2 and tried something similar yourself in Exercises 2.8, 2.18, and 2.19. It's time this technique had an official introduction. Definition 3.1 describes it thoroughly, and thereafter I explain why the technique makes sense.

Definition 3.1 (technique for measuring symmetry). The following process extracts from any physical object a group that describes the symmetry in that object.

1. Identify all the parts of the object that are similar, and give each such part a different number. (The rectangle had four alike corners, so we numbered them. We could have instead numbered the four sides, because the left is the same as the right and the top is the same as the bottom.)

2. Consider the actions that you can perform with your hands that may rearrange the numbered parts, but that leave the object taking up the same space it did originally. (If the object is too large or small for your hands, imagine holding a perfectly scaled model instead.) This collection of actions forms a group.

3. If you want to visualize the group, explore and map it as we did in Chapter 2 with the rectangle, shown in Figures 2.2 through 2.7.

Reviewing how Chapter 2 applied this technique to the rectangle puzzle should help clarify this definition. Figure 3.1 does just that.

Let's take a moment to consider this technique so that we can see the logic behind it. Step 1 of Definition 3.1 numbers the object's parts so that we can track the manipulations permitted in Step 2. That is, it enables us to distinguish all the object's different states from one another, and to see how they relate. Note that new each state is a new arrangement of the object's similar parts, a *re*-arrangement. The numbering allows us to clearly describe each of these rearrangements, which would otherwise be indistinguishable from each other.

The heart of the technique lies in Step 2, which gives both freedom and restriction: You should look for *all* manipulations of the object *provided that* it takes up the same space. The freedom (in fact, the command) to find *all* manipulations ensures that we get a *complete* description of the object's symmetry; to do so may require some exploration, as it did in the case of the rectangle. The restriction to manipulations that take up no new space ensures that our exploration pays attention to the object's shape. Recall that symmetry is not only a set of similar parts, but also their arrangement. (A starfish is more than just five alike arms in a pile!) The restriction in Step 2 ensures that this whole technique respects the *arrangement* of the object's parts.

Step 2 defines a group that describes the symmetry in the original object. Step 3 is optional, but we perform it in order to *see* the group, since we are studying group theory

3.1. Groups of symmetries

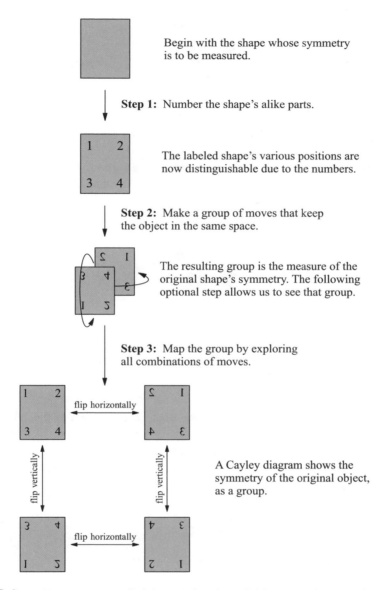

Figure 3.1. In Chapter 2, we applied the technique in Definition 3.1 to the rectangle. A summary of the steps we took is shown here as an example. Clearer illustrations of the rectangle's valid moves and its map appear in Figures 2.3 and 2.7, respectively.

visually. Although we have already experienced this technique in Chapter 2 when we studied the rectangle puzzle, now it is time to employ it for something more useful.

3.1.1 Shapes of molecules

Chemists need to classify the molecules they study by shape, because those shapes impact the molecules' behavior. They use group theory to do so. This gives us a chance to apply the technique in Definition 3.1 to something practical. Figure 3.2 represents a molecule of

Figure 3.2. A molecule of Boric acid, B(OH)₃

Boric acid; the red ball is boron, the blue oxygen, and the green hydrogen. Let's measure its symmetries as a chemist would, and find the group that describes them.

Step 1 of the technique in Definition 3.1 insists that we label the molecule's alike parts—in the case of Boric acid, this means the three "arms." I have done so in Figure 3.3.

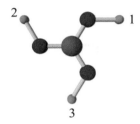

Figure 3.3. The same molecule as in Figure 3.2, but now with each of the three identical arms numbered, to distinguish them, as in Step 1 of Definition 3.1

Step 2 of Definition 3.1 requires us to ask what manipulations of this molecule are available. As Step 2 instructs, we imagine holding a scale model of the molecule and twisting it about. One of the first things to notice is that rotating one-third of a full circle in either direction is a valid move, because the result occupies the same space as the shape we started with; only the numbering has changed. (To help your mind's eye, see Figure 3.4.) On the other hand, we can find manipulations that are not admissable. For example, a horizontal flip was legal in the rectangle puzzle (Figure 2.3, page 14), but is not legal here. Figure 3.5 shows that a horizontal flip results in a shape that sits differently in space than the original.

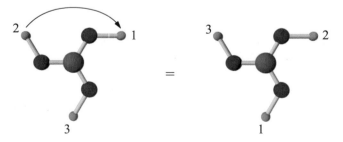

Figure 3.4. Rotating the numbered molecule from Figure 3.3 clockwise one-third of a full rotation rearranges the numbering, but the molecule occupies the same space as before. Thus a one-third turn is part of the group describing this molecule, as per Definition 3.1.

3.1. Groups of symmetries

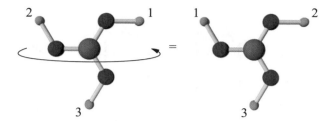

Figure 3.5. Flipping the numbered molecule from Figure 3.3 over horizontally situates the shape differently in space. (The arms do not reach in the same directions as originally.) Thus a horizontal flip is *not* part of the group describing this molecule, as per Definition 3.1.

If we conducted a full exploration of this group, as we did when mapping out the rectangle puzzle, we would generate a cyclic group with three actions, shown in Figure 3.6. It is satisfyingly natural that a molecule with three prongs like a propeller has a three-step cyclic symmetry group. The term "cyclic" is not used casually here; it is actually a technical term, although its meaning should be mostly obvious from Figure 3.6. We will talk more about cyclic groups in Chapter 5.

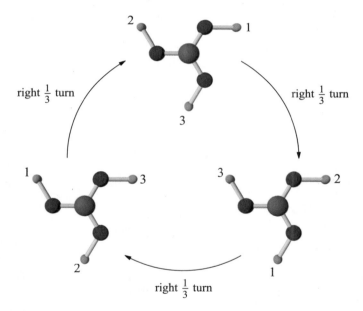

Figure 3.6. A complete map of the group obtained by exploring the symmetry of the labeled molecule from Figure 3.3. If we make this diagram abstract, as we did when creating Figure 2.9, we obtain the top left diagram in Figure 2.10, page 22, the cyclic group C_3.

3.1.2 Crystallography

Of course, chemists study more than one molecule at a time. One of the easiest ways for a chemist to learn about solids at the atomic level is to study solids whose atoms are naturally well organized. Solids whose atoms arrange themselves in a regular, repeating pattern are called crystals. The study of them is called crystallography.

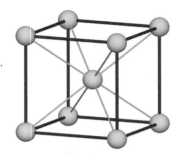

Figure 3.7. One cube from a crystal whose type chemists have named *body-centered cubic*. Each white ball represents an atom (of lithium or sodium, for instance) and each line a bond between atoms. The crystal structure formed by repeating many of these through space is shown in Figure 3.8.

There are hundreds of three-dimensional crystal patterns, but let us take one example. The atoms of several common elements (including sodium and lithium, for example) arrange themselves in a grid of cubes when they form a solid, with an extra atom at the center of each cube. This is called a *body-centered cubic* arrangement, and is shown in Figure 3.7. Many of these cubes together form a crystal, as shown in Figure 3.8. The cubes sit next to one another, and over and under one another, and in front of and in back of one another, in all three dimensions, extending indefinitely. Where cubes touch faces, they share the four atoms of the face between them.

The technique from Definition 3.1 can even be used to classify the symmetry of patterns like this one. I will not do so complex an example here, but most of that complexity comes from just one aspect of the crystal: that it seems to go on forever. Chemists study-

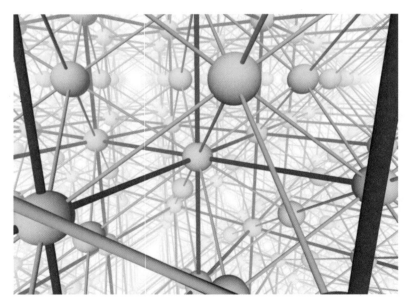

Figure 3.8. The cube from Figure 3.7 repeated to fill three-dimensional space. The different colors of the lines have no chemical significance; they make it easier to distinguish the edges of each cube from the diagonal lines through the cubes' centers.

3.1. Groups of symmetries

ing crystals do indeed treat them as patterns that repeat without end, and thus the groups describing them are infinite. Perhaps your curiosity is now aroused, and you're wishing that I *would* do an example. That's good! Although I do not classify the symmetry of a crystal, in the following section I give an example of classifying the symmetry of infinite repeating patterns in one dimension. The interesting aspect of the example remains: How do we deal with infinite objects, and what are infinite groups like?

The World Wide Web contains excellent resources for further investigation of the group theory of molecules and crystals. One particular site that helped generate some of the examples in this chapter is maintained by Jonathan Goss [10].

3.1.3 Art and Architecture

Crystals are patterns that repeat in three dimensions, but we can think of repeated patterns that are not three-dimensional. Such repeating patterns appear in several areas of art. First, let's consider simply patterns that repeat in only one direction, along a strip. Perhaps you have seen such a pattern embroidered on the hem of a piece of clothing, or the edge of a tablecloth. You may have seen a piece of jewelry composed of the same type of links chained together. The cornices of some buildings are sculpted in a one-dimensional repeating pattern called a *frieze*. Mathematicians have adopted this term, and call all patterns that repeat in one dimension *frieze patterns* and the groups that describe them *frieze groups*. A simple frieze pattern is shown in Figure 3.9. It may not be intricate or beautiful enough for you to want on your clothes or your house, but a simple first example is best. Although Figure 3.9 only shows five leaves, the ellipses at either end of the diagram indicate that one should imagine the pattern extending infinitely left and right.

Figure 3.9. A frieze pattern, which can be extended indefinitely to the left or right.

Applying the technique from Definition 3.1 to an infinite object will be a new experience for us. Step 1 of that technique asks us to number all the object's alike parts, which can be a challenge if there are infinitely many! I number only those parts of the pattern that are present in Figure 3.9, and use a numbering scheme that makes it obvious how you could continue numbering in either direction if more of the frieze pattern were drawn. But what are the alike parts of the pattern, which I must number? It might be tempting to say that the leaves are the alike parts, but that is only half of the story. Each leaf is also itself symmetrical, its left side like its right side. So the parts we need to number are actually both *sides* of each leaf. Figure 3.10 shows a numbering that starts with 1 at the right side of the center leaf and counts to the right. In mirror image fashion, it uses negative numbers to go from the left of the center leaf to the left. If the vine were extended and more leaves shown in either direction, it is clear how we could continue numbering them.

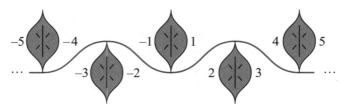

Figure 3.10. The frieze pattern from Figure 3.9, now with all alike parts labeled, as Step 1 of Definition 3.1 requires.

Step 2 from Definition 3.1 instructs us to find the manipulations of the frieze that rearrange the numbered parts but keep the shape occupying the same space. One such action we have seen twice before—a horizontal flip. Just as Figure 2.3 showed that a horizontal flip is a valid move for the rectangle, and Figure 3.5 showed that it is not a valid manipulation of an atom of Boric acid, so Figure 3.11 shows that it is a valid move for this frieze pattern. The second manipulation shown in Figure 3.11 has not yet appeared in this book; it is called a glide reflection. To perform a glide reflection, move the pattern to the right and then flip it vertically, as shown by the arrow in that figure. An upper leaf moves to the right and then flips down to end up where the adjacent lower leaf had been.

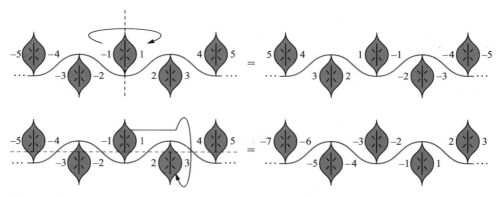

Figure 3.11. The legal moves for the leaf frieze shape from Figure 3.9, according to Step 2 of Definition 3.1.

This glide reflection is the first time we have had any manipulation that involved *moving* an object in any direction. This is because only with an infinite object is it possible to move it and still have it look the same when the move is completed. Because the pattern repeats infinitely along one axis, moves in that direction may not alter the space it occupies. (Note that another legal manipulation is to move the center leaf twice as far, to take the place of the right leaf, with no vertical flip involved. But this is the same as two sequential glide reflections, so by including a glide reflection, we will end up generating this move as well.)

If we carefully map the interactions among these two moves, as Step 3 of Definition 3.1 recommends, we obtain a Cayley diagram of the group describing the symmetry in this frieze pattern. Like the frieze pattern itself, the Cayley diagram is infinite and thus we can only see a portion of it at one time, and must use ellipses to suggest how it could be extended. We will not go through the details of the mapping out of such a diagram

3.1. Groups of symmetries

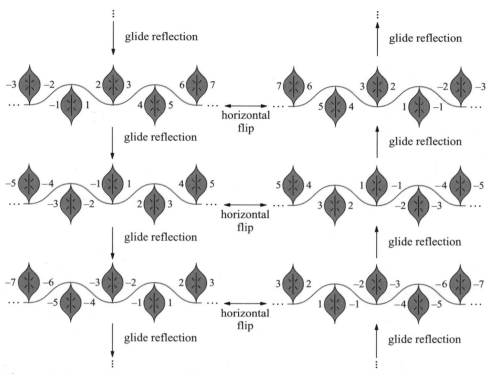

Figure 3.12. A portion of the infinite Cayley diagram of the group of symmetries of the frieze pattern in Figure 3.9.

here, as we did with the rectangle in Chapter 2. Figure 3.12 shows the resulting diagram, the abstract version of which is shown in Figure 3.13.

The study of group theory has yielded a very interesting fact about frieze patterns. The symmetry of any frieze pattern can be described by one of seven different infinite groups.[1] That is, although you may spend a lifetime creating new and different frieze

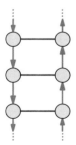

Figure 3.13. The abstract version of Figure 3.12. All the friezes and names of actions have been removed, leaving only nodes representing different states, and two kinds of arrows connecting them. The pattern of such connections, however, remains unchanged from Figure 3.12 to here.

---

[1] Actually, some of the frieze groups are isomorphic to one another, a concept we will learn in Chapter 8. But the frieze groups that are isomorphic (i.e., have the same structure) are still different in the actions that comprise them, and thus different in their visual appearance.

patterns, the *symmetry* in each of those patterns can only be of one of seven different types. Each such type is described by a group, one of which we just saw in Figures 3.12 and 3.13. Exercise 3.11 asks you to diagram the other six frieze groups, given six different vine-and-leaf patterns that exemplify them.

This frieze example shows that the technique from Definition 3.1 is more broadly applicable than it may at first have seemed—it can even apply to infinite objects! We used it to measure the symmetries in a one-dimensional repeating pattern, and mentioned earlier that chemists use it to classify the three-dimensional repeating patterns that occur naturally in crystals. Just as frieze patterns can be described by one of seven frieze groups, so crystal patterns can be described by one of 230 crystallographic groups [19]. As you may therefore suspect, we can also classify two-dimensional repeating patterns with the same technique, yielding one of what are appropriately called the 17 wallpaper groups.

Two-dimensional repeating patterns, like those found in wallpaper, will not be investigated in depth here. Although the technique from Definition 3.1 can be applied to such patterns in principle, it takes a great deal of patience and a very large piece of paper. Such diagrams have several different types of arrows, and extend infinitely in two directions, like the wallpaper patterns they describe. They can be very tangled.

But notice the specific import generators have in visualization: a Cayley diagram for a group relies on having a small set of generators to use as arrows. A Cayley diagram is a map of the group, and the arrows are the roads. So on the one hand, if the set of actions we use as arrows does not generate the whole group, then there will be some unreachable nodes, and the diagram will not be connected. On the other hand, including an arrow for *every* action in a group would typically clutter diagrams beyond readability. (In fact, in infinite groups it simply could not be done!) A set of generators provides a few arrow types that still connect all the nodes; this is what makes Cayley diagrams possible.

3.2 Groups of actions

This book introduced groups with two examples, Rubik's Cube and the rectangle. The example applications we've seen so far in this chapter are like Rubik's Cube and the rectangle in that they deal with physical objects with inherent symmetry. Therefore each application so far has seen use of Definition 3.1 to find and illustrate the groups describing the symmetry of some physical object.

But the definition of a group we learned in Chapter 1 is not restricted only to the symmetry of physical objects. Any collection of actions satisfying the rules from Defintion 1.9 is a group. Indeed, some important examples of group theory (and some fun ones) do not involve the symmetry of a physical object at all. This second section of the chapter briefly shows two such examples.

3.2.1 Dancing

Several traditional styles of country dancing involve predefined steps called *figures* that move couples around the dance floor. In square dancing and contra dancing, the dancers learn these steps by name and practice following a caller who orders them to perform specific figures in time with the music. In both these kinds of dances, couples stand in pairs forming a box, as shown in Figure 3.14.

3.2. Groups of actions

Figure 3.14. The arrangement of two couples in a square, ready to begin a square dance or a contra dance.

Dancing a figure rearranges the dancers. If they correctly obey the caller, every dance ends with the dancers back to the locations in which they began the dance. Figure 3.15 shows the effects of six example contra dance figures. The particular steps are not shown, only the effect the figure has on the dancers' arrangements.

The collection of all such figures turns out to form a group. That is, all the criteria from Definition 1.9 are satisfied by the collection of contra dance figures. Furthermore, that group has the same structure as the group describing the symmetries of a square (D_4). In other words, the Cayley diagrams of both groups have the same shape, but with different names on the nodes and arrows. When two groups have the same structure, we say they are *isomorphic*, a topic we will study in Chapter 8.

Readers interested in learning more about the group theory in dance figures can begin with the Exercises in Section 3.4.4 of this chapter. Further reading is available on the World Wide Web; the author has found the website of Larry Copes [5] and an article by Ivars Peterson [15] particularly informative.

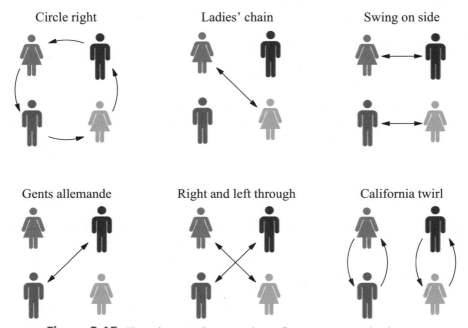

Figure 3.15. How six example contra dance figures rearrange the dancers.

3.2.2 Polynomial roots

By now the reader is certainly convinced that group theory shows up in diverse situations. But it would be a great disservice to the history of mathematics if I did not mention one more application, the very reason that group theory was invented! In the nineteenth century, two young mathematical prodigies, Neils Abel and Évariste Galois, solved a mathematical problem that had stood unsolved for centuries. It has come to be called "the unsolvability of the quintic." It is one of the great discoveries in mathematics, and when you come to Chapter 10 of this book, you will be ready to read about it in some detail. It rests on the fact that the solutions to polynomial equations have a certain relationship to one another. They form a group.

Galois did not call the patterns he noticed groups; later mathematicians who extended his work gave them that name. But he was the first to notice them and study them. Galois found groups when studying the roots of polynomials, but since his day mathematicians have found a multitude of other contexts in which groups arise, some of which you have just read about.

As in contra dancing, the actions in the groups Galois studied are not manipulations of a physical object, but instead rearrangements of a collection of things. Instead of dancers, Galois was rearranging the solutions to polynomial equations. The remarkable discovery he made was that by examining the group of those rearrangements, you could tell how difficult it was to solve the original equation. Some groups, he proved, correspond to equations that are impossible to solve.[2]

This brief mention of the work of Galois, skipping many details (including Abel's contribution), is included here for two reasons. First, it gives you a bit of the historical origins of the subject, which are themselves group theory's first application. Second, you now have something to look forward to as your study of group theory progresses; the end of this book describes some famous, deep, and beautiful mathematics. The intervening chapters prepare you for it.

3.3 Groups everywhere

This chapter has applied group theory to chemistry, visual art, dancing, and mathematics. From these applications, it is clear that group theory arises in diverse contexts, yet these examples are just a few from among the many applications of group theory. There is neither space nor need to cover all the applications of group theory here, but I mention some in passing, so that readers interested in a specific application know where to look for more information.

We saw two ways that groups are useful to chemists—molecular shape and crystal structure. Another way chemists use group theory is for studying and enumerating isomers, using Pólya's technique. A paper of Kennedy, McQuarrie, and Brubaker [12] gives a good overview of this technique with references to sources with more detail. Theoretical physics has used group theory when looking at fundamental particles [4, 17].

Section 3.2.2 revealed that the birth of group theory was in the solution to a famous mathematical problem from algebra. Since then, several other branches of mathematics

---

[2]Exactly what "impossible" means in this context is described in Chapter 10.

have made use of groups as well; I mention a few examples. Group theory has tight connections to linear algebra (which can describe spatial manipulations, like some of our group actions) and number theory (in which we will dabble in Section 8.3). Topology, the abstract study of shapes and spaces, with applications in physics, astronomy, and elsewhere, uses groups to categorize spaces and shapes. The textbook *Group Representations in Probability and Statistics* covers uses of group theory in probability and statistics [7].

Subgroups and cosets (topics from Chapter 6) can be used to design error-correcting codes, which help minimize error in transmitted messages. Every time you download a file from the Internet, your computer uses codes like these to keep your download uncorrupted by minor noise or static on the network lines.

David Benson's book *Music: A Mathematical Offering* [2] contains a chapter on the application of group theory to music. It uses many different results from group theory, from introductory through advanced. The technique mentioned above that chemists use to count isomers, Benson uses to enumerate pitch classes with certain properties.

I could continue with further examples from poetry, juggling, sudoku, and more. But the point has been made. Symmetry is everywhere, and group theory makes an essential contribution to the study of symmetry. Our eyes can spot symmetry, but the measuring technique from this chapter takes us beyond just spotting it. Cayley diagrams show that an object's symmetries are not a disordered collection, but relate in what are often intricate and beautiful ways. Group theory studies those relationships mathematically and uses them to solve diverse problems.

3.4 Exercises

3.4.1 Basics

Exercise 3.1. Consider the six contra dance figures shown in Figure 3.15. Which of them, if any, when repeated twice in succession, return the dancers to their original positions? Which, if any, do not return the dancers to their original positions until the move is repeated three times? Four? Do any figures require more repetitions than four to return the dancers to their original positions?

Exercise 3.2. Consider the horizontal flip shown on top of Figure 3.11. What is the effect of repeating that action twice in succession? How is the glide reflection shown in the same figure different?

Exercise 3.3. Figure 3.4 shows that if you rotate a boric acid molecule clockwise one-third of a full turn, it takes up the same space as it did originally. This clockwise rotation generates the Cayley diagram shown in Figure 3.6. How would the diagram be different if I had generated it with a counterclockwise rotation instead? Would the group size or structure change?

Exercise 3.4. Name three physical manipulations you could do to the cube in Figure 3.7 that would rearrange its parts but leave it taking up the same space, as in step 2 of Definition 3.1.

3.4.2 The symmetry of molecules

Exercise 3.5. Apply the technique from Definition 3.1 to a molecule of ethylene, shown below. (Include the optional Step 3.) Where have we seen the resulting group before and what is its name?

The chemical formula for ethylene is C_2H_4; the dark atoms indicate carbon and the light blue atoms indicate hydrogen. A scale model of this molecule would lie flat on a table; it has no depth.

Exercise 3.6. Apply the technique from Definition 3.1 to a molecule of sulfur chloride pentafluoride, shown below. As in the previous exercise, perform the optional Step 3. We have not named the resulting group so far in this book; can you find it in *Group Explorer*?

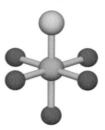

The chemical formula for sulfur chloride pentafluoride is SF_5Cl; the yellow atom indicates sulfur, the white indicates chlorine, and the magenta atoms indicate fluorine.

Exercise 3.7. Repeat the previous exercise on a molecule of benzene, as shown below.

The chemical formula for benzene is C_6H_6; the dark atoms indicate carbon and the light blue atoms indicate hydrogen. A scale model of this molecule would lie flat on a table; it has no depth.

Exercise 3.8. Repeat the previous exercise on a molecule of eclipsed ferrocene, $Fe(C_5H_5)_2$, as shown below.

3.4. Exercises

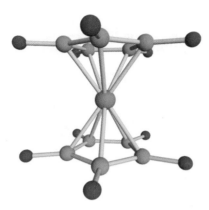

Exercise 3.9. Your work in the exercises in this section has shown you several symmetry groups, of various sizes. Is there any relationship between symmetry group size and how much symmetry the corresponding object displays?

Exercise 3.10. What physical objects with which you are familiar have some symmetry? Choose one and apply the technique in Definition 3.1 to create and diagram the group describing its symmetry.

3.4.3 Repeating patterns

Exercise 3.11. As mentioned in this chapter, there are exactly seven groups that describe all possible symmetries for frieze patterns. The diagram of one such group appears in Figures 3.12 and 3.13.

This exercise has six parts, (a) through (f), each asking you to apply the technique in Definition 3.1 to a frieze that exemplifies one of the other six frieze groups. The friezes you need are shown below.

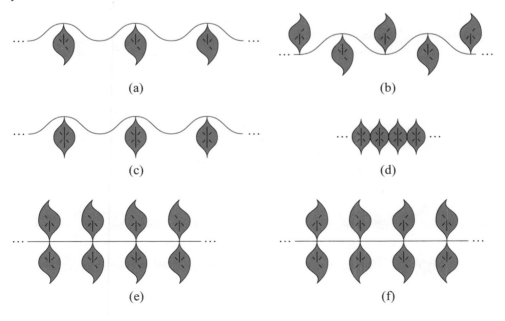

If you found the example in the chapter an insufficient springboard for classifying infinite groups, refer to the solutions in the Appendix for the solution to part (d), one of the more difficult parts.

Exercise 3.12. For each of the following letter sequences, think of it as a frieze pattern. Match each letter sequence with the leaf frieze from the previous exercise that has the same symmetry group.

(a) LLLLLLLL

(b) MMMMMMMM

(c) MWMWMWMW

(d) HHHHHHHH

(e) SSSSSSSS

3.4.4 Dancing

Exercise 3.13. Take just the contra dance figure called "Circle right" from those shown in Figure 3.15.

(a) Create a Cayley diagram for the group generated by this action alone (i.e., Circle right and any number of repetitions of it).

(b) Does this group contain any of the other contra dance figures from Figure 3.15?

Exercise 3.14.

(a) Repeat Exercise 3.13 for the figure "Ladies' chain."

(b) Repeat Exercise 3.13 for the figure "Gents allemande."

(c) Repeat Exercise 3.13 for the group generated by the two figures "Ladies' chain" and "Gents allemande" together.

Exercise 3.15.

(a) If you had to generate all the contra dance figures in Figure 3.15 using only combinations of "Circle right" and another contra dance figure of your choice, which would you choose? Show that your answer is correct.

(b) Is it possible to generate all the contra dance figures in Figure 3.15 using only combinations of "Ladies chain" and one other contra dance figure of your choice? If so, which contra dance figure should you choose? If not, how many more contra dance figures beyond "Ladies' chain" would you need?

(c) How few contra dance figures can you start with, and still generate all the contra dance figures in Figure 3.15 from their combinations?

Exercise 3.16.

(a) When you generate a group by combining contra dance figures as in Exercise 3.15, can you generate any contra dance figures that are not shown in Figure 3.15?

(b) How big is the group of contra dance figures generated from those in Figure 3.15?

(c) Diagram the entire group.

4

Algebra at last

So far you have been learning about groups in a way that is unique to this book. Our unofficial definition of a group, Definition 1.9, is not how mathematicians define a group. Though there are many benefits to the approach I have taken (and will continue using later), it would be an incomplete education in group theory that did not acquaint you with the official definition of a group, the one you would find in any *other* group theory book.

That standard definition comes with its own natural visualization technique, called a multiplication table. Since we are focusing on visualization, multiplication tables will be our inroad to the study of the standard definition of a group. But there is some background material we must cover first, in Sections 4.1 and 4.2.

4.1 Where have all the actions gone?

In the Cayley diagrams we made in Chapter 2, actions in a group showed up in the diagram as arrows. But in a typical Cayley diagram there were only arrows for *some* of a group's actions, specifically the generators. Multiplication tables will fill an important gap in our study by allowing us to see *all* the actions in a group and how they interact with each other.

To see why Cayley diagrams do not illustrate all of a group's actions, recall Rule 1.8 on page 6, which stipulated that any combination of actions is also an action in its own right. So, for example, even though Chapter 2 introduced the rectangle puzzle as having two actions (horizontal and vertical flips), it is also legal to flip first horizontally and then vertically. Furthermore, this combination can be thought of as *one single action*. But the Cayley diagram for the rectangle puzzle (Figure 2.7 on page 18) does not include arrows for this combined action. It only illustrates the two generators that were introduced with the puzzle, horizontal and vertical flips. This subtle point was addressed in Exercise 2.13: generators receive special treatment in Cayley diagrams, because they are represented by the diagram's arrows. I encourage you to reexamine Figure 2.7 briefly to verify this fact and your understanding of it.

4. Algebra at last

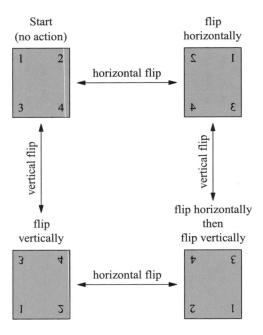

Figure 4.1. An annotated copy of Figure 2.7 from page 18. Above each rectangle in the figure appears a phrase describing an action that leads there from the starting point.

Although Cayley diagrams do not show us every action in a group directly, they do display the full set of actions indirectly. Every *location* in a Cayley diagram such as Figure 2.7 naturally correlates with the *action* that leads to that location. Therefore the full set of actions is implicitly present in the diagram as the locations to which they lead. Let us alter the diagram in Figure 2.7 to explicitly demonstrate this point. Figure 4.1 is a copy of Figure 2.7 in which each rectangle configuration has been labeled with the name of an action that leads there from the starting configuration. This makes the correspondence between the locations and the actions that lead to them clear. Take a careful look at Figure 4.1 now and compare it to Figure 2.7. Note that the starting configuration is associated with the action "none," because to get there from the start, simply do nothing!

Removing the rectangles and leaving only the annotations, as in Figure 4.2, has the effect of making the diagram no longer be about configurations, but about actions. If you prefer, it has changed from a diagram about *nouns* to one about *verbs*. Figure 4.2 does not include arrow labels because they would be redundant. To distinguish "vertical flip" arrows from "horizontal flip" ones at a glance, the vertical arrows are dashed. We can infer which arrow is which by the actions they connect: Because the solid arrow points from the start to "horizontal flip," it must symbolize the horizontal flip action, and thus the dashed arrow symbolizes the vertical flip.

Diagrams like Figure 4.2 will become very useful to us in the next section, so let us take a careful look at that diagram before proceeding. The following definition summarizes the transformation that turned Figure 2.7 into Figure 4.2. As you read it, look over Figures 2.7, 4.1, and 4.2 to be sure that you see how each part of the definition plays out in that example.

4.1. Where have all the actions gone?

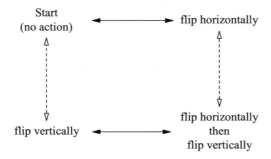

Figure 4.2. A simplified copy of Figure 4.1. The rectangles themselves have been removed, and replaced by the actions that led to them. The labels on the arrows have become superfluous, and were therefore removed. The two types of arrows are now distinguished by dashed vs. solid lines.

Definition 4.1 (diagram of actions). The following three steps transform a Cayley diagram into one that focuses on the group's actions.

(i) Choose any node and call it the "start." A natural choice for starting node often exists, as it did in the rectangle puzzle.

(ii) Relabel each remaining node in the diagram with a path that leads there from the start.

(iii) Remove the arrow labels, and distinguish arrows in some other way (e.g., by color or using solid lines vs. dashed lines).

Let us call the result of such a transformation a ***diagram of actions***.

The transformation described in Definition 4.1 can be done to any Cayley diagram, even a purely abstract one, such as the diagram of the Klein 4-group in Figure 2.9 on page 21. (Recall that the lines in Figure 2.9 lack arrowheads, for simplicity and to indicate bidirectionality. We'll call them arrows nonetheless.) We will need to refer to the different colored arrows in some way, in order to speak of paths in the diagram. Let's write R for red arrow and B for blue. Applying the transformation in Definition 4.1 results in Figure 4.3. As a test of your understanding of Definition 4.1, check that Figure 4.3 is the correct result of transforming Figure 2.9 into a diagram of actions.

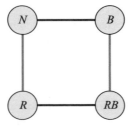

Figure 4.3. The result of labeling nodes according to "actions" (in this case the colors of the diagram's arrows) in the abstract Cayley diagram from Figure 2.9 on page 21. The group shown is the Klein 4-group V_4, which we first encountered in Chapter 2.

4.2 Combine, combine, combine

The diagram of actions for a group presents new information to us, information about the relationships among the actions in the group. In fact, we will soon see that diagrams of actions, like Figure 4.3, can be used as a simple kind of calculator. They calculate the effect of combining any two of the group's actions.

It may seem strange to say that Figure 4.3 has actions in it, since it is an abstract diagram. Of course, we expect mathematics to be abstract at times, but concrete circumstances help ground our understanding. So recall that we saw concrete examples in Sections 2.2 through 2.5 that fit the abstract pattern in Figure 4.3. Here in this abstract case, the "actions" are just following red and blue arrows in the diagram (called actions R and B respectively). To see how any two actions combine, simply begin at the starting node (N for none) and perform the two actions one after the other. This is how we can use Figure 4.3 as a calculator.

For example, we can verify that "R followed by B" has the same effect as "B followed by R." Use the diagram to follow the red (R) arrow from the start node, and then the blue (B) arrow. This ends us at the RB node. Now verify for yourself that B followed by R leads from the start to the same place. Thus those two different phrases, "R followed by B" and "B followed by R," really describe equivalent actions, because they have the same final result.

It is common to express this relationship using notation that comes from algebra. (In fact, group theory is usually studied in a very algebraic way, as this chapter will continue to show, and hence falls under the mathematical subject area called abstract algebra.) To say "R followed by B" we write RB as if we were multiplying two variables in algebra; to say "B followed by R" we will just write BR. Then the sentence

"R followed by B" is equivalent to "B followed by R"

can be written clearly and concisely as

$$RB = BR.$$

This observation is a very simple one, but Figure 4.3 can also calculate more complex relationships among the group's actions. Consider, for instance, the lengthy sequence of actions $RRRBRRBBB$. We can use common conveniences of algebraic notation to write this same sequence of actions more concisely, as $R^3BR^2B^3$, again as if we were multiplying variables in algebra. Figure 4.3 can simplify this lengthy sequence as well. Beginning at the start, perform the given sequence of actions—follow the red arrow three times, then the blue once, the red twice, and finally blue three times. You should end up at the node labeled R, which gives us the equation

$$R^3BR^2B^3 = R.$$

We have used the diagram of actions as a calculator to do a multiplication problem; we multiplied R^3 times B times R^2 times B^3 and found it to equal R. Any diagram of actions (as in Definition 4.1) can be used as a calculator in this same way.

(Perhaps you're wondering whether it's okay to simplify $R^3BR^2B^3$ further and write it as R^5B^4. Find out for yourself whether those two expressions mean the same thing by trying the technique just described.)

4.3. Multiplication tables

Now it's becoming clear why group theory is part of algebra—these equations actually look like algebra. But there is an important difference to keep in mind. Although we write RB as if it is a multiplication of R and B, we use "multiplying" metaphorically—we're really combining the actions by doing them sequentially. I will continue to use the term "multiplication" in this loose way throughout this book. Using common names like multiplication and addition as placeholders for abstract operations is common in mathematics.

The equation $R^3BR^2B^3 = R$ (and each of an endless supply like it that we could come up with) is a fact about how the actions in the group combine. The method we used for finding that equation shows us that we don't need to worry about long, complicated sequences of actions like $R^3BR^2B^3$, because no matter how lengthy a sequence of actions we write down, our diagram of actions will always help us simplify it to an equivalent action from among R, B, RB, or N.

4.3 Multiplication tables

Knowing how to use a diagram of actions as a calculator for a group is like having a secret decoder ring—you can read facts from a diagram of actions because you know the secret. But the purpose of visualization is not to keep secrets; it is to make things clear. Let's therefore create a table containing all the information about how group actions combine so that we no longer need to know the secret of how to read that information from a Cayley diagram.

Such a table, which shows how every pair of group actions combine, is called a multiplication table. Recall how a gradeschool multiplication table looks; if you've forgotten, see Figure 4.4. We wish to make a multiplication table with group actions in place of numbers, and in which "multiplication" means combining two actions one after the other, as above.

Continuing with the group shown in Figure 4.3 as our example, the four actions we need to be concerned with are N, R, B, and RB. So we make a table analogous to Figure 4.4, but with N, R, B, and RB as row and column headings; see Figure 4.5. We can fill the table by calculating the combinations of various pairs of actions, using Figure 4.3. For example, in the R row and the B column, we write the result of combining R and B,

	1	2	3	4	5	6
1	1	2	3	4	5	6
2	2	4	6	8	10	12
3	3	6	9	12	15	18
4	4	8	12	16	20	24
5	5	10	15	20	25	30
6	6	12	18	24	30	36

Figure 4.4. A standard elementary-school multiplication table, in this case only showing how to multiply the numbers 1 through 6.

	N	R	B	RB
N				
R				
B				
RB				

Figure 4.5. Beginning to create a multiplication table for the group V_4, whose Cayley diagram appears in Figure 4.3

which is RB. In the RB row and the RB column, we write the result of combining RB and RB, which is N. In performing the computations leading to these answers, Figure 4.3 is our calculator. Completing the whole table in this manner results in Figure 4.6.

Because N stands for "no action," combining it with another action has no effect, e.g., $NB = B$. For this reason, the first row and column of Figure 4.6 are just multiplication by N, and thus simply repeat the headings. This makes the row and column headings superfluous. It is therefore possible to omit those headings, and write the table as on the right of Figure 4.6. The row and column headings have become part of the table itself, its first row and column. This convention is used by *Group Explorer* because it makes the table simpler. But the headings can be a useful reference, so later multiplication tables in this book include them.

	N	R	B	RB
N	N	R	B	RB
R	R	N	RB	B
B	B	RB	N	R
RB	RB	B	R	N

N	R	B	RB
R	N	RB	B
B	RB	N	R
RB	B	R	N

Figure 4.6. On the left, a completed version of the multiplication table begun in Figure 4.5. Each entry can be computed using Figure 4.3 as a calculator. On the right, the same table with row and column headings removed, because they appear as the first row and column of the remaining table anyway.

Multiplication tables are a potent new visualization technique because they reveal patterns in the way the actions in the group combine, something Cayley diagrams do not explicitly do. To make such patterns visually obvious, it helps to color the cells of the multiplication table systematically. Assigning a different color to each action in the group enables us to color each cell according to the action that appears there. Figure 4.7 shows six multiplication tables colored in this way. The six groups depicted in the tables are the same depicted by the six Cayley diagrams in Figure 2.10 on page 22.

For now, the patterns in Figure 4.7 are of purely visual interest, but we will study them more in chapters to come. What remains in this chapter is to use multiplication tables as an inroad to the standard definition of a group.

4.3. Multiplication tables

Cyclic group C_3 (or $\mathbb{Z}_3$)

	e	a	a^2
e	e	a	a^2
a	a	a^2	e
a^2	a^2	e	a

Symmetric group S_3

	e	r	r^2	f	fr^2	fr
e	e	r	r^2	f	fr^2	fr
r	r	r^2	e	fr^2	fr	f
r^2	r^2	e	r	fr	f	fr^2
f	f	fr	fr^2	e	r^2	r
fr^2	fr^2	f	fr	r	e	r^2
fr	fr	fr^2	f	r^2	r	e

Direct product group $C_3 \times C_3$

	(e,e)	(a,e)	(a^2,e)	(e,a)	(a,a)	(a^2,a)	(e,a^2)	(a,a^2)	(a^2,a^2)
(e,e)	(e,e)	(a,e)	(a^2,e)	(e,a)	(a,a)	(a^2,a)	(e,a^2)	(a,a^2)	(a^2,a^2)
(a,e)	(a,e)	(a^2,e)	(e,e)	(a,a)	(a^2,a)	(e,a)	(a,a^2)	(a^2,a^2)	(e,a^2)
(a^2,e)	(a^2,e)	(e,e)	(a,e)	(a^2,a)	(e,a)	(a,a)	(a^2,a^2)	(e,a^2)	(a,a^2)
(e,a)	(e,a)	(a,a)	(a^2,a)	(e,a^2)	(a,a^2)	(a^2,a^2)	(e,e)	(a,e)	(a^2,e)
(a,a)	(a,a)	(a^2,a)	(e,a)	(a,a^2)	(a^2,a^2)	(e,a^2)	(a,e)	(a^2,e)	(e,e)
(a^2,a)	(a^2,a)	(e,a)	(a,a)	(a^2,a^2)	(e,a^2)	(a,a^2)	(a^2,e)	(e,e)	(a,e)
(e,a^2)	(e,a^2)	(a,a^2)	(a^2,a^2)	(e,e)	(a,e)	(a^2,e)	(e,a)	(a,a)	(a^2,a)
(a,a^2)	(a,a^2)	(a^2,a^2)	(e,a^2)	(a,e)	(a^2,e)	(e,e)	(a,a)	(a^2,a)	(e,a)
(a^2,a^2)	(a^2,a^2)	(e,a^2)	(a,a^2)	(a^2,e)	(e,e)	(a,e)	(a^2,a)	(e,a)	(a,a)

Direct product group $C_2 \times C_2 \times C_2$

	000	100	010	110	001	101	011	111
000	000	100	010	110	001	101	011	111
100	100	000	110	010	101	001	111	011
010	010	110	000	100	011	111	001	101
110	110	010	100	000	111	011	101	001
001	001	101	011	111	000	100	010	110
101	101	001	111	011	100	000	110	010
011	011	111	001	101	010	110	000	100
111	111	011	101	001	110	010	100	000

Quasihedral group with 16 elements

	a	b	c	d	e	f	g	h	i	j	k	l	m	n	o	p
a	a	b	c	d	e	f	g	h	i	j	k	l	m	n	o	p
b	b	c	d	e	f	g	h	a	j	k	l	m	n	o	p	i
c	c	d	e	f	g	h	a	b	k	l	m	n	o	p	i	j
d	d	e	f	g	h	a	b	c	l	m	n	o	p	i	j	k
e	e	f	g	h	a	b	c	d	m	n	o	p	i	j	k	l
f	f	g	h	a	b	c	d	e	n	o	p	i	j	k	l	m
g	g	h	a	b	c	d	e	f	o	p	i	j	k	l	m	n
h	h	a	b	c	d	e	f	g	p	i	j	k	l	m	n	o
i	i	l	o	j	m	p	k	n	a	d	g	b	e	h	c	f
j	j	m	p	k	n	i	l	o	b	e	h	c	f	a	d	g
k	k	n	i	l	o	j	m	p	c	f	a	d	g	b	e	h
l	l	o	j	m	p	k	n	i	d	g	b	e	h	c	f	a
m	m	p	k	n	i	l	o	j	e	h	c	f	a	d	g	b
n	n	i	l	o	j	m	p	k	f	a	d	g	b	e	h	c
o	o	j	m	p	k	n	i	l	g	b	e	h	c	f	a	d
p	p	k	n	i	l	o	j	m	h	c	f	a	d	g	b	e

Alternating group A_5

Figure 4.7. Multiplication tables of the same small, finite groups whose Cayley diagrams appear in Figure 2.10 on page 22.

4.4 The classic definition

Group theory is usually not introduced the way this book introduces it. It is only after covering several chapters of fundamentals that most textbooks reveal that every group can be seen as a collection of actions, the way this book introduces them. Most textbooks introduce groups by stating the official definition of a group, Definition 4.2 on page 51. It uses very mathematical language, which has the advantage of being very precise but the disadvantage of being difficult to penetrate, especially for newcomers. Because that standard definition lends itself more easily to constructing multiplication tables than does our definition (Definition 1.9 on page 7), we can take advantage of our recent acquaintance with multiplication tables to ease its introduction.

We have been calling the objects in a group "actions" because our definition requires a group to be a collection of actions (with certain additional restrictions). But the standard definition of a group is not based on actions. So we need to use more general terminology; we'll call the objects in a group the ***elements*** of the group. This is a mathematical term; mathematicians call collections of objects "sets" and the things in them "elements."

Along with this terminology comes some useful notation. To say that N is one of the elements of the group V_4, the standard notation is

$$N \in V_4.$$

The symbol $\in$ looks like an "e" in reference to the word "element." This is pronounced "N is an element of V_4." This terminology and notation is used not only in Definition 4.2 but in the remainder of this book and throughout mathematics in general.

In addition to this new terminology, let us also revive some old terminology. The previous section taught us that multiplication tables show how to combine any two group elements into one. Mathematicians call a method for combining objects an "operation," a word familiar to high-school algebra students. Such students learn the "order of operations," that multiplication and division must be done before addition and subtraction. These are the four basic operations of arithmetic—each takes two numbers and combines them in some way to produce a new number (e.g., $2 + 3 = 5$ or $6.3 \div 7 = 0.9$). Because each of the operations mentioned here combines *two* things into one, we can more specifically call them ***binary operations.***

The combining of group elements, as depicted in a multiplication table, is also a binary operation. We say it is a binary operation *on* the elements of the group, meaning that it can combine any two of those elements to yield another one. We've been writing this binary operation using multiplicative notation (R combined with B written as RB). Other common ways to denote group operations include additive notation $R + B$ and other creative symbols such as $R * B$ and $R \cdot B$. As long as the writer is consistent, the particular symbol used matters little.

These two new terms, element and operation, are central to Definition 4.2, and knowing them better prepares us to understand that definition. Interested readers can find more formal definitions of set and operation in nearly any textbook on discrete mathematics, or in a more traditional introductory group theory textbook.

Every multiplication table depicts a binary operation; it shows it to us in full. That's what multiplication tables do—they have two axes (across and down), because the op-

4.4. The classic definition

	1	2	3	4
1	1	2	3	4
2	2	2	1	3
3	3	1	4	4
4	4	3	4	2

	1	2	3	4
1	1	2	3	4
2	2	4	1	3
3	3	1	4	2
4	4	3	2	1

Figure 4.8. Can you figure out which of these two multiplication tables depicts the binary operation for a group? To learn how, read Sections 4.4.1 and 4.4.2 and then try the exercises in Section 4.5.3 on page 55.

eration is binary. But because the multiplication tables we're studying depict *groups,* they will reflect the defining characteristics of a group from Definition 1.9. Not just any square grid with symbols in it is the multiplication table for a group. In fact, two distinct features distinguish those tables that depict groups from those that do not. Sections 4.4.1 and 4.4.2 cover those features, and enable us better to understand Definition 4.2. But if you like thinking ahead, try figuring out which of the two multiplication tables in Figure 4.8 depicts a group. (And what's wrong with the one that does not?)

4.4.1 Associativity

Recall the example from page 44 in which we simplified the sequence $R^3BR^2B^3$. Although we simplified it to one basic operation, R, we can actually write lots of less significant simplifications. Here are a few examples.

$$R^3BR^2B^3 = RBR^2B^3 \quad \text{(because } R^3B = RB\text{)}$$
$$R^3BR^2B^3 = R^3BB^3 \quad \text{(because } BR^2 = B\text{)}$$
$$R^3BR^2B^3 = R^3BB \quad \text{(because } R^2B^3 = B\text{)}$$

In the first simplification on the list, we're concentrating on the R^3B at the beginning of $R^3BR^2B^3$, and simplifying it alone. We might draw our attention to it with parentheses, like this.

$$(R^3B)R^2B^3 = (RB)R^2B^3$$

What is in parentheses on the left-hand side has been simplified to what is in parentheses on the right-hand side. Writing the parentheses didn't change the meaning of the equation at all; it just made it easier to understand. Parentheses are *permitted anywhere* but *required nowhere.*

Because group operations permit parentheses for clarity but neither require them nor have their meaning changed by parentheses, we say that group operations are *associative.* That is, you may *associate* any portion of a multiplied chain of elements together without changing the meaning. The two other examples above are rewritten parenthesized here.

$$R^3(BR^2)B^3 = R^3(B)B^3$$

$$R^3B(R^2B^3) = R^3B(B)$$

Not all binary operations are associative. For instance, consider subtraction of whole numbers.
$$4 - (2 - 5) \neq (4 - 2) - 5$$
Because the placement of parentheses in this chain of subtractions changes the result, subtraction is not associative. When subtracting, parentheses matter, unlike when adding, or when combining group elements.

Combining *actions* in sequence is always associative, and thus group operations are always associative. This is the first of two special properties every group operation must have; the second is discussed below. Together they form the heart of Definition 4.2. Several exercises at the end of this chapter will help you get to know how multiplication tables for groups display both associativity and inverses.

4.4.2 Inverses

Rule 1.6 in Definition 1.9 requires each group element to be reversible. That is, given any group element, you should also be able to find its opposite, which we call its *inverse*. If you perform an action and then perform its inverse, the end result is the same as if you had done nothing at all. In Rubik's Cube, for instance, the inverse of a clockwise turn for a face is a counterclockwise turn of the same face. Doing the one and then the other has no impact on the state of the cube; it is equivalent to having taken no action.

In order to write this principle down using algebraic notation, we need a symbol that means "no action." In the multiplication tables I made for V_4, I used N for "no action;" often group theory textbooks use 0, 1, or e instead. This special element of a group is called the *identity* element. In this book I will usually write e for the identity. (Do not confuse this use of the lowercase letter e with its other common use in mathematics. The Euler number $e = \lim_{n \to \infty} \left(1 + \frac{1}{n}\right)^n \approx 2.271828$ is used for computing compound interest, radioactive decay, etc.)

Having this symbol enables us to write equations saying that every element of a group has an inverse. As an example, consider the group S_3, whose multiplication table appears in Figure 4.7. Using the notation in that multiplication table, with e as the identity, we can write equations showing each element of S_3 has an inverse. Here are six equations, one for each element of S_3. The first equation, for example, shows that r can be undone by r^2—so r^2 is the inverse of r.

$$r(r^2) = e$$
$$(r^2)r = e$$
$$ff = e$$
$$(fr)(fr) = e$$
$$(fr^2)(fr^2) = e$$
$$ee = e$$

In this group, elements such as f and fr are their own inverses, but two are not, r and r^2. Instead, they are one another's inverses. All of these equations could have been written without parentheses, due to associativity, but I chose to add them for clarity.

4.4. The classic definition

The group V_4 is a more unusual example, because every element in it is its own inverse. (Think of the rectangle puzzle—any flip twice changes nothing.)

$$NN = N$$
$$RR = N$$
$$BB = N$$
$$(RB)(RB) = N$$

The fact that every element of a group has an inverse is the second of two key facts in Definition 4.2, the first being associativity, covered in the previous section. This chapter's exercises will also give you a chance to examine the impact of this fact on the appearance of multiplication tables. But now, we are ready to read and analyze the classic definition of a group.

4.4.3 Putting it all together

Definition 4.2 (group). A set G is a *group* if the following criteria are satisfied.

1. There is a binary operation $*$ on G.

2. That operation is associative — i.e., for any elements $a, b, c \in G$, the equation $(a * b) * c = a * (b * c)$ is true.

3. There is an identity element $e \in G$. That is, for any other element $g \in G$, the equations $e * g = g$ and $g * e = g$ are true.

4. Every element $g \in G$ has an inverse, written g^{-1}, and the equations $g * g^{-1} = e$ and $g^{-1} * g = e$ are true.

Part (1) of this definition essentially says "a group is a multiplication table." Part (2) restricts attention only to those multiplication tables that display associative operations, as explained in Section 4.4.1. Part (4) says that every element has an inverse (as in Section 4.4.2) and it depends upon the notion of an identity element (also from Section 4.4.2), required by Part (3).

Notice that a new notation has been introduced—g^{-1} means the inverse of the group element g. So, for instance, in V_4, we could write

$$R^{-1} = R,$$

and in S_3, we could write

$$r^{-1} = r^2.$$

In any group, $e^{-1} = e$. Can you see why the definition makes this true?

The discussions leading up to this point have made it clear that Definitions 1.9 and 4.2 are very related. But are they really the same? That is, if Definition 1.9 says that something is a group, will Definition 4.2 agree? Or vice versa? The previous two sections constitute an informal argument for why the answer to the first question must be yes. But we will return to this and definitively answer the second question also when we cover permutation groups in Chapter 5.

4.4.4 Past, Present, Future

This chapter has provided an important new outlook on our subject matter. We can now think of groups using either of two equally correct paradigms, as a collection of actions or as a set with a binary operation. Each has its own advantages, and sometimes a group theoretic fact is more clear from one viewpoint than from the other. Thus it is good to have both at our disposal as we proceed into the next few chapters, which get into more serious study of groups and unearth some of the famous theorems of this branch of mathematics. But at present, it is time to exercise the new viewpoint on group theory this chapter has given us.

4.5 Exercises

4.5.1 Basics

Exercise 4.1. Consider the lightswitch group shown in Figure 2.8. Let L stand for the action of flipping the left switch and R stand for the action of flipping the right switch.

(a) Which of the following equations are true and which are false in this group?

$$LRRRR = RRL \qquad L = RR$$
$$LR = RLRLRL \qquad R^8 = R^{100}$$

(b) Let N stand for the non-action (leaving the switches untouched). Which of the following equations are true?

$$(LNR)^2 = LNR \qquad NN = N$$
$$RL = N \qquad R^4 = N$$
$$(LNR)^3 = R^3 L^3 \qquad LRLR = N$$

(c) What is the smallest positive power of R that equals N?

(d) What is the smallest positive power of L that equals N?

(e) What is the smallest positive power of RL that equals N?

(f) What is the smallest positive power of LR that equals N?

Exercise 4.2.

(a) Apply the transformation in Definition 4.1 to the lightswitch Cayley diagram in Figure 2.8.

(b) Create a multiplication table for the lightswitch group.

Exercise 4.3. Each part below describes a set with a binary operation on it. For each one, determine whether it is commutative and whether it is associative.

(a) the addition operation on the set of all whole numbers

(b) the subtraction operation on the set of all whole numbers

4.5. Exercises

(c) the multiplication operation on the set of positive real numbers

(d) the division operation on the set of positive real numbers

(e) the exponentiation operation on the set of positive whole numbers (that is, the operation written a^b, but typed $a \wedge b$ on many calculators)

4.5.2 Creating tables

Exercise 4.4. The Cayley diagrams for two groups are shown here, the cyclic group C_5 on the left and the Quaternion group Q_4 on the right.

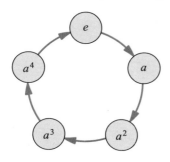
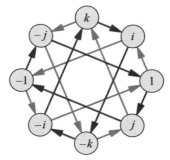

(a) The red arrow in the diagram for C_5 represents multiplication by what element?

(b) What is $a^3 \cdot a$ in C_5?

(c) What is $a^3 \cdot a \cdot a$ in C_5?

(d) If 1 is the identity element, then what do red arrows in the diagram for Q_4 represent? What do the blue arrows represent?

(e) What is i^2? What is $j \cdot i$?

(f) What is $i \cdot j \cdot j$?

Exercise 4.5. Using the Cayley diagrams from Exercise 4.4, answer the following questions.

(a) How do you use the diagram of C_5 to multiply $x \cdot a^2$ in C_5, for any element x?

(b) How do you use the diagram of Q_4 to multiply $x \cdot k$ in Q_4, for any element x?

Exercise 4.6. Create a multiplication table for each of the following Cayley diagrams.

(a) C_5, as shown on the left of Exercise 4.4. Use the template given here.

	e	a	a^2	a^3	a^4
e					
a					
a^2					
a^3					
a^4					

(b) Q_4, the quaternion group with eight elements, as shown on the right of Exercise 4.5. Use the template given here.

	1	i	j	k	-1	$-i$	$-j$	$-k$
1								
i								
j								
k								
-1								
$-i$								
$-j$								
$-k$								

(c) A_4, the alternating group with twelve elements:

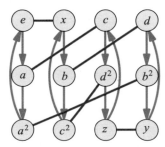

Recall that lines without arrowheads are bidirectional—it is as if it has arrowheads going both ways. Make your own table, using e as the identity.

Exercise 4.7. It is possible to suggest the full multiplication table for an infinite group by showing just part of it. Fill in the following partial table for the operation of addition on the set of all whole numbers; the ellipses indicate the table continues infinitely in all directions.

		-3	-2	-1	0	1	2	3	$\cdots$
$\vdots$	$\ddots$				$\vdots$				$\iddots$
-3					-3				
-2					-2				
-1					-1				
0	$\cdots$	-3	-2	-1	0	1	2	3	$\cdots$
1					1				
2					2				
3					3				
$\vdots$	$\iddots$				$\vdots$				$\ddots$

4.5. Exercises

Exercise 4.8. Exercises 2.4 through 2.8 of Chapter 2 asked you to draw Cayley diagrams for three groups. Use the diagrams you drew to make multiplication tables for those same groups. Note that if your diagram is not yet a diagram of actions, you may need to apply the transformation in Definition 4.1.

Exercise 4.9. Exercises 2.18 and 2.19 of Chapter 2 asked you to find the pattern describing the sequence of Cayley diagrams for the "n-gon puzzle." I mentioned in that exercise that the family of groups describing such puzzles are called the *dihedral groups*. You will study them in detail in Chapter 5, and this exercise previews some of that material.

Find the pattern describing the sequence of multiplication tables for those same groups. You might consider the following steps.

(a) Create multiplication tables from the Cayley diagrams for triangle, square, and regular pentagon puzzles.

(b) Discern a pattern and describe it. This will be your conjecture about n-gons for $n > 5$.

(c) Return to the group library in *Group Explorer* and investigate the same groups as in Exercise 2.19. (If you do not remember which ones they are, you can use the search feature again and look up "triangle," "square," etc.)

   (i) Do your multiplication tables from part (a) agree with those in *Group Explorer*? Note that *Group Explorer* may have used different names for the elements of the group than you did, and may have chosen a different order for the rows and columns, but the *pattern* in the table should be the same. You can rename elements and reorder the rows and columns in *Group Explorer*'s multiplication tables to help with the comparison.

   (ii) Does your conjecture about the pattern of multiplication tables for n-gons with $n > 5$ hold up against all the data you can find in *Group Explorer*? (This includes groups other than the three you just examined.)

(d) Can you give a convincing reason why the conjecture you made ought to be true?

4.5.3 Almost tables

Exercise 4.10. Consider the following multiplication table that displays a binary operation.

	e	A	B
e	e	A	B
A	A	e	e
B	B	e	e

(a) Explain succinctly why the binary operation is not associative. Can you write your answer as one equation?

(b) Does the operation have inverses?

Exercise 4.11. Consider the following multiplication table that displays a binary operation.

	3	2	1
3	3	2	1
2	2	2	1
1	1	1	1

(a) Explain succinctly why the binary operation does not have inverses. Can you write your answer as one equation?

(b) Is the operation associative?

Exercise 4.12. Consider the following multiplication table that displays a binary operation.

	1	x	y
1	1	x	y
x	x	y	y
y	y	y	y

(a) Does this operation have inverses? Justify your answer.

(b) Is the operation associative? Justify your answer.

Exercise 4.13. For each multiplication table below, explain why it does *not* depict a group.

(a)

	e	1	2	3	4
e	e	1	2	3	4
1	1	4	e	3	2
2	2	3	4	e	1
3	3	e	1	4	3
4	4	2	1	2	e

(b)

	e	a	b	c
e	e	a	b	c
a	a	a	b	c
b	b	a	b	c
c	c	a	b	c

(c)

	e	a	b	c
e	e	a	b	c
a	a	e	b	a
b	b	c	e	b
c	c	c	a	e

(d)

	e	f	g
e	f	e	g
f	e	g	f
g	g	f	e

Exercise 4.14. The following multiplication table does not depict a binary operation on the set $\{e, x, y\}$. The reason is part of the definition of a binary operation; we would say that this binary operation lacks *closure*. Can you spot the problem and explain it in your own words?

4.5. Exercises

	e	x	y
e	e	x	y
x	x	y	s
y	y	s	x

Exercise 4.15. Why can the same element not appear twice in any row of a group's multiplication table? Does this restriction also apply to columns?

Exercise 4.16. Exercises 2.14 through 2.17 on page 24 asked you to translate the rules from Definition 1.9 into criteria about diagrams. The goal was to create criteria for judging whether a diagram was a Cayley diagram—i.e., whether it represented a group.

This exercise will show that the answers to Exercises 2.14 through 2.17 are not sufficiently restrictive. That is, there are diagrams that satisfy those criteria but are not Cayley diagrams for any group. Consider the two diagrams shown below, neither of which is a valid Cayley diagram.

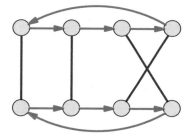
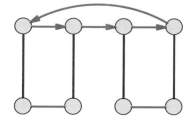

(a) Does each diagram meet all the criteria that were your answers to Exercises 2.14 through 2.17?

(b) Try to convert each of these diagrams into a multiplication table. What problem arises in each case?

(c) Can you explain what is wrong with these diagrams that caused the problems you encountered in part (b)?

(d) Create another diagram satisfying the criteria of Exercises 2.14 through Exercises 2.17, yet having the same problem as the two diagrams above.

This exercise shows a slight discrepancy between the actions-based definition of group (Definition 1.9, illustrated using Cayley diagrams) and my algebraic one (Definition 4.2, illustrated using multiplication tables). This discrepancy will be cleared up in Section 6.1.

Exercise 4.17. Explain why a Cayley diagram must be connected. That is, why must there be a path from every node to every other node?

Exercise 4.18. When creating a multiplication table for a group, if you try to include two different identity elements, what goes wrong? What does this lead you to conclude about groups?

4.5.4 Small groups

Exercise 4.19. Complete each of the following multiplication tables so that it depicts a group. There is only one way to do so, if we require that 0 be the identity element in each table. Then search *Group Explorer*'s group library to determine the names for the groups the tables represent.

(a)

	0	1
0		
1		

(b)

	0	1	2
0			
1			
2			

(c)

	0
0	

(d)

	0	1	2	3
0				
1		2		
2				
3				

(e)

	0	1	2	3
0				
1		3		
2				
3				

Exercise 4.20. The following table can be completed in more than one way, and still have the result depict a group. Find all possible such completions of the table, again using 0 as the identity element. How many did you find? Search *Group Explorer*'s group library to determine the names for the groups each of your resulting tables represents.

	0	1	2	3
0				
1		0		
2				
3				

Exercise 4.21. From Exercise 4.19 part (a) you can conclude that there is only one pattern for a group containing two elements. This is because the only difference between the multiplication table you computed and that of any other group with two elements will be the names of those elements. So the pattern of interactions among elements (or colors if we were to color the cells of the table) would be no different.

(a) How many patterns are there for groups containing three elements?

(b) Containing one element?

(c) Containing four elements?

Chapter 9 attacks the general question, "How many groups there are with n elements?"

4.5.5 Table patterns

These exercises preview Section 5.2, about an important family of groups called the *abelian* groups.

Exercise 4.22. We saw earlier in this chapter that in the group V_4, the equation $RB = BR$ is true. In fact, for any two elements $a, b \in V_4$, the equation $ab = ba$ is true. That is, the order in which you combine elements does not matter. Consider each group whose multiplication table appears in Figure 4.7 (except A_5, whose details are too small to see). For which of those groups does the order of combining elements matter?

Exercise 4.23. Groups in which the order of multiplication of elements does not matter are called *commutative* or *abelian*. Look through the groups in *Group Explorer*'s group library, starting with the smallest, until you find one that is noncommutative. What is the name of the smallest noncommutative group?

Exercise 4.24. What visual pattern do the multiplication tables of commutative groups exhibit?

4.5.6 Algebra

Exercise 4.25. To go along with the other algebraic notation we've seen in this chapter, there is also an algebraic notation for generators. For instance, the group C_5, which appears in the first few exercises of this chapter, is generated by the element a. The standard notation for this is $C_5 = \langle a \rangle$. The $\langle a \rangle$ means "what you can generate from a," and so the equation $C_5 = \langle a \rangle$ is saying "C_5 is the group generated from a." From Figure 4.3, we can write $V_4 = \langle R, B \rangle$, saying that R and B together generate V_4.

Show your understanding of this new notation by filling in the blanks below using however many elements are necessary to generate the group. Use as few elements as possible.

(a) From the Cayley diagram in Exercise 4.4, we see that $Q_4 = \langle$ _____ $\rangle$.

(b) From the Cayley diagram in part (c) of Exercise 4.6, we see that $A_4 = \langle$ _____ $\rangle$.

(c) There is more than one way to generate most groups. Find a different (yet still correct) answer to each of the previous two questions.

Exercise 4.26. Use the multiplication tables you constructed in Exercise 4.6 to determine the inverses for each element of each of the three groups from that problem.

(a) In the cyclic group C_5, the inverses are

$$e^{-1} = \underline{},$$
$$a^{-1} = \underline{},$$
$$(a^2)^{-1} = \underline{},$$
$$(a^3)^{-1} = \underline{}, \quad \text{and}$$
$$(a^4)^{-1} = \underline{}.$$

(b) In the quaternion group Q_4, the inverses are

$$1^{-1} = \underline{}, \qquad i^{-1} = \underline{},$$
$$j^{-1} = \underline{}, \qquad k^{-1} = \underline{},$$
$$(-1)^{-1} = \underline{}, \qquad (-i)^{-1} = \underline{},$$
$$(-j)^{-1} = \underline{}, \quad \text{and} \quad (-k)^{-1} = \underline{}.$$

(c) In the alternating group A_4, the inverses are

$$e^{-1} = \underline{}, \qquad a_1^{-1} = \underline{}, \qquad c_1^{-1} = \underline{},$$
$$x^{-1} = \underline{}, \qquad a_2^{-1} = \underline{}, \qquad c_2^{-1} = \underline{},$$
$$y^{-1} = \underline{}, \qquad b_1^{-1} = \underline{}, \qquad d_1^{-1} = \underline{},$$
$$z^{-1} = \underline{}, \qquad b_2^{-1} = \underline{}, \quad \text{and} \quad d_2^{-1} = \underline{}.$$

(d) In general, how do you use a multiplication table to find an element's inverse?

4.5. Exercises

Exercise 4.27. Inverses can be used to solve equations. In the group C_5, to solve $a^2x = a$ for x, I can proceed as in high school algebra:

$$a^2x = a$$
$$(a^2)^{-1}a^2x = (a^2)^{-1}a \quad \text{multiply both sides by } (a^2)^{-1}$$
$$x = (a^2)^{-1}a \quad \text{cancel } a^2 \text{ with its inverse}$$
$$x = a^4 \quad \text{use Exercises 4.6 and 4.26 to compute the answer}$$

Computing $(a^2)^{-1}a$ in C_5 gives $x = a^4$.

Try solving each of these equations in C_5.

(a) $a^3x = a^2$

(b) $a^4a^2x = a$

(c) $ax(a^3)^{-1} = e$

Exercise 4.28.

(a) If I have the equation $a^2x(a^2)^{-1} = a$ to solve as in the previous exercise, can I cancel the a^2 and the $(a^2)^{-1}$? Why or why not? (Hint: Is the result you get by canceling actually a solution to the equation?)

(b) If I have a similar equation, but in the group Q_4 from Exercise 4.6, $ixi^{-1} = j$, can I cancel the i and i^{-1}? Why or why not?

(c) Your answers to parts (a) and (b) should be different. What makes them different? Hint: Apply what you learned from the exercises in Section 4.5.5.

Exercise 4.29. Consider the equation $b_1 \cdot t \cdot a_2 = y$ in the group A_4; I want to solve for t. The previous exercise is a warning that I cannot simply proceed as follows.

$$\cancel{b_1^{-1}} \cdot \cancel{a_2^{-1}} \cdot \cancel{b_1} \cdot t \cdot \cancel{a_2} = b_1^{-1} \cdot a_2^{-1} \cdot y$$
$$t = b_1^{-1} \cdot a_2^{-1} \cdot y = x$$

What should I do instead?

Exercise 4.30. Solve these equations for t.

(a) In Q_4, $jitk^{-1} = -kj$.

(b) In A_4, $t(b_2)^2 = xyz$.

(c) In S_3, $rtf = e$. (See Figure 4.7 for a multiplication table for S_3.)

Exercise 4.31. Let's say you have a group G with identity element e. Take any three elements a, b, and c in G.

(a) What does the equation $ab = e$ say about the relationship between a and b?

(b) If both $ab = e$ and $ac = e$, can you use algebra to show that $b = c$?

(c) Can an element in a group have two different inverses?

Exercise 4.32. The set of integers (all positive and negative whole numbers, and zero) is often written as $\mathbb{Z}$. Use Definition 4.2 to answer each of the following questions about $\mathbb{Z}$.

(a) Is it a group using ordinary addition as the operation?

(b) Is it a group using ordinary multiplication as the operation?

(c) Are the even integers a group using ordinary addition as the operation?

(d) The even integers are sometimes written $2\mathbb{Z}$, because they can be obtained by multiplying every integer by 2. If we think of $3\mathbb{Z}$, $4\mathbb{Z}$, and in general any $n\mathbb{Z}$ in the same way, for what integers n is the set $n\mathbb{Z}$ a group using ordinary addition as the operation?

Exercise 4.33. The rational numbers (often written $\mathbb{Q}$) are the set of fractions $\frac{a}{b}$, where a and b are integers (but $b \neq 0$). For example, $\frac{1}{2}$, $\frac{-6}{11}$, and $\frac{50}{3}$ are all rational. Any integer, including zero, is rational, because you can just divide it by 1. For example, 10 is the rational number $\frac{10}{1}$.

Use Definition 4.2 to answer each of the following questions about $\mathbb{Q}$.

(a) Is it a group using ordinary addition as the operation?

(b) Is it a group using ordinary multiplication as the operation?

(c) Call $\mathbb{Q}^+$ the positive rational numbers (only those greater than zero). Is $\mathbb{Q}^+$ a group under ordinary addition?

(d) Is $\mathbb{Q}^+$ a group under ordinary multiplication?

(e) Call $\mathbb{Q}^*$ the nonzero rational numbers (all positive and negative ones, only leaving out zero). Is $\mathbb{Q}^*$ a group under ordinary addition?

(f) Is $\mathbb{Q}^*$ a group under ordinary multiplication?

(g) Why are groups like $\mathbb{Q}$, $\mathbb{Q}^+$, and $\mathbb{Q}^*$ difficult to visualize using multiplication tables and Cayley diagrams?

5
Five families

We have learned two powerful group theory visualization techniques. Cayley diagrams show us groups as collections of actions, and multiplication tables show them to us as binary operations. This chapter uses both to give a well-rounded introduction to five famous families of groups. Along the way a variety of new concepts will also arise.

We will begin by meeting the cyclic groups, for many reasons the perfect place to start our tour. Not only are cyclic groups the simplest kind of symmetry groups, but meeting them first will make the rest of the chapter clearer: Cyclic groups show up in all other groups as what we call *orbits,* and we will learn a new visualization technique based on orbits, called a *cycle graph.* Cyclic groups will also help us understand the groups we meet thereafter, the abelian groups and the dihedral groups. Both these families of groups can be built by combining cyclic groups using product operations, which we will preview in this chapter and study fully in Chapter 7.

Last, we will meet the symmetric and alternating groups, which will show the connection between groups and the idea of rearranging a set of items. Many examples and exercises in this book have dealt with rearrangements of things, which mathematicians call *permutations.* The chapter culminates with two beautiful results from group theory: the symmetries of the five Platonic solids and an illustrated proof of Cayley's theorem, as promised at the end of Chapter 4.

In some ways, this chapter is a gateway from introductory material to deeper material. Of course, everything in it (including the proof of Cayley's theorem) is done visually.

To see how familiarity with the five families this chapter introduces will pay off, take a moment to briefly consider each of the following questions.

(1) Is there more than one group with ten elements?

(2) Is there a group in which no element (other than the identity) is its own inverse?

(3) Do all groups describe the symmetries of some object?

Think how helpful it would be, when trying to answer such questions, to have a large, well-organized library of groups in your memory. The tour this chapter gives exposes you to many new groups, organizes them into memorable categories, and highlights their key features visually.

5.1 Cyclic groups

5.1.1 Things that spin

Chapter 3 taught us how to describe the symmetries of physical objects using groups. The most basic family of groups, the cyclic groups, describe objects that have only rotational symmetry. The symmetry group for the molecule of Boric acid we analyzed (Figure 3.6, page 29) is a cycle of three actions. What makes cycles simple is that they only have one kind of action. There's only one thing you can do to the molecule $B(OH)_3$ that respects its symmetry—rotate it. You can rotate it backwards or rotate it several times, but those are consequences of one fundamental rotation, in this case of 120 degrees clockwise.

The group of symmetries of the Boric acid molecule has three elements because the molecule has three arms. Similar shapes with more arms are described by cyclic groups with more elements. There may not be molecules with these shapes, but we can imagine them. Also, some everyday objects have only rotational symmetry—consider the pinwheels and propellers in Figure 5.1. Notice the similarities between the simple analysis of the propeller's symmetries in Figure 5.2 and that of the Boric acid molecule in Figures 3.4 and 3.5.

Figure 5.1. A six-bladed propeller and an eight-foiled pinwheel, whose symmetry groups are the cyclic groups with six and eight elements, respectively.

Because the cyclic group describing the $B(OH)_3$ molecule has three elements, it is commonly called C_3. The groups of symmetries for the propeller and the pinwheel in Figure 5.1 would therefore be C_6 and C_8, respectively. You may hear C_3 called "the cyclic group of order 3," because *order* is the group theory term for size, or number of elements. When speaking of the whole family of cyclic groups, or of a nonspecific member of it, it is common to write C_n, meaning that n could stand for any positive whole number.

The most common way to name the elements in C_n is to call the identity 0, the clockwise rotation 1, the clockwise rotation done twice 2, and so on, up to $n - 1$. We do not have an action called n in C_n because an n-bladed propeller rotated n times returns to its original position, so n rotations are really no different than zero rotations. Each number therefore signifies how many times to repeat the one fundamental action, named 1.

5.1. Cyclic groups

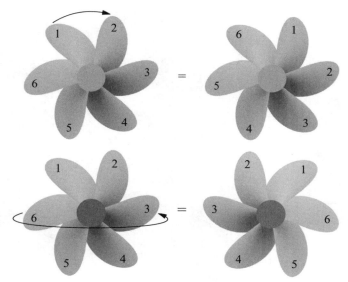

Figure 5.2. Propellers have only rotational symmetry. Rotating them clockwise (as shown in the top row) keeps them occupying the same space, but flipping them over (as in the bottom row) does not. The lower right propeller rotates backwards compared to the other three, and thus doesn't sit the same in space.

Figure 5.3 shows Cayley diagrams for C_3, C_5, and C_n using this naming convention. The rightmost diagram in Figure 5.3 makes use of an ellipsis to show that the Cayley diagram for C_n is always a cycle for any value of n.

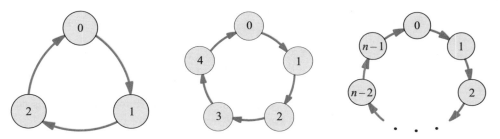

Figure 5.3. Cayley diagrams for the cyclic groups C_3, C_5, and C_n

5.1.2 Multiplication tables and modular addition

You may also see a cyclic group C_n referred to as $\mathbb{Z}_n$. This alternate naming convention comes from the common mathematical use of $\mathbb{Z}$ to refer to the integers. Since we think of C_n as the first n nonnegative integers, many books and papers call it $\mathbb{Z}_n$ instead.

In Exercise 4.6, you created a multiplication table from the Cayley diagram of C_5, also known as $\mathbb{Z}_5$. Using numbers as names for actions, as described above, that multiplication table becomes the left one in Figure 5.4. Although much of that table looks like ordinary addition (e.g., $0+3 = 3$ and $2+2 = 4$), not all of it does (e.g., $2+3 = 0$ and $3+3 = 1$).

The official name for the operation in C_5 is **modular addition,** or less formally, **clock addition.** It works just the way we count on a clock—because the numbers wrap around

	0	1	2	3	4
0	0	1	2	3	4
1	1	2	3	4	0
2	2	3	4	0	1
3	3	4	0	1	2
4	4	0	1	2	3

	0	1	2	3	4	5	6	7	8	9	10	11	12	13	14
0	0	1	2	3	4	5	6	7	8	9	10	11	12	13	14
1	1	2	3	4	5	6	7	8	9	10	11	12	13	14	0
2	2	3	4	5	6	7	8	9	10	11	12	13	14	0	1
3	3	4	5	6	7	8	9	10	11	12	13	14	0	1	2
4	4	5	6	7	8	9	10	11	12	13	14	0	1	2	3
5	5	6	7	8	9	10	11	12	13	14	0	1	2	3	4
6	6	7	8	9	10	11	12	13	14	0	1	2	3	4	5
7	7	8	9	10	11	12	13	14	0	1	2	3	4	5	6
8	8	9	10	11	12	13	14	0	1	2	3	4	5	6	7
9	9	10	11	12	13	14	0	1	2	3	4	5	6	7	8
10	10	11	12	13	14	0	1	2	3	4	5	6	7	8	9
11	11	12	13	14	0	1	2	3	4	5	6	7	8	9	10
12	12	13	14	0	1	2	3	4	5	6	7	8	9	10	11
13	13	14	0	1	2	3	4	5	6	7	8	9	10	11	12
14	14	0	1	2	3	4	5	6	7	8	9	10	11	12	13

Figure 5.4. Multiplication tables for the groups C_5 and C_{15}

in a cycle, so does the addition. If at 10:00pm, you start to watch a three-hour movie, you wouldn't say it will end at 13:00pm, but rather at 1:00am. That's how 10 and 3 are added *modulo* 12, or *mod* 12. Counting past 12 does not lead to 13, but rather back to 1. The only difference between a typical wall clock and the Cayley diagram for C_{12} is that the number at the top will be 12 on the clock, but 0 in the Cayley diagram. This difference is not significant, but just a matter of convention.

Similarly, in C_5, when we add $3 + 4$, we cannot get 7 because C_5 contains only 0, 1, 2, 3, and 4. Instead, we count 4 steps from 3, because that's the meaning of $3 + 4$, wrapping around the "clock" when necessary, $3 \to 4 \to 0 \to 1 \to 2$. To be sure this makes sense, try adding 3 and 2 in C_5; your result should be 0. If your mind's eye is not used to envisioning five-hour clocks, use the one in Figure 5.3 for convenience.

Modular addition has a nice visual effect on multiplication tables; each row is different from the row above it only by being cycled one cell to the left. In colored multiplication tables, this extends the color spectrum in the top row diagonally through the table, creating a rainbow effect; see the table for C_{15} on the right of Figure 5.4. That table also exemplifies the general pattern for multiplication tables for cyclic groups. Notice the diagonal strip of 14s, from the lower left corner to the top right. Above that strip, the table obeys the laws of ordinary addition, because none of the sums exceeds 14. Below that strip, addition mod 15 shows its colors. Figure 5.5 illustrates this pattern in the general case, C_n.

5.1.3 Orbits

We started our study of families of groups with C_n partly because of its simplicity and partly because cyclic groups are fundamental to the study of group theory in several ways. Shortly we will meet the abelian groups, and thereafter the dihedral groups, which can

5.1. Cyclic groups

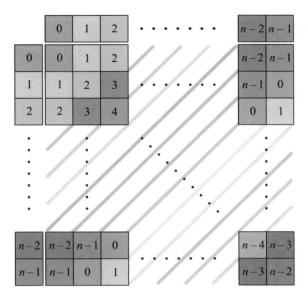

Figure 5.5. The pattern of multiplication tables for cyclic groups. The ellipses indicate linear sequences and the colored diagonal lines connect strips that are numbered (and therefore colored) the same.

be built up in straightforward, regular ways from cyclic groups. But first, let's learn how to find cyclic groups inside all other groups.

Look back at the first Cayley diagrams in this book, in Figure 2.10 on page 22. In the top right diagram, of S_3, consider just the blue arrows, which represent the action r. Starting at the identity e, the blue arrows lead in a cycle around the outside of the diagram, tracing out a copy of C_3 inside S_3. The standard term for this cycle is the *orbit* of the element r. Orbits are usually written with braces to indicate that the elements are to be treated together as a set; this one is therefore $\{e, r, r^2\}$.

Every element in a group can trace out an orbit. In the same diagram for S_3, imagine starting from e and repeatedly following the f arrows. A shorter orbit is generated, $\{e, f\}$, because $f^2 = e$.

The orbits of r and f we can clearly see in the diagram as it is shown in Figure 2.10; r's orbit is the outermost ring and f's is a single vertical two-way arrow. But other orbits exist. For instance, we can compute the orbit of fr^2 by doing just what its name suggests—follow the f arrow and then the r arrow twice—see where you end up, and repeat as necessary. Trace out this path in the diagram; you should find that the orbit of fr^2 is $\{e, fr^2\}$. So the orbit of fr^2 is not visually obvious in the diagram of S_3 in Figure 2.10, but it is an orbit nonetheless. Chapter 6 will show that there are many ways to structure a Cayley diagram, and S_3 can be represented in a way that makes the orbit of fr^2 obvious.

Even those orbits that are clear in a Cayley diagram do not always look circular. Consider the diagram of $C_3 \times C_3$, also in Figure 2.10. The orbit of (e, a) is the three-element cycle along the top of the diagram. Though it is a copy of the group C_3, in the diagram it is laid out linearly, not circularly.

Element	Orbit
r	$\{e, r, r^2\}$
r^2	$\{e, r^2, r\}$
f	$\{e, f\}$
fr	$\{e, fr\}$
fr^2	$\{e, fr^2\}$

Figure 5.6. The orbits for the group S_3, shown on the left in a table, and on the right as the group's cycle graph.

If we consider each element of S_3 in turn and compute its orbit, we come up with the list shown on the left of Figure 5.6. (Note that the sets $\{e, r, r^2\}$ and $\{e, r^2, r\}$ are the same; we are not concerned with the order in which the elements appear in an orbit.) We cannot compute an orbit for the identity element, since it has no arrow to lead us anywhere.

5.1.4 Introducing cycle graphs

From lists like the one in Figure 5.6, we can create a diagram connecting the elements that appear in an orbit together. Such diagrams show how the group is made up of the orbits of its elements. Since each orbit is a cycle, the resulting diagram is called a *cycle graph*. The cycle graph for S_3, constructed from the list in Figure 5.6, is shown in that same figure.

We will use cycle graphs throughout the rest of this chapter to see how the families of groups we meet are built on cycles. However, since cyclic groups themselves are comprised of only one cycle, their cycle graphs are rather unrevealing compared to the cycle graph of S_3. In fact, they look no different from the groups' Cayley diagrams, except that by convention cycle graphs do not use arrowheads. Compare the Cayley diagrams in Figure 5.3 with the cycle graphs for the same three groups, in Figure 5.7.

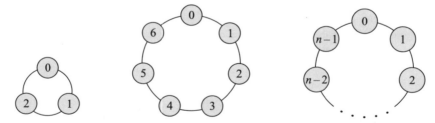

Figure 5.7. Cycle graphs for the cyclic groups C_3, C_5, and C_n, the same groups whose Cayley diagrams appear in Figure 5.3.

5.2 Abelian groups

Cyclic groups lead naturally to abelian groups in two ways. First, all cyclic groups are abelian, and the abelian groups retain some of the important simplicities that cyclic groups have. Second, we will see that there is a natural way to construct abelian groups by piecing together cyclic ones. This fact is called the Fundamental Theorem of Abelian Groups, which this chapter will preview, leaving the full details until Chapter 8.

5.2. Abelian groups

Named after Neils Abel, one of the founders of group theory, **abelian** groups are those in which the order in which one performs the actions is irrelevant. That is, if a and b are any two actions in an abelian group, then the action a followed by the action b yields the same result as b followed by a. The algebraic way to say this is that a *commutes with* b, and so abelian groups are often referred to as **commutative**. The equation $ab = ba$ says that a and b commute; a group is abelian when that equation is true no matter which two elements from the group a and b stand for.

5.2.1 Noncommutativity in Cayley diagrams

To see the impact of commutativity on the appearance of a Cayley diagram, let's imagine we have the Cayley diagram of some abelian group G. Say the diagram's red arrows represent multiplication by a and its blue arrows represent multiplication by b. (Recall the convention from Exercise 4.6 that arrows represent multiplication *on the right*.) In such a diagram, ab is the combination of a red arrow and then a blue one, in that order, while ba is the combination of a blue and then a red arrow. Commutativity stipulates that $ab = ba$, and so from any node in the Cayley diagram, if you follow a red arrow and then a blue one, you must arrive at the same point you would have if you had first followed the blue arrow and then the red one. Figure 5.8 illustrates this visually.

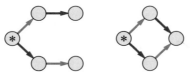

Figure 5.8. On the left is a pattern that never appears in Cayley diagrams for abelian groups: from the node marked ∗, following a red and then blue arrow does not reach the same node as following a blue and then red. The pattern on the right will always appear instead.

The patterns in Figure 5.8 give us an easy way to spot abelian groups using Cayley diagrams: Ensure that every pair of arrows leaving a node closes to a diamond shape, as on the right of Figure 5.8. Whether or not the pattern appears laid out exactly in a diamond shape is unimportant—it's the pattern of connections that matters. An example will make this more clear.

Figure 5.9 shows Cayley diagrams for two slightly different groups. Look carefully through the two diagrams in Figure 5.9 and see if you can find an instance of the left pattern from Figure 5.8 in either one. If you find an instance of that pattern, you know

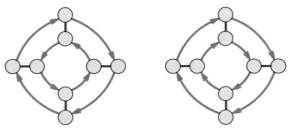

Figure 5.9. Cayley diagrams for two very similar groups, D_4 and $C_2 \times C_4$. Test these groups for commutativity by finding instances of the patterns in Figure 5.8.

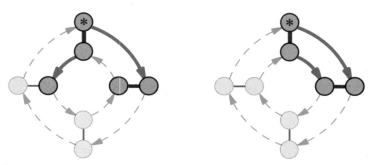

Figure 5.10. The diagrams from Figure 5.9 reappear here, highlighted to show instances of the patterns in Figure 5.8. The left diagram highlights an instance of the pattern that makes it non-abelian: if a stands for red and b for blue, then $ab \neq ba$ because from $*$, ab leads to a different destination than ba does. The right diagram highlights the analogous paths, but the pattern indicates commutativity.

the corresponding group is not abelian. Recall that connections without arrowheads in a Cayley diagram count as arrows in both directions. When you have finished, examine Figure 5.10 and compare your answer to the explanation in its caption.

The straight lines in Figure 5.8 become smooth curves in Figure 5.10, but the pattern of connections is the same. Note that the $*$ element could have been placed anywhere and similar patterns highlighted; Cayley diagrams embody a great deal of symmetry.[1] In Chapter 8 we will see how the visual patterns from Figure 5.8 lay the foundation for the Fundamental Theorem of Abelian Groups.

5.2.2 Commutative multiplication tables

Abelian groups are easier to spot from their multiplication tables than from their Cayley diagrams. Consider Figure 5.11, sketching a partial multiplication table, cut in half diagonally from top left to bottom right. Only the rows and columns for the elements a and b are shown, and consequently the cells for both ab and ba, which are directly across the diagonal dividing line from one another. The equation $ab = ba$ requires that these two cells must contain the same group element. That is, they must mirror one another, with the diagonal dividing line acting as the mirror.

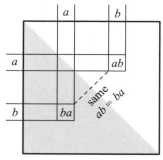

Figure 5.11. The equation $ab = ba$, which defines abelian groups, causes their multiplication tables to be symmetric across the diagonal shown here.

---

[1] We will define this symmetry specifically in the next chapter, specifically in Definition 6.1.

5.2. Abelian groups

	e	r	r^2	f	fr^2	fr
e	e	r	r^2	f	fr^2	fr
r	r	r^2	e	fr^2	fr	f
r^2	r^2	e	r	fr	f	fr^2
f	f	fr	fr^2	e	r^2	r
fr^2	fr^2	f	fr	r	e	r^2
fr	fr	fr^2	f	r^2	r	e

	(0,0)	(0,1)	(0,2)	(0,3)	(1,0)	(1,1)	(1,2)	(1,3)
(0,0)	(0,0)	(0,1)	(0,2)	(0,3)	(1,0)	(1,1)	(1,2)	(1,3)
(0,1)	(0,1)	(0,2)	(0,3)	(0,0)	(1,1)	(1,2)	(1,3)	(1,0)
(0,2)	(0,2)	(0,3)	(0,0)	(0,1)	(1,2)	(1,3)	(1,0)	(1,1)
(0,3)	(0,3)	(0,0)	(0,1)	(0,2)	(1,3)	(1,0)	(1,1)	(1,2)
(1,0)	(1,0)	(1,1)	(1,2)	(1,3)	(0,0)	(0,1)	(0,2)	(0,3)
(1,1)	(1,1)	(1,2)	(1,3)	(1,0)	(0,1)	(0,2)	(0,3)	(0,0)
(1,2)	(1,2)	(1,3)	(1,0)	(1,1)	(0,2)	(0,3)	(0,0)	(0,1)
(1,3)	(1,3)	(1,0)	(1,1)	(1,2)	(0,3)	(0,0)	(0,1)	(0,2)

Figure 5.12. The symmetry of the table on the right tells us it is abelian. The left table does not have the required diagonal symmetry.

Since this same pattern holds for any two elements a and b, the whole upper right of the table (the white triangle) is mirrored in the lower left (the gray triangle). If you folded the table in half along the diagonal dividing line, pairs of identical elements would touch. Consider the two example multiplication tables in Figure 5.12; it is easy to spot which of the two is abelian by looking for this kind of diagonal mirroring.

5.2.3 Intricate cycle graphs

Abelian groups that are not cyclic often have very interesting cycle graphs. Take a moment to inspect those in Figure 5.13. Let's take a closer look at the simplest of them, the top left one. It shows that $C_3 \times C_3$ is composed of four three-step orbits, and Figure 5.14 shows where in the Cayley diagram for $C_3 \times C_3$ we can find these orbits.

The top two orbits are fairly self-explanatory; they are the orbits of the red and blue arrows, respectively. The other two orbits were constructed from sequences of red and blue arrows, to visit nodes the first two orbits did not reach.

This cycle graph is an example of a general pattern. In Section 8.4 you will encounter a famous theorem that will enable you to prove that for any prime number p, the cycle graph of $C_p \times C_p$ will always have $p + 1$ orbits with p elements in each, all touching at the identity.

But now it is time to leave the gentle realm of commutative groups and dip into some of the rich complexity non-abelian groups can offer. The smallest step into that realm leads to the dihedral groups. You have already seen the dihedral group D_4 on the right of Figure 5.9, and from that figure you can tell that although D_4 is not abelian, it is very close.

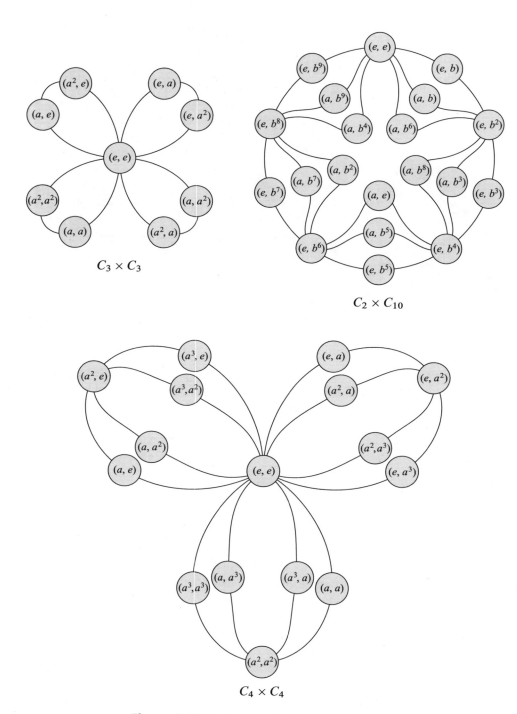

Figure 5.13. Cycle graphs for three abelian groups

5.2. Abelian groups

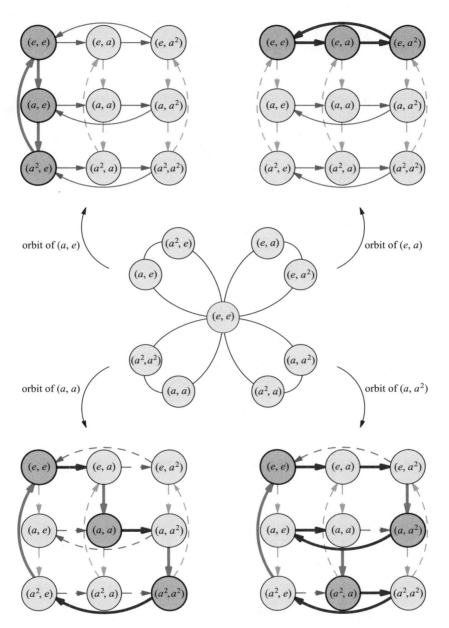

Figure 5.14. In the center is the cycle graph of $C_3 \times C_3$, and next to each of the four orbits is a copy of the Cayley diagram for $C_3 \times C_3$ with the corresponding orbit highlighted. These Cayley diagrams are structured like the original in Figure 2.10 on page 22.

5.3 Dihedral groups
5.3.1 Things that flip (and spin)

While cyclic groups describe objects that have only rotational symmetry, dihedral groups describe objects that have both rotational and bilateral symmetry. Objects have *bilateral symmetry* if they look the same when flipped over (usually in a specific direction, such as horizontally). Figure 5.2 showed that a propeller does not have bilateral symmetry—flipping it over changes the space it occupies. The easiest geometric examples of objects with both rotational and bilateral symmetry are regular polygons, several of which are shown in Figure 5.15. "Regular" means that all their sides and angles are equal.

Figure 5.15. Regular polygons, from three- through seven-sided: triangle, square, pentagon, hexagon, and heptagon.

A polygon with n sides is called an n-gon, and the dihedral group that describes the symmetries of a regular n-gon is written D_n. All the actions in C_n are also actions in D_n because rotating a regular polygon preserves the space it occupies. But because regular polygons also permit flipping, we expect there to be more elements in D_n than in C_n, and indeed there are. While C_n contains n actions, D_n contains twice as many, $2n$. Let us see what these n new actions do.

Recall that C_3 described the symmetries of the three-armed molecule of Boric acid, as in Figure 3.6. Let's compare this to the symmetries of a regular three-sided polygon, an equilateral triangle. (If you did Exercise 2.18 this will seem familiar. If not, you may want to create for yourself a numbered triangle to make it easier to follow along.) We've seen that rotations respect the shape of the Boric acid molecule or a propeller or a pinwheel, but horizontal flips do not (Figures 3.4, 3.5, and 5.2). Figure 5.16 shows that this changes with an equilateral triangle; both rotations and flips respect its shape.

If we use only the rotation from Figure 5.16, the resulting group will be the copy of C_3 we already know to be in D_3. If we call that clockwise rotation r, then our copy of C_3 is the orbit of r. Let us call the flip shown in Figure 5.16 f, and see what happens

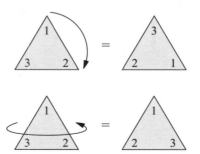

Figure 5.16. Both rotations and horizontal flips respect the shape of the equilateral triangle. The one shown here has its corners labeled to show the effects of those moves, as in Exercise 2.18 (page 24).

5.3. Dihedral groups

when we use it together with the orbit of r. From the starting position, if we flip the triangle over, we reach a new position, one outside the orbit of r. From this position, using r reaches two other new positions. These three new positions bring the total to six. Applying f to any of these new positions returns us to the orbit of r, and therefore the complete list of actions in D_3 is e, r, r^2, f, rf, and $r^2 f$. Indeed, this is the process you went through and carefully recorded in a Cayley diagram if you did Exercise 2.18.

5.3.2 Cayley diagrams for D_n

The leftmost Cayley diagram in Figure 5.17 shows D_3 as just described. If that diagram seems familiar (even apart from Exercise 2.18), it is because D_3 has already appeared in this book, but under the name S_3 (Figure 2.10, page 22). Both names are common, and you will learn the origin of the name S_3 at the end of this chapter.

Dihedral groups come in many sizes, but as a family they share important structural characteristics. Cayley diagrams for other dihedral groups resemble the one for D_3, as exemplified by the center diagram in Figure 5.17, showing D_5. The rightmost diagram in Figure 5.17 shows the pattern for an arbitrary D_n.

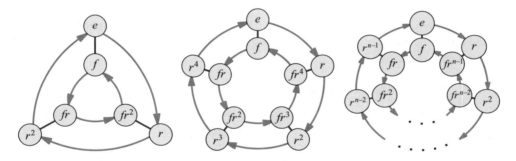

Figure 5.17. Cayley diagrams for the dihedral groups D_3 (left) and D_5 (center), as well as the general scheme for Cayley diagrams of D_n (right).

The outer ring in each of these diagrams is the orbit of r, and is a copy of the cyclic group C_n. The inner ring is a cycle of the same size, but pointing counterclockwise instead of clockwise. To see the reason for this reversal, imagine yourself holding a triangle in the initial configuration shown on the left of Figure 5.16. A friend stands facing you, and you hold the triangle up between the two of you, so that you see the front face and your friend sees the back. You can swap viewpoints with your friend by doing the f action to the triangle. When you see the triangle in its initial configuration, your friend sees it in its flipped configuration, the bottom right of Figure 5.16.

Your performing action r to the triangle will be different than your friend's doing it. When your friend turns the triangle clockwise, it will look counterclockwise to you. It is the same if you temporarily take the role of your friend by flipping the triangle, rotating it clockwise, and flipping back: the overall effect is a counterclockwise rotation. To say this algebraically, $frf = r^{-1}$.

The inner ring of the diagrams in Figure 5.17 represents your friend's point of view, the one you can take on by performing action f, and put off again the same way. Thus the

	e	r	r^2	r^3	r^4	f	fr	fr^2	fr^3	fr^4
e	e	r	r^2	r^3	r^4	f	fr	fr^2	fr^3	fr^4
r	r	r^2	r^3	r^4	e	fr^4	f	fr	fr^2	fr^3
r^2	r^2	r^3	r^4	e	r	fr^3	fr^4	f	fr	fr^2
r^3	r^3	r^4	e	r	r^2	fr^2	fr^3	fr^4	f	fr
r^4	r^4	e	r	r^2	r^3	fr	fr^2	fr^3	fr^4	f
f	f	fr	fr^2	fr^3	fr^4	e	r	r^2	r^3	r^4
fr	fr	fr^2	fr^3	fr^4	f	r^4	e	r	r^2	r^3
fr^2	fr^2	fr^3	fr^4	f	fr	r^3	r^4	e	r	r^2
fr^3	fr^3	fr^4	f	fr	fr^2	r^2	r^3	r^4	e	r
fr^4	fr^4	f	fr	fr^2	fr^3	r	r^2	r^3	r^4	e

Figure 5.18. Multiplication table for the dihedral group D_5

f actions connect the inner and outer rings. The r action rotates in the opposite direction in the inner ring because the inner ring corresponds to a viewpoint from the other side of the triangle, where clockwise and counterclockwise have been switched.

5.3.3 Multiplication tables for D_n

This separation of the elements of D_n into two categories—the inner and outer rings—is also visible in multiplication tables. Consider the table for D_5 in Figure 5.18. It is not difficult to mentally divide this table into quarters by splitting it in half horizontally and again vertically; in fact, the colors of the cells encourage you to do so.

This is because the headings in the table have been organized with those from the Cayley diagram's outer ring listed first (e through r^4) and its inner ring listed next (f through fr^4). This makes the upper left quarter of the table show only the interactions among members of the outer ring. It's a copy of the multiplication table of C_5, because the outer ring of the Cayley diagram is a copy of C_5.

This table will be analyzed further in the next section, and in Exercise 5.35.

5.3.4 A preview of Chapter 7

Consider the overall color pattern of Figure 5.18. The top left quarter uses the same colors as the bottom right, and the bottom left matches the top right. This color pattern clusters the elements that do not involve the flip f in two quadrants of the table, and the elements that do involve f in the other two quadrants. If we call them "flip" and "non-flip" elements for short, the clustering makes the following facts clear.

> Any non-flip times a non-flip is a non-flip.
> Any non-flip times a flip is a flip.
> Any flip times a non-flip is a flip.
> Any flip times a flip is a non-flip.

5.3. Dihedral groups

Figure 5.19. Multiplication table for D_5 with annotation that clusters elements according to whether they involve a flip. The resulting two-by-two pattern matches the multiplication table for the group of order two, shown on the right.

Annotating the table to illustrate this pattern results in Figure 5.19, a simple two-by-two table overlayed on the original ten-by-ten table. This two-by-two table is actually the multiplication table for the only group of order two.

The color pattern in the multiplication table for D_5 revealed within that table the structure of a smaller group, C_2. Shrinking a group this way is called taking a *quotient*, and we have only previewed it here. Group quotients and the complementary notion of group products are central to the study of groups, and Chapter 7 will cover both notions thoroughly. We'll learn that a dihedral group D_n can be obtained by combining the groups C_n and C_2 using what's called a *semidirect product*, and the quotient in Figure 5.19 is reversing that product operation.

5.3.5 Cycle graphs for D_n

Before we move beyond dihedral groups, let's examine their cycle graphs, to solidify our understanding of the roles of the elements in D_n. Figure 5.6 shows that the orbit of r in D_3 contains three elements, and each element outside that orbit is in a little orbit consisting only of itself and the identity. For larger dihedral groups than D_3, this pattern continues. D_n consists of an r orbit of size n, and n other elements each in an orbit of size two. Let us see why this is so.

Any element of D_n that's not in the orbit of r can be written as fr^m for some number m between 0 and $n-1$. (If you can't recall why this is so, look back at Figure 5.17.) I will explain why doing the action fr^m twice has the same result as doing nothing. Recall the description from Section 5.3.2 of you and a friend holding a triangle. That example taught us that performing the action frf is the same as performing the action r in reverse. It did so by having us realize that surrounding r with f's was like taking on the role of our friend, who faces the opposite direction. The same argument applies to surrounding r^m with f's,

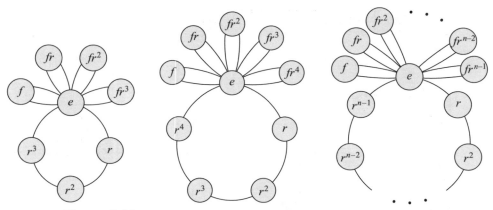

Figure 5.20. Cycle graphs for D_4, D_5, and the general pattern for D_n

and so $fr^m f$ is the same as r^m in reverse. Consequently, $fr^m f$ followed by r^m returns the triangle to its original position. Algebraically, we can either write it as $fr^m f = r^{-m}$ or as $fr^m fr^m = e$. Both signify that fr^m done twice returns to the original state.

This explanation did not depend on the fact that the regular polygon in question was a triangle; it could have been a pentagon, square, or any other regular polygon. So in any D_n, every element outside the orbit of r will have its own two-step orbit. This makes the cycle graph of D_n look like one ring of size n with many two-element orbits connected to it at the identity. Examples for D_4, D_5, and the general pattern for D_n appear in Figure 5.20.

Now it's time to move on from D_n, which has been our first exposure to a family of non-abelian groups. As you recall from Figure 5.9, the dihedral groups are structurally very close to being abelian, and are therefore not very complex. The following section opens the floodgates, letting in the full measure of the complexity of group theory.

5.4 Symmetric and alternating groups

Many groups we've seen in this book have been collections of ways to rearrange things. Chapter 1's exercises had you rearranging coins on a table and paintings on walls. It also talked about Rubik's Cube, which we can think of as rearranging colored stickers. In Chapter 3, we numbered the parts of symmetric objects so that manipulations of those objects could be analyzed as rearrangements of the numbers. Rearrangements, which mathematicians call *permutations,* are intimately tied to group theory in two ways. First, we will meet two important families of groups, the symmetric and alternating groups, that can be built from permutations in a natural way. Then we will see a visual proof of Cayley's theorem, which says that permutations are a powerful enough tool that from them we can build *any* group.

5.4.1 Permutations

A ***permutation*** is an action that rearranges a collection of things. Permutations can describe the shuffling of cards in a deck, a rearrangement of the letters in a word, or the way you sort the CDs on your shelf. Mathematicians usually refer to permutations of small positive integers, partly just to make permutations easy to write down.

5.4. Symmetric and alternating groups

Figure 5.21. Three example permutations of the numbers 1, 2, 3, and 4. For example, the leftmost one permutes the elements by moving each one step to the right, but moving 4 all the way to the left. The central permutation does not alter the positions of 1 or 4, but swaps the other two numbers.

Following this convention, let's assume we have four things we can rearrange, 1, 2, 3, and 4. A few example ways to permute these numbers are shown in Figure 5.21. There are many notations mathematicians use to represent permutations, but we will use the notation in Figure 5.21.

5.4.2 Permutation groups

Permutations cooperate well with the original requirements we put on groups, in Definition 1.9 on page 7. If we take all permutations of a certain collection of things (such as the numbers 1, 2, 3, and 4), we clearly have a predefined, unvarying list of actions (satisfying Rule 1.5), each of which is unambiguous (satisfying Rule 1.7). Furthermore, the work of any two permutations done in succession can just as easily be done by one permutation, the combination of the two (satisfying Rule 1.8). Figure 5.22 exemplifies how any two permutations can be combined. Lastly, every permutation is reversible (satisfying Rule 1.6), as Figure 5.23 shows by example. So there is a good reason why so many examples in this book have involved rearranging things: permutations are natural group-building tools.

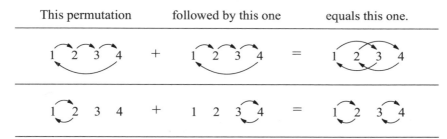

Figure 5.22. Two examples of combining two permutations to form a new one. In the top row, the first permutation moves 1 to 2, which the second permutation moves to 3, so their combination must move 1 to 3. In the bottom row, the first permutation moves 1 to 2, which the second permutation leaves untouched, so in combination, they move 1 to 2. Try following the paths of 2, 3, and 4 through each pair of combined permutations yourself.

For instance, Exercise 1.4 asked you to count all possible ways to rearrange three paintings on three walls. Exercise 2.5 built on Exercise 1.4, asking you to you draw a Cayley diagram for that group, and Exercise 4.8 asked you to make its multiplication table. In this chapter we've seen that this group has the name D_3, but in earlier chapters we called it by its more common name, S_3. The S stands for symmetric, because the group of all permutations of a given size is called a ***symmetric group.*** So S_n represents the group of all permutations of n things.

We've seen a diagram of S_3, but have not yet encountered other members of the symmetric group family. The smallest members of the family, S_1 and S_2, you will analyze yourself in Exercise 5.18, but the larger S_n groups are both complex and intriguing. Their

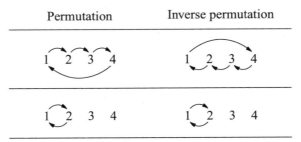

Figure 5.23. Two examples of permutations alongside their inverses. Try combining each permutation shown here with its inverse, in the way Figure 5.22 combines permutations. The result will always be the permutation that moves no numbers.

size increases quickly; the order of S_n is n factorial (written $n!$, meaning the product of all whole numbers from 1 to n). Thus S_4 has order $1 \cdot 2 \cdot 3 \cdot 4 = 24$ and S_5 has order $1 \cdot 2 \cdot 3 \cdot 4 \cdot 5 = 120$. Cayley diagrams for groups as large as S_5 can be tangled and hard to draw, but the Cayley diagram for S_4 is still small enough to admit a very pleasing arrangement, shown in Figure 5.24. As you will read in the next section, it is no coincidence that it looks cube-like.

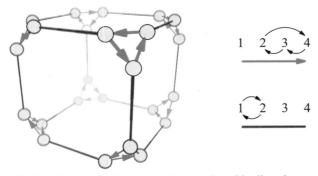

Figure 5.24. A Cayley diagram for S_4 arranged on a cube with clipped corners (usually called a "truncated cube"). The legend on the right shows which permutations in S_4 the arrows represent. Node labels were omitted to eliminate clutter.

Although the collection of all permutations of n items forms a group, creating a group of permutations does not *require* taking *all* permutations of a given size. It is often possible to form a group from just some of the permutations from S_n. One famous way is to take exactly half of the elements of S_n, creating what is called an **alternating group**. Not just any half will do; random choosing will rarely result in a group. But if you take every element in S_n and square it, that collection of squared elements will be exactly half of S_n, an alternating group called A_n. The creation of A_3 from S_3 is shown in Figure 5.25. As you recall from Exercise 4.21 there is only one structure for groups of order 3, so A_3 has the same structure as the cyclic group C_3.

5.4.3 The Platonic solids

There are only five three-dimensional shapes all of whose faces are regular polygons that meet at equal angles. We call them the Platonic solids, and you can see renderings of

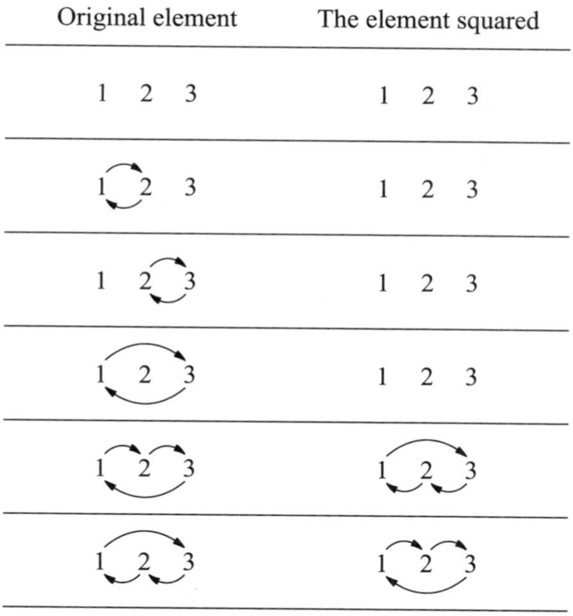

5.4. Symmetric and alternating groups

Original element	The element squared
1 2 3	1 2 3
1↔2 3	1 2 3
1 2↔3	1 2 3
1⌒3 2	1 2 3
1↔2↔3	1↔2↔3
1↔2↔3	1↔2↔3

Figure 5.25. The left column shows all the elements of S_3 as permutations, and the right column shows all their corresponding squares. Because four of the six elements square to the identity (the permutation that moves nothing around, the "no action"), the right column contains only three distinct elements. Those three elements comprise the group A_n.

them in Figure 5.26. I bring them up now to show their connection to the symmetric and alternating groups.

Recall the technique Chapter 3 taught for classifying the symmetries of any three-dimensional object. We applied it to molecules and dancers and other things, and we could apply it to the Platonic solids. Exercise 5.25 asks you to try this for the tetrahedron, and helps you out with a diagram. Applying the technique to the other Platonic solids, especially the icosahedron or dodecahedron, is a bookkeeping challenge. This work has already been done, and we will see the results.

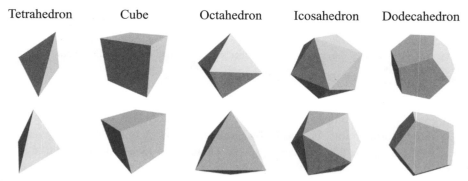

Figure 5.26. The five platonic solids, the only three-dimensional objects that are comprised of faces made of the same regular polygon repeated, connected to one another at the same angle throughout. Each solid is shown from two different angles, to help indicate depth.

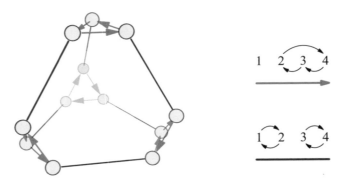

Figure 5.27. A Cayley diagram for A_4 arranged on a truncated tetrahedron. The legend on the right shows which permutations in A_4 the arrows represent. As in Figure 5.24 and Figures 5.28 and 5.29, node labels are omitted to eliminate clutter.

The symmetry group for the tetrahedron is A_4, half of all permutations of four items. We can lay out a Cayley diagram for A_4 in a way that makes its connection to the tetrahedron clear; see Figure 5.27. The cube and octahedron have the same symmetry group, S_4. This is why it is no coincidence that the Cayley diagram for S_4 in Figure 5.24 looked cube-like. It is possible to arrange the diagram like an octahedron by using arrows that refer to different permutations; compare Figures 5.24 and 5.28. The dodecahedron and icosahedron also share a symmetry group, A_5, of order 60. Its Cayley diagram can therefore be tailored to suggest either shape, as shown in Figure 5.29.

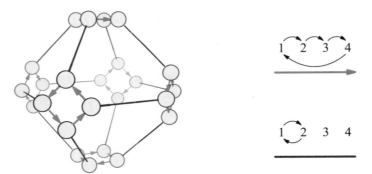

Figure 5.28. A Cayley diagram for S_4 arranged on a truncated octahedron. (Compare to Figure 5.24, a truncated cube.)

So groups of permutations are important in group theory. We saw that permutations can be used to form groups very easily, by simply taking all permutations of a given size. We called these groups the symmetric groups S_n, and then took exactly half of each S_n to form the alternating groups A_n. Then we saw that some members of the S_n and A_n families describe the symmetries of some beautiful shapes, the five Platonic solids. But beyond even all this, there remains a yet more significant way that permutations relate to group theory. Cayley's theorem states that relationship.

5.4. Symmetric and alternating groups

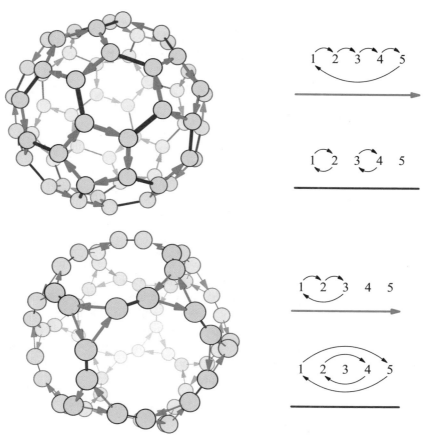

Figure 5.29. Two Cayley diagrams for A_5, one arranged on a truncated icosahedron and the other on a truncated dodecahedron. To make this possible, the arrows represent different elements in the one diagram than in the other.

5.4.4 Cayley's Theorem

At the beginning of Section 5.4, I stated that permutations can be used to construct any group. This statement effectively says that all of group theory can be found in permutations, a fact known as Cayley's theorem. As a first step toward proving this theorem, let's see how every Cayley diagram and multiplication table expresses a collection of permutations.

Consider the Cayley diagram for S_3 on the left of Figure 5.30. The nodes in the diagram have been numbered 1 through 6 to make it easy to talk about permutations of them using familiar notation. Each color of arrows in the diagram can be interpreted as a permutation: The red arrows move 1 to 2, 2 to 3, 3 to 1, 4 to 6, and so on. The permutation representing this is on the top right of Figure 5.30. The blue arrows in the diagram for S_3 interchange pairs of nodes, and the permutation representing this appears below the red one. To create these permutations from the diagram, I've just written the numbers from the diagram's nodes in order in a row, 1 to 6, and connected them with arrows the same way the diagram connects them.

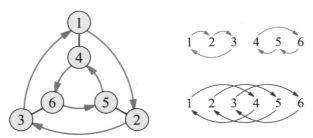

Figure 5.30. A Cayley diagram for the group S_3, with nodes numbered 1 through 6 to facilitate analyzing how the arrows permute the elements (shown to the right)

Cayley's theorem says that these two permutations embody the relationship among the six elements of S_3. That is, if we create a group from the two permutations in Figure 5.30, we will find that the combinations of those permutations with one another form a six element group with the same structure as S_3. Furthermore, the red permutation will behave in that group the same way that r behaves in S_3, and the blue permutation will behave like f in S_3. Two groups with the same structure are called *isomorphic,* a term we will use loosely for now, and study in Chapter 8. We have not yet proven any of this, but we are about to. First, let us see how it relates to multiplication tables.

In Figure 5.30 I created permutations from the arrows in a Cayley diagram, which represent right multiplication. What part of a multiplication table shows us right multiplication? Consider the multiplication table for the group V_4 in Figure 5.31, with elements renamed 1 through 4 to make it easier to relate to permutations.

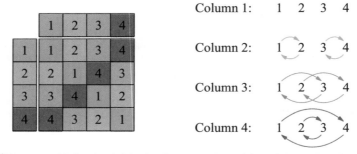

Figure 5.31. A multiplication table for the group V_4, with nodes numbered 1 through 4 to facilitate analyzing how the arrows permute the elements. The permutation for each arrow color is shown on the right.

Each cell contains the result of its row heading times its column heading. So one column in a multiplication table contains the results of multiplying each row heading by the same column heading. For instance, the column labeled 3 in the multiplication table in Figure 5.31 contains the results of each group element multiplied on the right by 3. That column contains a complete record of what right multiplication by 3 means in V_4. Notice that the column is simply a *reordering* of the numbers 1 through 4, a permutation. (Your answer to Exercise 4.15 explains why each column in a multiplication table is always a reordering of the elements.) That column turns a 1 into a 3, a 2 into a 4, a 3 into a 1, and a 4 into a 2. We can represent it by the following permutation.

5.4. Symmetric and alternating groups

So each column in a multiplication table is a permutation of the row headings. The four permutations represented by the four columns of the multiplication table in Figure 5.31 are on the right of that same figure.

Notice that we could apply the procedure exemplified in Figure 5.30 to any Cayley diagram, and we could apply the procedure exemplified in Figure 5.31 to any multiplication table. The important difference is that from a multiplication table we do not obtain simply a few permutations we could use to build a group by combining them with each other, but rather the full set of permutations that form the group. This difference arises because the column headers of a multiplication table contain every group element, whereas the arrows in a Cayley diagram usually only depict a set of generators. Cayley's theorem states that in both cases, the resulting group will have exactly the same structure as the original. It is now time to see why Cayley's Theorem is true. We begin with a proof based on the visual computations in Figures 5.30 and 5.31, then illustrate that proof in Figure 5.32.

Theorem 5.1 (Cayley). *Every group is isomorphic to a collection of permutations.*

Proof. We have just seen that from the columns of any group's multiplication table, we can create a permutation for each element of the group, as Figure 5.31 exemplifies. We can also make a multiplication table out of those permutations. This proof explains why such a multiplication table must behave the same as the original.

To make it easy to refer to them, let's say the group's elements are numbered 1 through n, with 1 standing for the identity, just as in Figure 5.31. We also need a way to refer to the permutations we created, one created for each element of the original group as in Figure 5.31. So let's write p_1 for the permutation for column 1, p_2 for the permutation created from column 2, and so on. Let's compare two multiplication tables, the original one containing the numbers 1 through n, and a new one containing the permutations p_1 through p_n.

Imagine we're looking at the entry in the row for p_i and the column for p_j in the new table, and the cell contains p_k. That is, the table is saying $p_i \cdot p_j = p_k$. I will now give evidence that these permutations faithfully represent the elements from which we created them; that is, that the original table must say that $i \cdot j = k$.

Consider how the permutations treat the identity element from the original group. Because p_k represents the "multiply by k" column in the table, applying p_k to 1 means multiplying 1 by k. To say this with an equation, $p_k(1) = 1 \cdot k = k$. Similarly, doing $p_i \cdot p_j$ to 1 means multiplying 1 by i and then j, as follows.[2]

$$p_i \cdot p_j(1) = 1 \cdot i \cdot j = i \cdot j$$

Because $p_i \cdot p_j = p_k$, we know that $p_k(1)$ and $p_i \cdot p_j(1)$ must result in the same answer. Because the former is k and the latter is $i \cdot j$, that means $i \cdot j = k$.

So any equation like $p_i \cdot p_j = p_k$ from the table of permutations is an imitation of an equation from the original group, $i \cdot j = k$. Thus the entire permutation group follows the same pattern as the original group, as the theorem states. □

---

[2]For the detail-oriented: I have just revealed my convention for function composition. Since I use $f \circ g$ to mean "f then g," I am using the convention $(f \circ g)(x) = g(f(x))$.

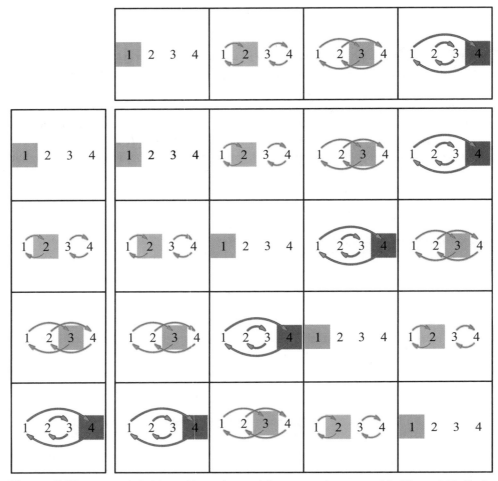

Figure 5.32. A multiplication table made up of the permutations created in Figure 5.31. Each cell highlights the destination to which the permutation sends 1, using the corresponding color from Figure 5.31, emphasizing what the colors of the arrows already showed: The two tables contain the same pattern.

This proof can be summarized as two steps: Create a permutation for each column in a group's multiplication table, and then inspect how those permutations treat the group's identity element. Figure 5.32 illustrates these two steps. It shows a multiplication table comprised of the permutations from Figure 5.31, and each cell of the table highlights the result of applying the permutation to the identity element 1. The correspondence between Figures 5.31 and 5.32 is clear: If I remove all but the highlighted elements from Figure 5.32, all that remains is the multiplication table from Figure 5.31.

5.4.5 Summary

Think how much you've learned in this chapter! You met five famous families of groups, together with the sizes, Cayley diagrams, and multiplication tables of many of them. You learned cycle graphs, permutations, and the proof of one of the most fundamental

5.5 Exercises

theorems in group theory. As promised at the outset, the library of groups you now know should give you a much better ability to answer questions like those on page 63. Try your new knowledge out on such questions in the following exercises, particularly those in Section 5.5.5.

5.5 Exercises

5.5.1 Basics

Exercise 5.1. If a group is generated by just one element, what kind of group is it?

Exercise 5.2.

(a) In the group C_5, compute $2 + 2$.

(b) In the group C_5, compute $4 + 3$.

(c) In the group C_{10}, compute $8 + 7$.

(d) In the group C_{10}, compute $9 + 1$.

(e) In the group C_3, compute $2 + 2 + 2 + 2 + 2 + 2$.

(f) In the group C_{11}, compute $10 - 8 + 1 - 7 + 6 + 5$.

Exercise 5.3. For each statement below, determine if it is true or false.

(a) Every cyclic group is abelian.

(b) Every abelian group is cyclic.

(c) Every dihedral group is abelian.

(d) Some cyclic groups are dihedral.

(e) There is a cyclic group of order 100.

(f) There is a symmetric group of order 100.

(g) If some pair of elements in a group commute, the group is abelian.

(h) If every pair of elements in a group commute, the group is cyclic.

(i) If the pattern on the left of Figure 5.8 appears nowhere in the Cayley diagram for a group, then the group is abelian.

Exercise 5.4.

(a) Use the Cayley diagram of the group D_5 in Figure 5.17 to compute $r \cdot f \cdot r$ in that group.

(b) Is the answer the same or different if you do the computation in the group D_3 instead?

(c) Is the answer the same or different if you do the computation in the group D_n instead?

(d) In this chapter, I introduced D_n as a group satisfying $frf = r^{-1}$. Use that equation to prove your answer to part (c).

Exercise 5.5. Compare the strengths and weaknesses of the three visualization techniques introduced in this book: Cayley diagrams, multiplication tables, and cycle graphs.

5.5.2 Understanding the families

Exercise 5.6. Sketch the following visualizations.

(a) a cycle graph for C_9

(b) a Cayley diagram for D_4

(c) a multiplication table for D_2

Exercise 5.7. Describe in words what each of the following visualizations looks like for C_{999}.

(a) Cayley diagram

(b) multiplication table

(c) cycle graph

Exercise 5.8. Describe in words what each of the following visualizations looks like for D_{999}.

(a) Cayley diagram

(b) multiplication table

(c) cycle graph

Exercise 5.9. What are the orders of the first ten symmetric groups, S_1 through S_{10}? What are the orders of their corresponding alternating groups, A_1 through A_{10}? Explain your answer for the order of A_1.

Exercise 5.10. The exercises for Chapter 3 asked you to create several Cayley diagrams. This chapter introduced a method for telling whether a group is abelian based on its Cayley diagram.

For each of the Chapter 3 exercises mentioned below, first determine whether each Cayley diagram from that exercise represents an abelian group. Then determine whether the group belongs to any of the five families introduced in this chapter, and if so, what the group's name is (e.g., D_4, S_3, etc.). Explain how you determine each of your answers.

(a) Exercise 3.5

(b) Exercise 3.6

(c) Exercise 3.7

(d) Exercise 3.8

(e) Exercise 3.11

(f) Exercise 3.13

(g) Exercise 3.14

(h) Exercise 3.16

Exercise 5.11. Explain why every cyclic group is abelian.

5.5. Exercises

Exercise 5.12. Why is it sufficient, when looking to see if a Cayley diagram represents an abelian group, to only consider the arrows? Why do we not need to examine every possible combination of paths?

Exercise 5.13.

(a) Create a cycle graph for the group V_4 using the multiplication table in Figure 5.31.

(b) Create a cycle graph for the group A_4 using the Cayley diagram in Exercise 4.6 part (c).

Exercise 5.14.

(a) Is there a dihedral group of order 7?

(b) If A_n has order 2520, what is n?

(c) If A_n has order m, what order does S_n have?

Exercise 5.15. For each part below, compute the orbit of the element in the group. Your answer will be a list of elements from the group that ends with the identity.

(a) The element r^2 in the group D_{10}

(b) The element 10 in the group C_{16}

(c) The element 25 in the group C_{30}

(d) The element 12 in the group C_{42}

(e) The element s in the group whose Cayley diagram is on the left below. (Assume the element a at the top left is the identity.)

(f) The element l in the group whose Cayley diagram is on the right below. (Assume the element a at the top is the identity.)

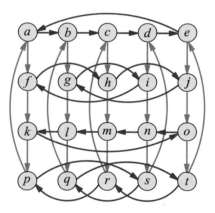
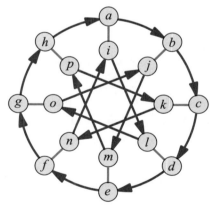

Exercise 5.16. Recall the notation for generators from Exercise 4.25. Use it to fill in the blanks below with however many elements are necessary to generate the group. Use as few elements as possible.

(a) $C_n = \{0, 1, \ldots, n-1\} = \langle \underline{\qquad} \rangle$

(b) $D_n = \{e, r, \ldots, r^{n-1}, f, fr, \ldots, fr^{n-1}\} = \langle \underline{\qquad} \rangle$

5.5.3 Small examples

Many of the smallest members of the families D_n, S_n, and A_n do not display the characteristic complexities of larger members of those families. The following exercises ask you to investigate the simplest (i.e., smallest) dihedral, symmetric, and alternating groups.

Exercise 5.17. Create multiplication tables for the smallest dihedral groups D_1, D_2, D_3, and so on, until you find the first non-abelian member of the family. Which is it, and how can you tell?

Exercise 5.18. Repeat Exercise 5.17 for the symmetric groups S_n. Use the permutation notation from this chapter.

Exercise 5.19. For each symmetric group whose multiplication table you created in Exercise 5.18, compute the elements of the corresponding alternating group, as in Figure 5.25. For each alternating group you compute, create

(a) a multiplication table,

(b) a Cayley diagram, and

(c) a cycle graph.

Exercise 5.20. Some of the smallest members of the families C_n, D_n, S_n, and A_n actually belong to more than one family, as long as we do not care about the names of the elements, but about the group structure. For instance, D_1 is a group with two elements, and its multiplication table has the same pattern as that of C_2, as shown here.

	e	f
e	e	f
f	f	e

	0	1
0	0	1
1	1	0

What other groups belong to more than one of the families we studied in this chapter? (Another way to read this question is, "Are there any groups in one of the families C_n, D_n, S_n, or A_n that are isomorphic to a group in another of those families?")

Exercise 5.21. For each of the following questions, either exhibit a group that answers the question in the affirmative or give a clear explanation of why the answer to the question is negative.

(a) Is there a cyclic group with exactly four generators? (Not that it takes four elements to generate the group, but that there are four different elements a, b, c, d in C_n and $C_n = \langle a \rangle = \langle b \rangle = \langle c \rangle = \langle d \rangle$.) Is there more than one such group?

(b) Is there a cyclic group with exactly one generator? Is there more than one?

Exercise 5.22. Open *Group Explorer* and sort the group library by order, with the smallest groups on top. Where in the list do you find the first group that's not in any of the families this chapter introduced? What is its name and size? How can you tell that it's not in any of the families you just learned?

5.5.4 Going a bit further

Exercise 5.23. This chapter gave propellers and pinwheels as examples of objects whose symmetries are described by cyclic groups, that is, objects with rotational symmetry only. What other objects fit in this category?

Exercise 5.24. This chapter gave regular polygons as examples of objects whose symmetries are described by dihedral groups, that is, objects with both rotational and bilateral symmetry, but no other symmetries. What other objects fit in this category?

Exercise 5.25. Analyze the symmetries of a tetrahedron using the technique from Definition 3.1, resulting in the Cayley diagram for its symmetry group. Here are a few hints to get you started.

First, if you need a tetrahedron, copy the following pattern onto a separate sheet, cut along the solid lines, and fold along the dotted lines. Keep the numbers on the outside of the figure as you fold, and the 1's should join to form the fourth vertex of the three-dimensional shape. You may need tape to hold that vertex together.

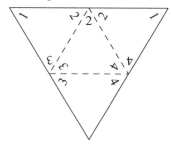

Second, decide on an orientation for the tetrahedron before you begin. For example, you might always keep one of its faces flat on the tabletop in front of you, and another facing up towards you. Obviously, as you manipulate the shape, it will change which face is on the tabletop or facing you.

Third, if you want your Cayley diagram to resemble the one in Figure 5.27, use the following two moves. One move is a 120-degree rotation of the tetrahedron about whatever vertex is pointing up, as shown on the left below. The other is a 180-degree rotation of the tetrahedron performed by putting two fingertips at the midpoints of opposite edges (edges that do not touch one another), as shown on the right below.

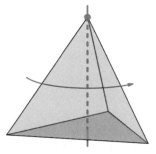
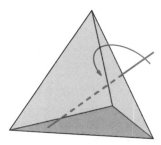

Exercise 5.26. As you know from the chapter, the symmetry group for the tetrahedron is A_4. We can think of it, as you saw in Exercise 5.25, as permuting four vertices. What physical features of the tetrahedron prevent its symmetry group from being all of S_4?

Exercise 5.27. As you know from the chapter, S_3 and D_3 are two different names for the same group. Yet no larger dihedral group is also a symmetric group. Give an argument based on the physical features of an n-gon for why this is so ($n \geq 4$).

Exercise 5.28. Section 5.2.3 describes what the cycle graph will look like for $C_p \times C_p$ if p is a prime number. Draw the cycle graph for $C_5 \times C_5$. (It is not necessary to label the elements.)

Exercise 5.29. Each part of this problem should help you answer the later parts.

(a) Is it true that any a in C_n will generate the whole group, i.e., $C_n = \langle a \rangle$? Why or why not? If not, for which elements a in which groups C_n is it false?

Hint: Although it is quite possible to solve this problem with what you know now, if it seems too tricky for the moment, come back to it after you've read Section 8.4.

(b) For how many different pairs a, b in D_3 can you write $D_3 = \langle a, b \rangle$? What if I change the D_3 to D_n?

(c) You may have noticed that $D_3 = \langle rf, r^2 f \rangle$, which for the moment I will rewrite as $\langle rf, r^{-1} f \rangle$. Does this same set of generators generate any other D_n? Which ones, and how do you know?

Exercise 5.30. In Exercise 4.32 you were introduced to the set $\mathbb{Z}$, which is a group under the operation of ordinary addition.

(a) Create a Cayley diagram for it.

(b) Create a cycle graph for it.

(c) Is it abelian?

(d) Are there any objects whose symmetries it describes?

Exercise 5.31. The group $\mathbb{Z}$ from the previous exercise is considered an infinite cyclic group. What do you think an infinite dihedral group would be like?

Exercise 5.32. In Exercises 4.32 and 4.33 you investigated some infinite groups of numbers. Are any of them abelian?

5.5.5 Going beyond

Exercise 5.33. The notation for generating a group, introduced in Exercise 4.25, can be extended to what are called "group presentations" by adding equations that describe how the elements relate. For example, the presentation for S_3 is

$$\langle r, f \mid r^3 = 1, f^2 = 1, frf = r^{-1} \rangle.$$

We can think of this as specifying a Cayley diagram; the r, f tells there will be two arrow colors (one for r and one for f) and the equations that follow tell us how those arrows tie together. The first two, $r^3 = 1$ and $f^2 = 1$, tell you the orders of the generators, and the last equation tells you how the generators relate. Such a presentation gives you all the information you need to construct a Cayley diagram for S_3, step-by-step.

5.5. Exercises

(a) Explain how the information from the presentation
$$\langle r, f \mid r^3 = 1, f^2 = 1, frf = r^{-1} \rangle$$
enables you to create each step of a Cayley diagram for S_3. That is, go through the step-by-step process and write down both the steps and how you concluded the information you needed at each step.

(b) Repeat part (a) for the presentation $\langle r, f \mid r^4 = 1, f^2 = 1, frf = r^{-1} \rangle$, which is only different in that the 3 changed to a 4. What group is this?

(c) What group has the presentation $\langle a \mid a^n = 1 \rangle$?

(d) What is the presentation for D_n?

(e) Sketch a Cayley diagram for the group $\langle a, b \mid a^4 = 1, b^4 = 1, a^2 = b^2, bab = a^{-1} \rangle$.

Exercise 5.34. This exercise gives you a different look at S_3 and S_4 by looking at them using a different set of generators than we have up until this point.

(a) The group S_3 can be generated by the following two permutations. Make a Cayley diagram for S_3 whose arrows represent these elements.

(b) The group S_4 can be generated by the following three permutations. Make a Cayley diagram for S_4 whose arrows represent these elements. Hint: It can be laid out as a truncated octahedron, similar to Figure 5.28, but with three arrow colors.

(c) What conjecture does this lead you to make about generating S_n?

Exercise 5.35. This exercise analyzes the multiplication table for D_5 in Figure 5.18, relating it to a Cayley diagram for that same group, shown in the center of Figure 5.17. It begins with a generic question that will help us relate multiplication to Cayley diagrams, and the remaining parts apply it to D_5.

(a) Consider an equation like $a \cdot b = c$ about the elements a, b, and c in a group. What part of a multiplication table for the group expresses the information $a \cdot b = c$? What part of a Cayley diagram for the group expresses the same information?

(b) The upper left quarter of the multiplication table for D_5 in Figure 5.18 corresponds to what portion of the Cayley diagram for D_5 in Figure 5.17? Justify your answer here using your answer from part (a).

(c) The lower left quarter of the multiplication table for D_5 in Figure 5.18 corresponds to what portion of the Cayley diagram for D_5 in Figure 5.17? Justify your answer here using your answer from part (a).

(d) Why do the diagonal stripes in the right half of the table in Figure 5.18 slant opposite to those in the left half of the table? This part's answer, too, can be justified using your answer from part (a).

Exercise 5.36.

(a) Is there a group in which no element (other than the identity) is its own inverse?

(b) The previous question was one of the three from page 63. Answer the other two as well.

(c) In every group, any two orbits touch at the identity. Is there any group containing two orbits that touch only at the identity? Is there any group containing two orbits that touch at the identity and at exactly one other place?

(d) Find a group with at least two elements in it, and only one solution to the equation $x^2 = e$, the solution $x = e$.

(e) Find a group that has exactly two solutions to the equation $x^2 = e$.

(f) Find a group that has more than two solutions to the equation $x^2 = e$.

(g) Find a group with at least two elements in it, and only one solution to the equation $x^3 = e$, the solution $x = e$, or explain why no such group exists.

(h) Find a group that has more than two solutions to the equation $x^3 = e$, or explain why no such group exists.

(i) Find a group that has exactly two solutions to the equation $x^3 = e$, or explain why no such group exists.

Exercise 5.37.

(a) There are two different groups of order 6. What are their names? Specify which, if any, are abelian.

(b) If there is only one group of a given order, to what family must it belong? Why?

(c) Find some values of n for which there is only one group of order n. Can you see a pattern among the numbers you found? (Chapter 7 will talk about the pattern that exists.)

Hint: The group library in *Group Explorer* may speed your search.

Exercise 5.38. Prove using algebra that if every element in a group has order 2, then the group is abelian.

Exercise 5.39. In finite groups, the orbit of an element can be defined as all the positive powers of that element; the orbit of a is $\{a^1, a^2, a^3, a^4, \ldots\}$, a sequence which cannot go on infinitely. Because the group is finite, eventually one of the elements encountered (maybe a^{20}, maybe a^{1000}, maybe later) must be a repetition of an element already on the list. Explain why this means that some power of a must equal e.

Exercise 5.40. The parts of this exercise ask you to explore the relationship among the elements in D_4, and how different layouts of its Cayley diagram can show those relationships in different ways. Begin with the Cayley diagram you created for D_4 in Exercise 5.6. Ensure that it follows the pattern given in Figure 5.17.

5.5. Exercises

(a) Make another copy of this diagram with one change: Reorder the elements of the inner ring so that the arrows representing r point clockwise in that ring, as they do in the outer ring. In order for it to still be a Cayley diagram of D_4, you must preserve the same pattern of connections. Therefore some of your f arrows will be stretched by this new layout, but try to stretch them as little as possible.

(b) Make another copy of the diagram, but this time arrange the nodes in two horizontal rows, the top row proceeding through the orbit of r starting from e, and the bottom row connected to the top row by four parallel f arrows.

(c) Make another copy of this two-row diagram, but this time arrange the bottom row so that each element fr^m is below the corresponding element r^m for every number m (between 0 and 3). The f arrows will no longer be parallel, but try to make it as organized as possible.

(d) For each of these three new ways of laying out the dihedral group D_4 (parts (a) through (c)), explain what the Cayley diagram for an arbitrary dihedral group D_n would look like if laid out similarly.

(e) In *Group Explorer*, open a Cayley diagram for the group D_4. It defaults to the pattern from Figure 5.17. How can you instruct *Group Explorer* to reorganize the nodes of the diagram in the same ways you did in this exercise? (You may need to refer to *Group Explorer*'s built-in help system for information on manipulating Cayley diagrams.)

5.5.6 Cayley's theorem

Exercise 5.41. The following applications will give you some hands-on experience with Cayley's theorem.

(a) Extract from the multiplication table of $C_2 \times C_4$ shown on the right of Figure 5.12 the eight permutations in S_8 that behave like a copy of $C_2 \times C_4$. You may want to make a copy of the table with the elements numbered from 1 to 8; the example in Figure 5.31 may help.

(b) Extract from the Cayley diagram of D_5 in Figure 5.17 the two permutations that describe its arrows, following the example of Figure 5.30. Here it may also be useful to number the elements, 1 to 10.

(c) Take the two permutations shown in Figure 5.30 and make a multiplication table from them and all their combinations. Organize your table and highlight what each permutation does to the identity element, as in Figure 5.32. The result should make it clear that the multiplication table you created is one for D_3, as shown in Figure 2.10 (under the name S_3).

Exercise 5.42. Cayley's theorem says that any group is isomorphic to a collection of permutations. If we restrict ourselves to only using permutations of three items (i.e., elements of S_3), what groups can we form? For instance, we can obviously create all of S_3 by taking every permutation of three elements, and we can obviously create the group C_1 by just taking the identity permutation from S_3. What other groups are possible?

Exercise 5.43. Repeating Exercise 5.42 for S_4 would be lengthy, so rather than ask you to search blindly, I will give you hints. All of the following groups can be built using collections of permutations of four things (i.e., elements of S_4). Find the appropriate collection of permutations for each group given.

(a) every group of order 1, 2, 3, or 4 (for a total of five groups)

(b) the group D_4

   Hint: Apply the technique from Chapter 3 to a square. Recall Exercise 2.8.

(c) the group A_4

Exercise 5.44. Although you can find a copy of C_6 in S_6 by simply taking the orbit of the permutation shown here,

that is not the most "efficient" way to embed C_6 in an S_n. The following permutation in S_5 also has order 6, and therefore its orbit is a copy of C_6 as well.

Thus we can fit C_6 in S_6 in an obvious way, or in C_5 with a little cleverness. So although the easiest way to embed C_n in a symmetric group is by taking a permutation that cycles the elements of S_n, for some n there is a way to embed C_n in a smaller symmetric group.

(a) For each n between 1 and 12, determine the smallest value of m such that C_n can be expressed in S_m. Can you find any pattern or determine any strategy for computing m from n?

(b) Does your answer change if instead of a copy of C_n, you must find in S_m a copy of D_n?

6

Subgroups

Though the chapter title may not suggest it, you've come to the exciting part of this book! Through five chapters, you've gained a lot of familiarity, learned the lay of the group theory land, and are now ready for in-depth study. Starting with this chapter, we'll be getting more analytical and learning more mathematical terminology, *but without giving up our visual roots*. Entering advanced realms doesn't mean leaving visualization behind; it can be just as helpful in advanced areas as in introductory ones, and often moreso. In fact, the degree to which visualization has helped me better understand the material in the latter half of this book is a large part of my motivation for writing it. This chapter is exciting because it starts to bring the power of visualization to bear on a few of the theorems and proofs in group theory, a trend that continues throughout the rest of the book.

Analyzing groups more deeply means asking questions about what kind of organizing structure they have, how they relate to one another, how to construct large ones by combining smaller ones, and how to dissect larger ones to expose the smaller ones inside. This chapter and those to come answer all these questions, beginning with looking inside groups to find smaller groups (subgroups). This chapter will be the foundation on which the following chapters of advanced study build. It will also deepen our understanding of the makeup of both the groups we know and our familiar means of visualizing them.

I begin laying that foundation with Section 6.1, which brings together the two paradigms this book has presented so far. We have seen how to view a group as a collection of actions and how to view it as an operation on a set. Cayley diagrams illustrate the former, action-based point of view and multiplication tables illustrate the latter, algebraic one. The following section looks at one of the most important things the algebraic point of view can teach us about Cayley diagrams, a concept called *regularity*. The remaining sections of this chapter use that concept to simplify their analysis of subgroups.

6.1 What multiplication tables say about Cayley diagrams

Multiplication tables are part of the algebraic view of groups, which encourages and enables us to use equations to describe relationships among elements of a group. As we know, such equations speak about Cayley diagrams by referring to their arrows. Consider the Cayley diagram for S_3, as shown on the left of Figure 5.17 on page 75. The equation $frf = r^{-1}$ is true in S_3 because the path frf is the same as the path r^{-1}, in the sense that following an f arrow, then an r arrow, then another f arrow accomplishes the same thing as following an r arrow in reverse.

This holds true *regardless of where in the Cayley diagram you begin*. Starting at the identity, frf and r^{-1} both lead to the same endpoint, r^2, but starting at any other element they both still lead to the same endpoint, though the particular endpoint will be different. We can illustrate the equation $frf = r^{-1}$ by lifting out just those two paths from the Cayley diagram for S_3, as shown in Figure 6.1.

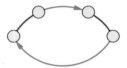

Figure 6.1. A Cayley diagram fragment representing the equation $frf = r^{-1}$. The top path, followed left-to-right, is frf, and the bottom path, followed left-to-right, is r^{-1}. This diagram indicates their equality by giving them common starting and ending points.

An algebraic equation is true not just about one portion of a Cayley diagram, but it is true *all across the diagram in the same way*. The pattern in Figure 6.1 appears not just in one part of a Cayley diagram for S_3, but at every point in the whole diagram. This is why I did not need to label the nodes in Figure 6.1; the pattern in that figure can fit anywhere in a Cayley diagram for S_3.

And that pattern is not the only one that permeates a Cayley diagram for S_3. Any other equation from S_3 can also be represented by a pattern that appears throughout every Cayley diagram for that group. Thus Cayley diagrams always have a uniform symmetry; every part of the diagram is structured like every other. We cannot have $frf = r^{-1}$ true at some part of a Cayley diagram and yet false in another part of the same Cayley diagram.

Definition 6.1 (regular). I call a diagram ***regular*** if it repeats every one of its internal patterns throughout the whole diagram, in the sense discussed above.[1] In particular, every Cayley diagram is regular; diagrams lacking regularity do not represent groups, and so we do not call them Cayley diagrams.

For instance, consider the arrows in the Cayley diagram in Figure 6.2 as representing actions and the numbers 0 through 7 as situations through which those actions navigate. Then in any of situations 4 through 7, if we call the red action r, then we can see that the equation $r^2 = 0$ is true. Yet in situations 0 through 3, things change; $r^2 \neq 0$ in any of those situations. The diagram is not regular, and thus it is not a Cayley diagram; it does not describe a group.

Regularity shows us how groups embody symmetry. We have used groups to measure symmetry, and said that Cayley diagrams faithfully represent groups; regularity is one of

---

[1] I use the term *regular* because the property I'm speaking of comes from [16], which classifies Cayley diagrams as those graphs whose automorphism groups have a subgroup that acts *regularly* on the diagram's nodes, a more technical way to describe this same property.

6.2. Seeing subgroups

the consequences of that faithful representation. Every part of a group looks like every other part. Groups themselves are full of symmetry.

6.1.1 Perfecting our unofficial definition

In Exercise 4.16, I asked you to inspect two Cayley diagrams, one of which is repeated in Figure 6.2. Exercises 2.14 through 2.17 asked you to come up with criteria for determining if a diagram represents a group as defined in Definition 1.9. The diagram in Figure 6.2 satisfies all of them! Therefore, from the viewpoint of the unofficial Definition 1.9, such a diagram would seem to describe a group. This is why I called that definition unofficial; for the sake of keeping Chapter 1 simple, I left out a technical point.

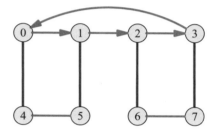

Figure 6.2. A diagram that satisfies many of the criteria for being a Cayley diagram, but violates one: equations about relationships among the generators change from one part of the diagram to another.

That technical point is what was stated above: a diagram is not a Cayley diagram (i.e., it does not describe a group) unless it is regular. I omitted any reference to regularity of diagrams in Chapter 1 because it requires speaking of equations, which it was too early to bring up. The informal argument in Section 4.4, that any group by Definition 1.9 is a group by Definition 4.2, glossed over this detail by nature of its informality. A rigorous argument would require making our informal definition mathematically precise; this could be done by incorporating regularity.

But there is no need to do so; we have gained what we need from the introduction of regularity, a better understanding of Cayley diagrams, in preparation for the coming sections. I will be using Cayley diagrams as our primary visualization tool for the rest of this book (occasionally also referring to multiplication tables when they are helpful). The following sections use this new idea of regularity to prove some essential theorems about subgroups.

6.2 Seeing subgroups

As we became acquainted with a wide variety of groups in Chapter 5, we talked a bit about their internal structure. Most importantly, we saw that every group has one or more cyclic groups inside it, called orbits. It's now time to take this idea beyond just cyclic groups.

Definition 6.2 (subgroup)**.** When one group is completely contained in another, the inner group is called a ***subgroup*** of the outer one. When a group H is a subgroup of a group G, the standard way to write this is $H < G$.

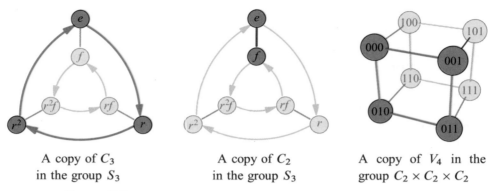

Figure 6.3. Three Cayley diagrams highlighting easy-to-spot subgroups.

All the orbits we saw in Chapter 5 are subgroups. For instance, the orbit of r in S_3, $\{e, r, r^2\}$, is a cyclic subgroup of order 3, a copy of C_3. We can write $\{e, r, r^2\} < S_3$, or if we're less concerned with formality, even $C_3 < S_3$. This is the first example shown in Figure 6.3. The second example in that figure, $\{e, f\} < S_3$, is also an orbit, but the third example is not, because the highlighted subgroup is not just a single cycle. The generator notation from Exercise 4.25 (page 59) gives us a handy way to describe these subgroups. For example $\{e, r, r^2\}$ is generated by r, and thus can be more succinctly written $\langle r \rangle$. The other two examples in Figure 6.3 are therefore $\langle f \rangle$ and $\langle 001, 010 \rangle$.

Every single group has some kind of subgroup, for the following two reasons. The identity by itself, $\{e\}$, is a copy of C_1 in any group, called the ***trivial subgroup.*** And technically every group is a subgroup of itself (e.g., $S_3 < S_3$), called the ***non-proper subgroup.***

The example subgroups in Figure 6.3 are easy to spot. We are familiar with these small groups from earlier chapters and can recognize their Cayley diagrams on sight, probably even if Figure 6.3 hadn't highlighted them for us. But not every subgroup in every Cayley diagram is visually obvious. Figure 6.4 shows two less obvious examples, yet they are by no means as visually obscure as subgroups can be. To visualize subgroups, we therefore need a method for bringing them out of hiding.

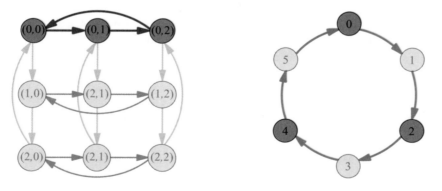

Figure 6.4. Two Cayley diagrams highlighting less obvious subgroups. The left diagram shows a copy of C_3 in $C_3 \times C_3$; the pattern is easy to spot, but it is not the familiar circular pattern of C_3. The right shows a copy of C_3 in C_6; each step in the copy of C_3 is *two* steps in the C_6 cycle.

6.3 Revealing subgroups

Any subgroup can be revealed by a suitable reorganization of the group's Cayley diagram. For example, recall Figure 5.29 on page 83, which shows two Cayley diagrams for A_5. One of these diagrams is organized to emphasize a cyclic subgroup of order five, and the other to emphasize a cyclic subgroup of order three. The five- and three-cycles are visible in the respective diagrams. This section explains how to reorganize Cayley diagrams to emphasize specific subgroups; I explain a similar strategy for multiplication tables in Exercise 6.26.

The language of generators makes it easier to discuss how Cayley diagrams and multiplication tables are organized. Consider the example $C_6 = \{0, 1, 2, 3, 4, 5\}$. This group can be generated several ways, including $C_6 = \langle 1 \rangle$ and $C_6 = \langle 5 \rangle$. Perhaps unexpectedly, we can also write $C_6 = \langle 2, 3 \rangle$. If indeed the fact $C_6 = \langle 2, 3 \rangle$ surprises you, I recommend taking a moment to verify that you can create every element of C_6 using just 2, 3, and addition mod 6. Neither 2 nor 3 alone generates C_6, but together they do. These different ways to generate C_6 give rise to correspondingly different ways to connect a Cayley diagram of C_6 with arrows, an idea you may remember from Exercise 2.9. Each is shown in Figure 6.5.

The left two diagrams in Figure 6.5 clearly represent a six-element cyclic group, but the right two diagrams disguise the cycle. The left two diagrams emphasize the non-proper subgroup $\langle 1 \rangle = C_6$, while each of the right two emphasizes both of the subgroups $\langle 2 \rangle$ and $\langle 3 \rangle$. Take a moment to spot the subgroups $\langle 2 \rangle$ and $\langle 3 \rangle$ in each of the right two Cayley diagrams in Figure 6.5.

The moral of the story is that we can make any subgroup visually prominent in a Cayley diagram by choosing generators for the group that *include the generators of that subgroup*. Sometimes it is also necessary to then choose a new layout for the nodes that untangles the resulting connections. This is how I brought out the subgroups $\langle 2 \rangle$ and $\langle 3 \rangle$: I generated C_6 as $\langle 2, 3 \rangle$, resulting in the third diagram in Figure 6.5. I then reorganized that diagram to look cleaner, resulting in the rightmost diagram in the same figure. Because there is sometimes more than one clear, useful layout, there is not necessarily one "perfect" or "correct" one; it may be a matter of preference.

Judicious choices of arrows and new layouts enable us to reveal hidden subgroups like $\langle 2 \rangle < C_6$. You can experiment easily with different ways to generate and organize Cayley diagrams using *Group Explorer*; refer to its documentation for details.

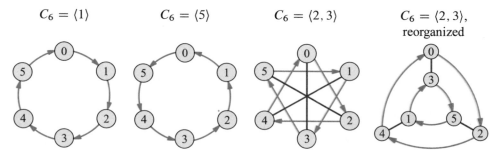

Figure 6.5. Four ways to connect a Cayley diagram of C_6, based on different choices of generators. Note how similar the rightmost diagram is to the familiar Cayley diagram of S_3. Can you spot the difference?

6.4 Cosets

The first step towards a deeper study of subgroups is to notice that identical copies of each subgroup appear throughout the rest of the group's Cayley diagram. For instance, consider the copy of C_3 in S_3, shown on the left of Figure 6.3. That same cyclic, three-element structure is repeated as the inner ring of the Cayley diagram (although it is reversed). Similarly, the copy of C_2 in S_3 shown in the middle of Figure 6.3 is repeated twice more throughout the diagram, as Figure 6.6 shows.

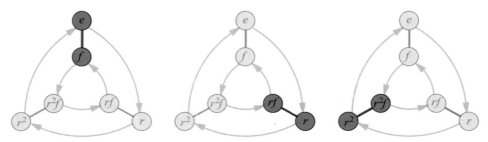

Figure 6.6. Copies of the subgroup $\langle f \rangle$ in S_3

But not all those copies are subgroups. A subgroup, because it is a group, must have an identity element. In Figure 6.6 only the leftmost copy $\{e, f\}$ contains e. Furthermore, multiplying two elements in $\{r, rf\}$, for example, does not yield another element in that set. For this reason also, it is not a group. So the sets $\{r, rf\}$ and $\{r^2, r^2f\}$ have a different name; they are called *cosets*, to indicate that they are copies of a subgroup structurally, but are not themselves groups.

Learning about subgroups and their cosets will reveal important structural properties about groups in general. Let's begin by making a few observations about cosets. I call them "observations" rather than theorems because they are simple enough that I can justify them with informal comments. A concept that shows up several times below is the regularity in every Cayley diagram, a concept introduced in Section 6.1. The first observation points out that cosets aren't specific to the example in Figure 6.6.

Observation 6.3. Every subgroup has cosets, and they cover every node of the group's Cayley diagram.

Let's see why this observation is true. In Figure 6.3, what allowed us to write $\{e, f\}$ as $\langle f \rangle$ and $\{000, 001, 010, 011\}$ as $\langle 001, 010 \rangle$ was that every group has a set of generators, even if that group is a subgroup of another. We can visualize how a subgroup is generated by exploring from the identity node in a Cayley diagram using just the generators for that subgroup.

But such exploration could be done from any node. We could explore using the generators for a subgroup H, but starting at any other element g in the larger group G. The regularity of the diagram means that such exploration will reveal the same pattern no matter where it begins. Thus such an exploration will trace out a structural copy of H beginning at g. So a Cayley diagram for G is full of copies of H touching every element of G.

Observation 6.4. We can also describe cosets algebraically. The copy of H based at a is named aH.

6.4. Cosets

The choice of the name aH is not arbitrary. Recall that to multiply ab in a Cayley diagram, start from the node a and follow the path for b. The meaning of aH in a Cayley diagram is similar: Start from the node a and follow *all* the paths in H. This describes the copy of H based at a, so the naming convention makes sense. Following that convention, the middle coset in Figure 6.6 is called $r\langle f \rangle$ and the right one is $r^2 \langle f \rangle$.

I have described how to compute a coset aH visually, but we can compute it algebraically as well. To do so, multiply the element a by the list of elements of H, with a the left element in each multiplication, as suggested by the name aH. An example of this computation for the coset $r \langle f \rangle$ is shown here.

$$r \langle f \rangle = r\{e, f\} = \{r \cdot e, r \cdot f\} = \{r, rf\}$$

For this reason all the cosets we have seen are actually called *left cosets,* because the multiplication is done on the left. I introduce the complementary notion of right cosets after two final observations.

Observation 6.5. Each coset can have more than one name.

For example, $r\langle f \rangle$ is the name given to the middle coset in Figure 6.6, because it is the copy of $\langle f \rangle$ based at r. However, it is just as true to say that it is the copy of $\langle f \rangle$ based at rf. As another example, consider the subgroup highlighted in $C_2 \times C_2 \times C_2$ in Figure 6.3 (call it H). The coset behind H could be described as $100H$, $101H$, $110H$, or $111H$, because exploration using the generators of H from any one of those elements reaches the same coset. The regularity in the *subgroup's* Cayley diagram is what makes this so.

Whenever a left coset aH contains an element b, we could just as easily have called the coset by the name bH. Since aH means the copy of H containing a and bH means the copy of H containing b, they refer to the same copy of H. Let's make this our final observation.

Observation 6.6. If b is in aH, then $aH = bH$.[2]

Thus in Figure 6.6, we can also call $r\langle f \rangle$ by the name $rf \langle f \rangle$, and $r^2 \langle f \rangle$ by the name $r^2 f \langle f \rangle$. The element you choose to use to name the coset is called the *representative*.

Let us put left cosets aside for a moment and find out about their counterparts, right cosets, which I have mentioned but not yet showed. From an algebraic point of view, you can probably guess how I will describe right cosets; a right coset Ha is computed just like a left coset, but the multiplication is done on the right. The right cosets of $\langle f \rangle$ in S_3 are therefore

$$\langle f \rangle r = \{e, f\}r = \{e \cdot r, f \cdot r\} = \{r, r^2 f\} \text{ and}$$
$$\langle f \rangle r^2 = \{e, f\}r^2 = \{e \cdot r^2, f \cdot r^2\} = \{r^2, rf\}.$$

Neither of these right cosets matches either of the left cosets of $\langle f \rangle$ we saw. Therefore they must look different in the Cayley diagram. Let's see how they look.

To compute a right coset visually, reverse the process for left cosets: Start with the subgroup $\langle f \rangle$ and follow the path for r from each element of that subgroup. The resulting set of destinations is the right coset $\langle f \rangle r$. Notice how this agrees with the algebraic computation; it multiplies each element of the subgroup on the right by r.

---
[2] A more formal argument justifying Observation 6.6 can be obtained by doing Exercise 6.15.

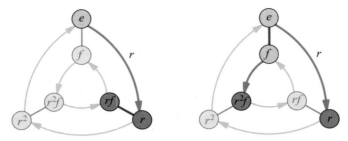

Figure 6.7. The left diagram shows the left coset $r\langle f\rangle$ in S_3, the nodes that f arrows can reach after the path to r has been followed. The right diagram shows the right coset $\langle f\rangle r$ in S_3, the nodes that r arrows can reach from the elements in $\langle f\rangle$.

Figure 6.7 compares the computations of $r\langle f\rangle$ and $\langle f\rangle r$ in a Cayley diagram. Figure 6.8 moves beyond that specific example, showing the distinction between left and right cosets in a Cayley diagram more generally. The reason that left cosets look like copies of the subgroup while the elements of right cosets are usually more scattered is that I adopted the convention that arrows represent right multiplication. If I had used the convention where arrows represent left multiplication, right cosets would have been copies of the subgroup and left cosets would have been scattered.

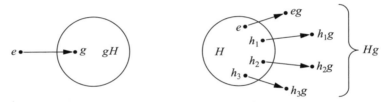

Figure 6.8. Each left coset gH is where H arrows can reach from g, which looks like a copy of H based at g, as in the left illustration. Each right coset Hg is the set of nodes to which the g arrows take the elements of H, as in the right illustration.

The subgroup $\langle f\rangle < S_3$ was a good first example because it showed us that left and right cosets are generally different. But because they are not always different, it is worth also seeing an example of when left and right cosets come out equal. Figure 6.9 illustrates

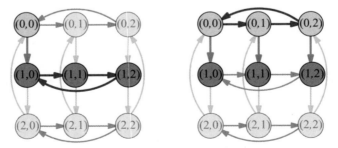

Figure 6.9. A left and right coset pair that are equal. Here $H = \langle(0,1)\rangle$ in the group $C_3 \times C_3$ and $g = (1,0)$; on the left we see the computation of gH and on the right Hg. The result is $\{(1,0),(1,1),(1,2)\}$ in both cases.

6.5. Lagrange's theorem

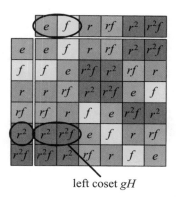
left coset gH

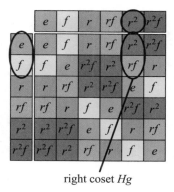
right coset Hg

Figure 6.10. The visual computation of left and right cosets in multiplication tables. The computations of gH and Hg are exemplified for $H = \{e, f\}$ and $g = r^2$ in the group S_3.

such a situation using the subgroup $H = \langle (0, 1) \rangle < C_3 \times C_3$. Although it only depicts the equality $gH = Hg$ for $g = (1, 0)$, we would find $gH = Hg$ to be true no matter which value of g we choose. Subgroups for which this is true are given the special name *normal*, and we will learn their importance in Chapter 7.

Computing left and right cosets using multiplication tables is also straightforward; I show how in Figure 6.10. I have ordered the row and column headings in the multiplication tables in that figure according to the subgroup $\langle f \rangle$ and its left cosets. The left coset $r^2 \langle f \rangle$ must appear in the r^2 row under the e and f columns, because those cells hold the results of $r^2 \cdot e$ and $r^2 \cdot f$. In any multiplication table, you can compute left cosets this way; the elements of gH are in the g row under the H columns. On the other hand, the right coset Hg can be computed in a similar way, but with the words "rows" and "columns" interchanged: Hg appears under the g column in the H rows.

6.5 Lagrange's theorem

Allow me to draw your attention to a pattern in all the examples of subgroups and their cosets that we've seen so far. Observation 6.3 stated that the left cosets of a subgroup cover the whole group. And in all the examples we've seen so far, not only is this true, but every element in the group has actually been in *exactly one* left coset. Figure 6.6 shows us not only that the left cosets of $\langle f \rangle$ cover all of S_3, but also that no element is in more than one of them; no two cosets overlap. This was also true of the right cosets of $\langle f \rangle$ we computed in the previous section, although we did not illustrate them.

Because the cosets of $\langle f \rangle$ do not overlap, I can consolidate all of Figure 6.6 into one Cayley diagram, with a color for each coset, as on the left of Figure 6.11. The right of Figure 6.11 shows a similar diagram, but for right cosets. The coloring schemes in these diagrams are possible because no element belongs to more than one coset, and thus no element requires more than one color.

When a collection is separated into categories, each item ending up in exactly one category, mathematicians call this a ***partition***. Figure 6.11 shows that the left cosets of $\langle f \rangle$ partition S_3 and the right cosets of $\langle f \rangle$ do also, but differently. A careful look at Figure 6.9 reveals that the cosets of $\langle (0, 1) \rangle$ also partition $C_3 \times C_3$.

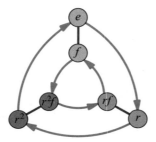 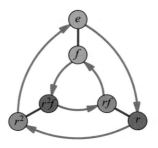

Figure 6.11. The left Cayley diagram for S_3 shows how the group is partitioned by the left cosets of the subgroup $\{e, f\}$; the right diagram shows how S_3 is partitioned by the right cosets of $\{e, f\}$.

The natural question is therefore whether this pattern we've noticed is a coincidence. Do the cosets of a subgroup always partition the group? The answer is given by the following theorem. I keep things simpler by dealing only with left cosets; Exercise 6.16 asks you to think through the same question for right cosets.

Theorem 6.7. *If H is a subgroup of G, then each element of G belongs to exactly one left coset of H.*

Proof. Suppose the element g of G appears to belong to two different left cosets, say aH and bH. In such a case, aH and bH must actually be two different names for the same coset (in the sense of Observation 6.5); here's why.

Since g is in aH, we can conclude that $gH = aH$, based on Observation 6.6. And since g is also in bH, we know $gH = bH$ for the same reason. The reason, then, that $aH = bH$ is that both are equal to gH. □

Theorem 6.7 teaches us to think of a group as being composed exclusively of non-overlapping copies of any subgroup (that subgroup's left cosets). We saw a specific example of this in Figure 6.11, and you can easily picture a similar division of $C_3 \times C_3$ suggested by Figure 6.9. The shapes of these two example partitions are different, but the abstract idea is the same in both cases. That idea, that cosets partition a group, is depicted a few ways in Figure 6.12.

Picturing groups this way reveals an important relationship between a group and each of its subgroups, expressed in the following important theorem, named after the Italian-

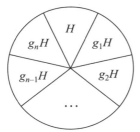 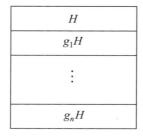 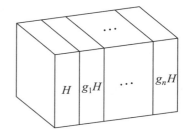

Figure 6.12. Several ways to picture the idea that cosets partition a group. The appearance of cosets in a diagram differs based on the layout of the diagram, so each of these illustrations represents the same idea, but for a different type of Cayley diagram layout.

6.5. Lagrange's theorem

born mathematician Joseph Louis Lagrange. In stating the theorem, I use the standard notation $|G|$ to mean the size (order) of the group G. Although there is a version of this theorem for infinite groups, in this section we assume G is finite.

Theorem 6.8 (Lagrange's Theorem). *If $H < G$, then the order $|H|$ of the subgroup divides the order $|G|$ of the larger group.*

Here I'm using the word ***divides*** in its common mathematical sense, as shorthand for "divides evenly into." In other words, we say that a smaller whole number divides a larger when the smaller number is one of the factors of the larger. Six divides twelve, but not twenty, because $20 \div 6$ leaves a remainder; the division does not come out "even."

Proof. Theorem 6.7 proved that the group G is partitioned into copies of H. So the size of G can be determined just by counting how many copies of H there are and multiplying that number by the size of each one, the number $|H|$. So if n is the number of left cosets (including H itself), we can write the equation

$$|G| = \overbrace{|H| + |H| + \cdots + |H|}^{\text{repeated } n \text{ times}} = n|H|.$$

□

The number n in the above proof comes up so frequently in the study of groups and their subgroups that it has a name.

Definition 6.9 (index). If $H < G$ then the ***index*** of H in G, written $[G : H]$, is how many times $|H|$ goes into $|G|$.

$$[G : H] = \frac{|G|}{|H|}$$

The index is the number of left cosets of H, if we count H itself as a coset. Most of the time when the word coset is used, it is understood that the subgroup itself is both a left and a right coset of itself, because $H = eH = He$.

You will notice that the index $[G : H]$ must also divide $|G|$, because the above equation could also be written as

$$|H| = \frac{|G|}{[G : H]}.$$

Let's make these ideas concrete by returning to the example of $\langle f \rangle < S_3$. The order of S_3 is 6, which we can see is the sum of the sizes of the three cosets of order 2. The index of $\langle f \rangle$ in S_3 is therefore 3, because it has three cosets.

$$[S_3 : \langle f \rangle] = \frac{|S_3|}{|\langle f \rangle|} = \frac{6}{2} = 3$$

Since Lagrange's Theorem has a famous mathematician's name attached to it, you may suspect that it has some interesting implications. Your suspicion would be correct! One important use of the theorem is that it significantly narrows down the possibilities for subgroups. For instance, in a group of order 8, you know not to bother looking for subgroups of order 3, 5, 6, or 7, because none of those numbers divides 8. More generally, in any group G, you know not to look for subgroups H with $|H| > \frac{|G|}{2}$, because numbers larger than half of $|G|$ cannot divide $|G|$. (The only exception here is the one non-proper subgroup $H = G$.) When you try the exercises in this chapter, keep these shortcuts in mind.

A natural and fruitful question is whether Lagrange's Theorem can be "turned around." The theorem states that if there is a subgroup H, then its order divides $|G|$. "Turning this around" means asking if for every n that divides $|G|$, we can find a subgroup $H < G$ of order n. The short answer is no, and Exercise 6.31 will guide you in demonstrating why. But some modified versions of this turned-around Lagrange's Theorem are both true and useful. Among them are the Sylow Theorems, which we will study in Chapter 9.

Be sure to try the exercises in Section 6.6.3 on Hasse diagrams, because Chapter 10 depends on an understanding of such diagrams. They help us understand the relationships among subgroups in general.

6.6 Exercises

6.6.1 Basics

Exercise 6.1. Which of these three diagrams is a Cayley diagram? Which satisfies all the requirements for being a Cayley diagram except for regularity?

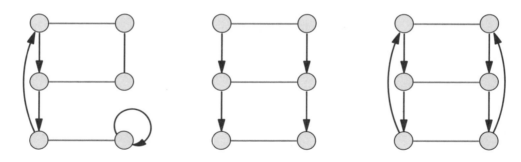

Exercise 6.2. Each of the diagrams below highlights some nodes. In each case, determine whether the highlighted set of nodes form a subgroup. You may need to reorganize some of the diagrams or add more arrows to make the answer obvious.

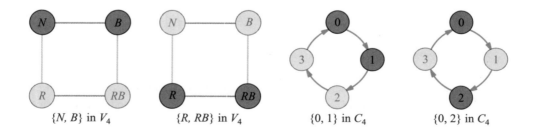

6.6. Exercises

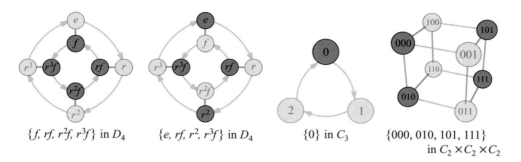

$\{f, rf, r^2f, r^3f\}$ in D_4 $\{e, rf, r^2, r^3f\}$ in D_4 $\{0\}$ in C_3 $\{000, 010, 101, 111\}$ in $C_2 \times C_2 \times C_2$

6.6.2 Understanding subgroups

Exercise 6.3. Make Cayley diagrams for the following groups with *every* non-identity element represented by an arrow.

(a) C_3

(b) V_4

(c) C_5

(d) S_3

Exercise 6.4. Which of the following subgroups does Lagrange's Theorem preclude from appearing as a subgroup of C_8? Which of them does Lagrange's Theorem preclude from appearing as a subgroup of D_5?

(a) C_2

(b) C_3

(c) V_4

(d) C_5

(e) S_3

Exercise 6.5. Find every subgroup for each of the following groups. Compute the order and index of each subgroup you find.

(a) V_4 Hint: There are five.

(b) C_5

(c) S_3 Hint: There are six.

(d) C_8

(e) D_4

(f) $C_3 \times C_3$

Exercise 6.6.

(a) What elements are in the left coset $r^m \langle f \rangle$ in D_n?

(b) What elements are in the right coset $\langle f \rangle r^m$ in D_n?

(c) How do these cosets look in a typical Cayley diagram for D_n?

Exercise 6.7.

(a) If e is the identity element, then what is $\langle e \rangle$?

(b) If a is in the subgroup $\langle b, c \rangle$, then what is $\langle a, b, c \rangle$?

(c) If a is not in the subgroup $\langle b, c \rangle$, and that subgroup's index is 2, then what is $\langle a, b, c \rangle$?

(d) If a is not in the subgroup $\langle b, c \rangle$, and that subgroup's order is 28, then what do you know about the order of $\langle a, b, c \rangle$?

Exercise 6.8. Consider a cyclic group C_n. If m is a number that divides n, then describe all the subgroups C_n has of order m.

Exercise 6.9. If a is the permutation that interchanges the numbers 1 and 2, but leaves all other numbers alone, then what is $[S_n : \langle a \rangle]$?

Exercise 6.10. Which of the following statements are true? Explain each of your answers.

(a) A group with 18 elements cannot have a subgroup of order 9.

(b) A group with 22 elements cannot have a subgroup of order 12.

(c) A group with 12 elements cannot have a subgroup of order 22.

(d) A group with 10 elements must have a subgroup of order 10.

(e) A group with 12 elements must have a subgroup of order 6. (See Exercise 6.31.)

Exercise 6.11. A particular special case of Lagrange's Theorem bears highlighting on its own. As we learned in Chapter 5, any element g in a group can be used to create a cyclic subgroup called the orbit of g. We write $|g|$ as shorthand for the size of that orbit (i.e., $|g| = |\langle g \rangle|$), and call it *the order of g*.

Explain why the order $|g|$ of any element must divide the order $|G|$ of the group.

Exercise 6.12. Answer each of the following questions and explain why your answers are correct.

(a) How many subgroups does G have when $|G|$ is prime?

(b) What is the orbit of an element in such a group G?

(c) How many groups of order p are there, when p is prime?

Exercise 6.13.

(a) What subgroup of A_4 is generated by the following two permutations?

(b) What subgroup of S_4 is generated by the following two permutations?

Exercise 6.14. Is the "turned-around" version of Lagrange's Theorem mentioned at the end of this chapter at least true about cyclic groups?

6.6. Exercises

Exercise 6.15. Leading up to Observation 6.6, I gave an informal explanation of why that observation is true. Let's make that explanation more formal. Recall that Observation 6.6 says that if b is in aH, then $aH = bH$. So assume in the group G that the element b is in the left coset aH, and answer the following questions.

(a) What do we know about how a and b connect in a Cayley diagram of G?

(b) Explain why every node that can be reached from b using the generators of H can also be reached from a using those same generators.

(c) Explain why every node that can be reached from a using the generators of H can also be reached from b using those same generators.

(d) Explain why correct answers to parts (b) and (c) prove that whenever b is in aH, then $aH = bH$.

Exercise 6.16. Correct answers to this exercise show that all the work in this chapter that was specific to left cosets could also be done for right cosets.

(a) The justification following Observation 6.3 spoke from the point of view of left cosets. Explain why Observation 6.3 is also true about right cosets. That is, why is every element of the group in some right coset?

(b) Explain what changes would need to be made to my questions and your answers in Exercise 6.15 to make them justify a version of Observation 6.6 about right cosets.

(c) If Theorem 6.7 had been about right cosets, what changes would you need to make to its proof? Feel free to make reference to your answers to parts (a) and (b).

(d) Why is the number of elements in Hg equal to $|H|$?

(e) If Theorem 6.8 had been about right cosets, what changes would you need to make to its proof? Again, you may make use of earlier answers.

Exercise 6.17. Consider the group $\mathbb{Z}$ from Exercise 5.30.

(a) If n is in $\mathbb{Z}$, then is $\langle n \rangle$ a subgroup of $\mathbb{Z}$? What does it contain?

(b) For which $n \in \mathbb{Z}$ is $\langle n \rangle = \mathbb{Z}$?

(c) If n and m are both in $\mathbb{Z}$, then when is $\langle n \rangle < \langle m \rangle$?

(d) Can $\langle n \rangle$ itself also have subgroups?

(e) What subgroup of $\mathbb{Z}$ is generated by 2 and 5?

(f) What subgroup of $\mathbb{Z}$ is generated by 4 and 6?

Exercise 6.18.

(a) What is the one coset of $\langle 2 \rangle$ in $\mathbb{Z}$?

(b) How many cosets does $\langle 3 \rangle$ have in $\mathbb{Z}$, and what are they?

(c) Is $\langle n \rangle$ a normal subgroup of $\mathbb{Z}$?

Exercise 6.19. Recall the group $\mathbb{Q}$ of rational numbers under the operation of addition, from Exercise 4.33 part (a).

(a) Compare its subgroup $\langle 2 \rangle$ to the subgroup $\langle 2 \rangle$ of $\mathbb{Z}$. Note any similarities or differences.

(b) Compare its subgroup $\langle 2, 3 \rangle$ to the subgroup $\langle 2, 3 \rangle$ of $\mathbb{Z}$. Note any similarities or differences.

Exercise 6.20. For each H and G given below, find all left cosets of H in G, then state the index $[G : H]$.

(a) $H = \langle 4 \rangle$, $G = C_{20}$

(b) $H = \langle 6 \rangle$, $G = C_{15}$

(c) $H = \langle f \rangle$, $G = D_4$

(d) $H = A_4$, $G = S_4$

Exercise 6.21. List all the subgroup relationships that exist among the following eight groups.

$$C_2 \quad C_3 \quad C_4 \quad C_6 \quad \mathbb{Z} \quad \mathbb{Q}$$

$\mathbb{Q}^+$, as in Exercise 4.33 $\quad$ the even integers, $\langle 2 \rangle$ in $\mathbb{Z}$

6.6.3 Hasse diagrams

Exercise 6.22. The collection of all subgroups of a group can be arranged in a *Hasse diagram*, which follows the following three rules.

1. At the top of the Hasse diagram we place the whole group, and at the bottom the trivial subgroup $\{e\}$.
2. Subgroups are arranged between, with larger ones higher up.
3. Vertical or diagonal lines connect smaller subgroups to those larger subgroups that contain the smaller ones.

Here are two examples, the Hasse diagram for all subgroups of C_5 on the left, and for S_3 on the right.

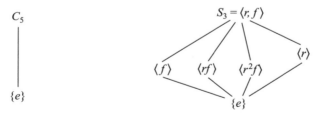

(a) Why are there no lines among any of the subgroups $\langle f \rangle$, $\langle rf \rangle$, $\langle r^2 f \rangle$, and $\langle r \rangle$ in S_3?

(b) What is the smallest subgroup of S_3 containing both f and rf?

(c) What is the smallest subgroup of C_5 with more than one element in it?

(d) Label each line in each of the above two Hasse diagrams with the index of the smaller subgroup in the larger. For example, you would label the lower-leftmost line in the S_3 diagram with 2, because $[\langle f \rangle : \{e\}] = 2$.

6.6. Exercises

Exercise 6.23. Make a Hasse diagram for each of the groups whose list of subgroups you found in Exercise 6.5.

Exercise 6.24. Make a Hasse diagram for the group C_{24}. Recall Exercise 6.8, whose answer will speed your search for subgroups.

6.6.4 Organizing visualizations

Exercise 6.25. In each of the following cases, draw a Cayley diagram for the given group that emphasizes the given subgroup. Choose a layout for the nodes that clusters together each left coset of the given subgroup.

(a) the subgroup $\langle 2 \rangle$ in C_4

(b) the subgroup $\langle 3 \rangle$ in C_9

(c) the subgroup $\langle 011, 110 \rangle$ of $C_2 \times C_2 \times C_2$, as diagrammed in Figure 6.3

(d) the subgroup $\langle r^2, f \rangle$ of $D_4 = \langle r, f \rangle$
Hint: Use arrows for r, r^2, and f.

(e) the subgroup A_4 of S_4
Hint: Take a Cayley diagram for S_4 from Chapter 5, label an element as the identity, and then square each element in the diagram visually (treating it as a path from the identity node). This quick, visual way to compute A_4 results in a diagram that must simply be reorganized to cluster A_4 apart from its one coset.

Exercise 6.26. Layouts for multiplication tables are determined by the ordering of the elements along the row and column headings. For most purposes, the order across the column headings should match the order down the row headings, and so just one order should be chosen.

It is usually helpful to order the elements in an organized way. If the group is cyclic, the natural order is the sequence of the cycle. Otherwise, choose a subgroup to list first, and then list its cosets in order thereafter, as was done in Figure 6.10.

Doing so emphasizes the subgroup by placing that subgroup's multiplication table in the upper left corner of the larger multiplication table. You can see this in Figure 6.10 and in the following table, organized by the subgroup $\langle r \rangle$ in S_3.

	e	r	r^2	f	r^2f	rf
e	e	r	r^2	f	r^2f	rf
r	r	r^2	e	rf	f	r^2f
r^2	r^2	e	r	r^2f	rf	f
f	f	r^2f	rf	e	r	r^2
r^2f	r^2f	rf	f	r^2	e	r
rf	rf	f	r^2f	r	r^2	e

Notice that the elements of the subgroup, too, are ordered using the same principles—because it is cyclic, they are in the order of the cycle; if the subgroup had not been cyclic, I would have chosen one of its subgroups to help me organize it, and so on.

(a) Organize a multiplication table of V_4 by an order-2 subgroup.

(b) Organize a multiplication table of C_8 by the subgroup $\langle 2 \rangle$.

(c) Organize a multiplication table of C_8 by the subgroup $\langle 4 \rangle$.

(d) Organize a multiplication table of C_9 by the subgroup $\langle 3 \rangle$.

Exercise 6.27. The previous exercises had you organize Cayley diagrams and multiplication tables in different ways. Are there also different ways to organize cycle graphs? If so, what significance do they have? Can they be used to emphasize specific subgroups? Why or why not?

6.6.5 Finding examples

When doing the exercises in this section, rely upon the library of groups you learned in Chapter 5.

Exercise 6.28. For each of the following questions, either find a group that answers the question in the affirmative or give a clear explanation of why the answer to the question is negative.

(a) Is there a non-cyclic group all of whose proper subgroups are cyclic?

(b) Is there a non-abelian group all of whose proper subgroups are abelian?

Exercise 6.29. For each of the following questions, either find a group that answers the question in the affirmative or give a clear explanation of why the answer to the question is negative.

(a) Is there a group of order 8 with a subgroup whose cosets partition the group into two different cosets (each coset therefore containing four elements)?

(b) Is there a group of order 8 with a subgroup whose cosets partition the group into eight different cosets (each element in its own coset)?

(c) Is there a group of order 8 with a subgroup whose cosets partition the group into just one big cluster (every element in one coset)?

(d) Is there a group of order 30 with a subgroup whose cosets partition the group into 20 different cosets?

(e) Is there an abelian group with a subgroup whose left and right cosets partition the group differently?

Exercise 6.30. For each integer n between 1 and 5, find a non-abelian group G with a subgroup H such that $|H| \geq 3$ and $[G : H] = n$.

Exercise 6.31. This exercise investigates subgroups of A_4. By Lagrange's Theorem, the only sizes possible for subgroups of A_4 are 1, 2, 3, 4, 6, and 12.

You can either do the work of this exercise using permutations, or using the following multiplication table for A_4. In it, the elements are colored according to their order: a, b, c, and d are of order 3, while the elements x, y, and z are of order 2. This coloring should prove helpful in answering the questions below.

6.6. Exercises

	e	x	y	z	a	b	c	d	a^2	b^2	c^2	d^2
e	e	x	y	z	a	b	c	d	a^2	b^2	c^2	d^2
x	x	e	z	y	b	a	d	c	c^2	d^2	a^2	b^2
y	y	z	e	x	d	c	b	a	b^2	a^2	d^2	c^2
z	z	y	x	e	c	d	a	b	d^2	c^2	b^2	a^2
a	a	c	b	d	a^2	d^2	b^2	c^2	e	x	z	y
b	b	d	a	c	c^2	b^2	d^2	a^2	x	e	y	z
c	c	a	d	b	d^2	a^2	c^2	b^2	z	y	e	x
d	d	b	c	a	b^2	c^2	a^2	d^2	y	z	x	e
a^2	a^2	b^2	d^2	c^2	e	y	x	z	a	c	d	b
b^2	b^2	a^2	c^2	d^2	y	e	z	x	d	b	a	c
c^2	c^2	d^2	b^2	a^2	x	z	e	y	b	d	c	a
d^2	d^2	c^2	a^2	b^2	z	x	y	e	c	a	b	d

(a) Describe all subgroups of order 1.

(b) Describe all subgroups of order 12.

(c) A subgroup of order 2 must be structurally the same as what group?
Express each order-2 subgroup of A_4 using generator notation.

(d) A subgroup of order 3 must be structurally the same as what group?
Express each order-3 subgroup of A_4 using generator notation.

(e) In part (a) of Exercise 5.37 we saw that there are only two groups of order 6, C_6 and S_3. Each of them has an element of order 2 and one of order 3. Thus we can look for subgroups of order 6 by pairing up order-2 elements with order-3 elements.
Express each order-6 subgroup of A_4 using generator notation.

(f) What is the significance of your result from part (e)?

7

Products and quotients

Chapter 6 looked inside groups to find subgroups, and so taught us something about the groups' internal structure. This chapter moves in the opposite direction, showing how groups can be assembled together to construct larger groups. We will learn two such construction processes, each a different kind of group product operation. Here I'm using the word "product" in its mathematical sense, meaning a kind of multiplication. We will also learn how certain subgroups can be the key to reversing such a process, facilitating a quotient operation that deconstructs any kind of product, revealing how a larger group may be constructed from smaller groups. This will enable us to picture clearly and simply the structure of many large groups, bringing otherwise large, complex objects within intellectual reach.

This chapter therefore teaches three processes: how to perform each of two product operations and how to perform a quotient operation. Each process is illustrated by several examples, and the exercises at the end of the chapter ask you to practice each of these processes. It is essential to "get your hands dirty" with such practice, because doing so strengthens your intuition for both the processes themselves and the groups that they construct and deconstruct.

I have organized this chapter's exercises to help you build your intuition one topic at a time. Each section in the chapter corresponds to a section in the exercises. So when you finish reading Section 7.1, if you feel you need to practice those concepts to strengthen your understanding before reading further, you can jump straight to the exercises in Section 7.6.1, which cover only the material from Section 7.1. Each section in the chapter follows this pattern, so that the material can be digested in bite-sized pieces.

We begin with the simpler of the two product operations we will study, the direct product.

7.1 The direct product

You have seen some direct product groups already in this book. Every group whose name contains the multiplication symbol × can be constructed using a process called the *direct*

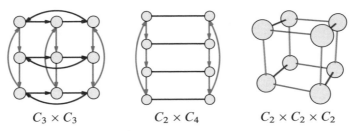

Figure 7.1. Three example direct product groups

product. The first such groups we saw were $C_3 \times C_3$ and $C_2 \times C_2 \times C_2$ in Figure 2.10, but we have seen others since then, such as $C_2 \times C_4$ in Figure 5.9. For convenience, I repeat the Cayley diagrams for these three groups in Figure 7.1. The $\times$ symbol is often pronounced "cross" or "times," so you might read $C_2 \times C_4$ as "cee two cross cee four." It is not a requirement that direct products only involve various C_n, but small examples usually do. In a few pages we will come to the larger example $S_3 \times C_2$.

Section 7.1.1 explains how to construct direct product groups and demonstrates the process on several examples. After that introduction to direct product groups, we will be prepared to learn how they will benefit our study of group theory in general (Section 7.1.3).

Before we learn the details of the direct product operation, let's pick up a few fundamentals by inspecting the examples in Figure 7.1. Those diagrams show us two reasons why the operation we are about to learn is a *product*, and uses the $\times$ notation. First, the diagram of $C_3 \times C_3$ is obviously a three-by-three (3×3) grid, making the name $C_3 \times C_3$ sensible. Second, $C_3 \times C_3$ is therefore obviously a group with nine elements, since after all, $3 \times 3 = 9$. These same two basic ideas also hold for $C_2 \times C_4$ (a two-by-four grid of eight elements) and $C_2 \times C_2 \times C_2$ (a two-by-two-by-two grid of eight elements), as we can see from the figure.

7.1.1 Constructing a direct product visually

I introduce the direct product construction as a visual procedure. From Cayley diagrams for groups A and B, we will create a Cayley diagram for the group $A \times B$ as described in Definition 7.1. Each step in the process is exemplified in Figure 7.2, using the groups C_2 and C_4 for A and B, respectively. The example therefore constructs the middle diagram from Figure 7.1. I later extend this method to cover multiplication tables, and from there we can gain an algebraic understanding of it.

Definition 7.1 (technique for constructing direct products using Cayley diagrams). To create a Cayley diagram of $A \times B$ from Cayley diagrams of A and B, proceed as follows.

1. Begin with the Cayley diagram for A.

    In the example in Figure 7.2, the group $A = C_2$ appears on the left.

2. Inflate each node in the Cayley diagram of A and place in it a copy of the Cayley diagram for B.

    In the middle of Figure 7.2, each of the two nodes of C_2 grew taller to contain a copy of a Cayley diagram for C_4.

7.1. The direct product

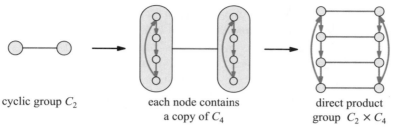

Figure 7.2. The construction of a Cayley diagram for the group $C_2 \times C_4$, using Cayley diagrams of C_2 and C_4, as described in Definition 7.1.

3. Remove the (inflated) nodes of A while using the arrows of A to connect corresponding nodes from each copy of B. That is, remove the A diagram but treat its arrows as a blueprint for how to connect corresponding nodes in the copies of B.

In the example in Figure 7.2, each C_4 has four nodes. Thus the top pair of nodes correspond to one another, and we connect them as a little copy of C_2. We do the same with the pair below them, and so on through all four pairs, resulting in a complete Cayley diagram for $C_2 \times C_4$. Because this step is more complicated than the first two, Figure 7.3 is devoted to illustrating this step alone.

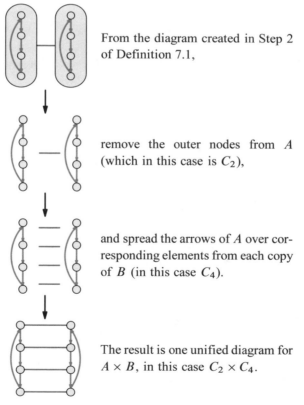

From the diagram created in Step 2 of Definition 7.1,

remove the outer nodes from A (which in this case is C_2),

and spread the arrows of A over corresponding elements from each copy of B (in this case C_4).

The result is one unified diagram for $A \times B$, in this case $C_2 \times C_4$.

Figure 7.3. An example execution of Step 3 of Definition 7.1, illustrated in greater detail than Figure 7.2 shows. As in Figure 7.2, in this example, $A = C_2$ and $B = C_4$.

The group $A \times B$ whose Cayley diagram results from the procedure in Definition 7.1 is called the ***direct product of*** A ***and*** B***,*** and we call A and B its ***factors.*** Note that it was essential to use different colors for A arrows than B arrows in Step 2, so that the merge in Step 3 created a valid diagram. Note also that it was helpful (though not strictly essential) to orient the copies of B perpendicular to the copy of A.

Notice that $C_2 \times C_4$ has copies of C_2 as its rows and copies of C_4 as its columns. The process described in Definition 7.1 uses two copies of C_4 and connects corresponding pairs of nodes from each copy, turning each pair into a copy of C_2. But the same diagram would have been achieved doing the product in the other order. Figure 7.4 shows that the direct product $C_4 \times C_2$ gives a group that is isomorphic to $C_2 \times C_4$. The $C_4 \times C_2$ structure is simply the $C_2 \times C_4$ structure turned on its side. It turns out that $A \times B$ always has the same structure as $B \times A$, for any groups A and B. We say that the direct product operation is *commutative,* and Exercise 8.36 asks you to prove it. For now, I use this fact without having given evidence for it.

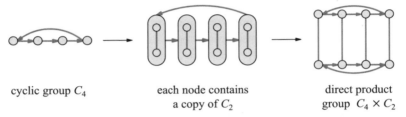

cyclic group C_4 each node contains a copy of C_2 direct product group $C_4 \times C_2$

Figure 7.4. Constructing a Cayley diagram for the group $C_4 \times C_2$, using Cayley diagrams of C_4 and C_2 and following the procedure in Definition 7.1. The result is a group that is equivalent to $C_2 \times C_4$, as constructed in Figure 7.2.

The nodes in the Cayley diagrams that the direct product process creates have no labels, but it would be helpful if they did so that we may refer to individual elements by name. There is a standard way to give names to the elements in a direct product group. High school algebra courses teach students to label points in the plane with names like $(1, 3)$ and $(-10, 6)$. These names are (x, y) pairs, the left number designating the point's x position and the right number its y position. In $A \times B$, we give each element a name that is a pair (a, b), with the a coming from A and the b coming from B. In the two simple examples we've seen so far, we can follow the high school algebra convention, with the first entry in the pair (the a) representing horizontal position, and the second entry (the b) representing vertical position, as shown in Figure 7.5.[1] In general, for a node in any direct product Cayley diagram, both components of its name come from Step 2 of Definition 7.1. A node gets the a component of its name from the inflated A node to which it belonged, and the b component of its name from the inner B node that it once was. In the simple examples in Figure 7.5, these correspond to columns and rows, respectively.

Given our recent work with subgroups, you may have noticed that C_2 is a subgroup of $C_2 \times C_4$; specifically, it is the subgroup $\langle (1, 0) \rangle$. Furthermore, the cosets of $\langle (1, 0) \rangle$ are clearly visible as the rows of $C_2 \times C_4$. They are both left and right cosets, which makes

---

[1] You could instead choose to do it the other way around, following the row-then-column convention of matrix algebra. The difference is superficial. In more complicated examples, it won't be possible to keep the horizontal/vertical distinction, but it's helpful when getting started.

7.1. The direct product

Steps 2 and 3 in the construction of $C_2 \times C_4$, with the element naming convention included

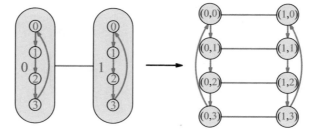

Steps 2 and 3 in the construction of $C_4 \times C_2$, with the element naming convention included

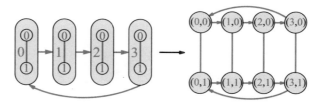

Figure 7.5. An element (a, b) in $A \times B$ gets its name from the a in A and b in B that gave rise to it in Step 2 of Definition 7.1. Colored numbers are used to make it easier to see that the names of the inflated A nodes go on the left of each pair, and the names of the inner B nodes on the right.

C_2 a normal subgroup. I mentioned normal subgroups in passing in Chapter 6, but they will be important throughout this chapter, so it's time for a formal definition.

Definition 7.2 (normal subgroup). A subgroup $H < G$ is called *normal* if each left coset of H is also a right coset of H (and vice versa). We indicate that H is a normal subgroup of G by writing $H \triangleleft G$.

We can therefore write the fact we just noticed as $\langle (1, 0) \rangle \triangleleft C_2 \times C_4$. Similarly, C_4 is normal in $C_2 \times C_4$ as the subgroup $\langle (0, 1) \rangle$ and its cosets are the columns of $C_2 \times C_4$. We will see that the factors in any direct product are normal subgroups (Exercise 7.12). That is, for any groups A and B, we will always have $A \triangleleft A \times B$ and $B \triangleleft A \times B$.

7.1.2 More direct product examples

Let's flesh out our understanding of the direct product by moving beyond the basic example of $C_2 \times C_4$, to the group $C_2 \times C_2 \times C_2$, which we first saw long ago in Figure 2.10. A three-dimensional direct product, such as $C_2 \times C_2 \times C_2$, can be thought of as two products taken successively, as in $(C_2 \times C_2) \times C_2$. The parentheses suggest we should first compute $C_2 \times C_2$, and then take the direct product of *that* group with C_2 again. This involves using the process in Definition 7.1 twice in succession, as illustrated in Figure 7.6. Notice that each time a diagram's nodes are inflated to contain copies of C_2, a new direction is used for the sake of clarity. The first inflation creates vertical C_2 copies and the second creates C_2 copies that extend deeper into the page, perpendicular to it. The result is a two-by-two-by-two grid, as the name $C_2 \times C_2 \times C_2$ suggests.

Thinking of the group as $(C_2 \times C_2) \times C_2$ suggests assigning the elements awkward names like $((0, 1), 0)$. The usual convention, however, is to use only one set of parentheses, naming group elements things like $(0, 1, 0)$. The first number in this three-part name is an element from the first C_2 factor, the second is from the second, and the third is from

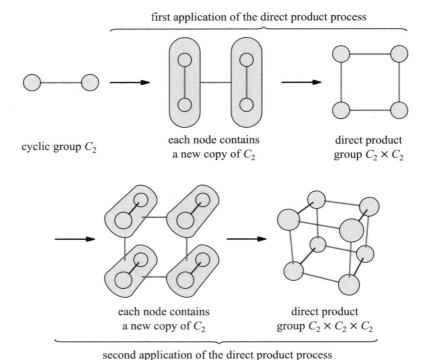

Figure 7.6. An example direct product with three factors, $C_2 \times C_2 \times C_2$. Constructing such a product involves applying Definition 7.1 twice, as marked in the figure.

the third. Similarly, for a group like $C_2 \times C_2 \times C_2 \times C_2$, elements will have names like $(1, 0, 1, 1)$, and so on for larger products. I occasionally shorten such names to make diagrams easier to read, as I did in Figure 2.10, dropping parentheses and commas altogether and turning $(0, 1, 0)$ into 010. This shortening is handy for making diagrams readable, as long as the names of the elements remain clear.

Our final example, $S_3 \times C_2$, involves a slightly larger, non-abelian group. It is not convenient to line up all the nodes in S_3 in one row, as I did in the previous example, so I render it as I have before, as on the left of Figure 7.7. When filling each node in S_3 with a copy of C_2 in the middle of that figure, I align the copies of C_2 to extend deeper into the page, as I did in the $C_2 \times C_2 \times C_2$ example. The result, on the right of Figure

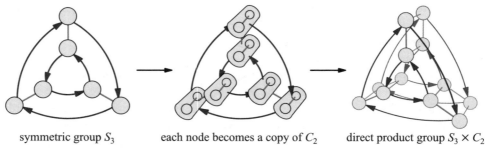

Figure 7.7. The direct product construction for $S_3 \times C_2$

7.1. The direct product

7.7, is our first non-abelian direct product. Following the naming convention from the previous section, elements in the direct product will have the following names.

Front copy of S_3: $(e, 0)$ $(r, 0)$ $(r^2, 0)$ $(f, 0)$ $(rf, 0)$ $(r^2 f, 0)$
Back copy of S_3: $(e, 1)$ $(r, 1)$ $(r^2, 1)$ $(f, 1)$ $(rf, 1)$ $(r^2 f, 1)$

7.1.3 Why direct products?

We have seen how to construct direct products, but of what value is this technique? It is one of several common, simple techniques for creating larger groups. If we understand the technique, we can understand some large groups more simply. This section explains how, using the fact that the factors in a direct product are independent. In a Cayley diagram for $A \times B$, following A arrows neither impacts nor is impacted by the location in group B.

For example, in the $C_4 \times C_2$ diagram I constructed in Figure 7.4, following a horizontal (C_4) arrow navigates to a new column, but stays in the same row. Another way to say it is that the green A arrows only impact the left component of each element's name (the A component). The orbit of the green arrow in $C_4 \times C_2$ is as follows.

$$(0, 0) \longrightarrow (1, 0) \longrightarrow (2, 0) \longrightarrow (3, 0) \longrightarrow (0, 0) \longrightarrow \cdots$$

Similarly, following a vertical (C_2) arrow changes the row and thus the second component of the element name, but stays in the same column and thus keeps the first component of the element name the same.

$$(0, 0) \longrightarrow (0, 1) \longrightarrow (0, 0) \longrightarrow \cdots$$

In a sense, the A arrows move you only in A and not in B, and the B arrows move you only in B but not in A. This independence of the two factors in a direct product leads to a new, useful way to think about direct product groups, which I now introduce.

We normally think of navigating a Cayley diagram by picturing each node as a place and the arrows as paths between places. You can picture yourself located at just one node at a time, and walking a path to get to another. As you do this imaginary walk through a group, what would it be like to stop and check a roadside map? Such a map could be a Cayley diagram for the group, with a "You are here" marker pointing to the node in which you stand, like one of the maps in Figure 7.8.

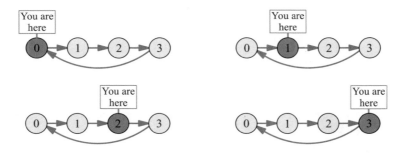

Figure 7.8. Four example roadside maps for a Cayley diagram of the group C_4, one map for each node

Navigating a direct product group can be thought of as navigating *two different Cayley diagrams simultaneously, and independently.* Figure 7.9 shows maps analogous to those in Figure 7.8, but each map in Figure 7.9 shows *two* groups and highlights a location in *each.* Perhaps rather than trying to imagine yourself in two places at once, imagine yourself remotely directing the movements of two others, sometimes instructing one person who's walking in C_4 which path to follow, and at other times instructing another person who's walking in C_2 to move. Your directing two people through two Cayley diagrams is a perfect analogy to your navigating the direct product group by yourself. Each node in $C_4 \times C_2$ is named for a C_4 node and a C_2 node, and we can interpret those names as horizontal and vertical positions of the original node in the $C_4 \times C_2$ grid, *or* as two different nodes in separate diagrams, one for C_4 and one for C_2. For instance, the element $(1, 0)$ in $C_4 \times C_2$ can mean either the node in the second row and first column in the $C_4 \times C_2$ diagram in Figure 7.5, or it can mean the second roadside map in the top row of Figure 7.9. Thus we have two equivalent ways to think of a direct product group like $C_4 \times C_2$. We can think of the direct product Cayley diagram, or of two Cayley diagrams used in tandem.

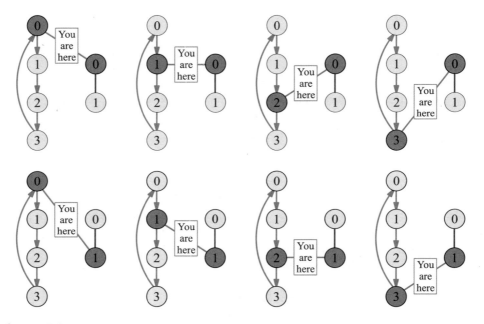

Figure 7.9. Each of these maps depicts a two-part location as you simultaneously navigate the two groups C_4 and C_2.

Let me make clear what I mean when I ask you to imagine directing two people who move independently. The motions of either person have no impact on either the position or the motion of the other person. A situation without such independence is shown in the rectangular Cayley diagram for D_4 in Figure 7.10. Vertical movements in that diagram can change horizontal positions, and thus the horizontal is not independent from the vertical. For instance, moving vertically from $(1, 0)$ does not result in $(1, 1)$, but rather in $(3, 1)$. In $C_4 \times C_2$, however, a vertical move from $(1, 0)$ would result in $(1, 1)$, causing no change in horizontal position. Although you could rearrange and rename the nodes of D_4 to fix

7.1. The direct product

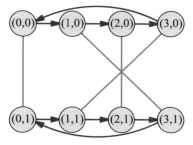

Figure 7.10. A rectangular layout of D_4, with elements named according to their row and column, as if it were a direct product group (which it is not).

this one problematic column, the tangles inherent in the structure of D_4 would create similar problems in other columns as you did so.

We saw our first example of two groups joined independently in Figure 2.8 on page 20. In the two-lightswitch group, each switch is independent of the other in the sense that flipping either one leaves the other unchanged. Alone, each switch is described by the group C_2, and so together, they are described by the group $C_2 \times C_2$, the more common name for which is V_4.

One of the benefits of viewing direct products as two groups to be navigated simultaneously is that some large groups become simpler to visualize. For instance, rather than draw the 56-element group $D_4 \times C_7$ as a rather messy 56-node Cayley diagram, we can instead draw two side-by-side Cayley diagrams, one for D_4 and one for C_7, and picture navigating them simultaneously and independently. Elements such as $(r^2, 5)$ represent two positions, one in D_4 and one in C_7, as shown in Figure 7.11. This involves drawing (and reading) many fewer nodes and arrows, and keeps us from having to look at any structure with more than eight nodes. Furthermore, not only is the 56-element group still faithfully represented (though less directly), but representing it as shown in Figure 7.11 helps us better understand the overall 56-element structure by seeing how it can be built by the direct product process. Thus direct products are one way of making some large groups easier to understand, by showing us how the larger group is built from smaller groups using a simple process.

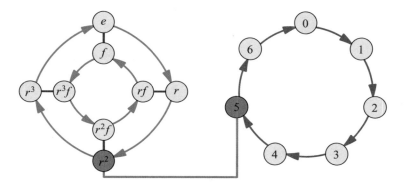

Figure 7.11. The element $(r^2, 5)$ from the group $D_4 \times C_7$, viewed as a pair of elements, one in a Cayley diagram from D_4 and one in a Cayley diagram of C_7.

7.1.4 The algebraic viewpoint

Both ways this book has taught us to approach group theory, from the actions-based viewpoint and from the algebraic one, are helpful in understanding group theoretic concepts. So let's try to understand direct product groups from the algebraic viewpoint as well, seeing them as a set of elements with a binary operation on it. Because multiplication tables show us groups in this way, knowing how to construct direct product multiplication tables is a good place to start. Fortunately, just as there is a visual way to create Cayley diagrams of direct product groups (Definition 7.1), there is also a visual way to create multiplication tables of direct product groups.

Definition 7.3 (technique for constructing direct products of multiplication tables). To create a multiplication table for the direct product group $A \times B$ from multiplication tables for the groups A and B, follow these three steps. Notice how analogous they are to the steps in Definition 7.1. Figure 7.12 illustrates them using $A = C_4$ and $B = C_2$ as an example.

1. Begin with a multiplication table for A.

   In Figure 7.12, the table for C_4 is shown with blue element names.

2. Inflate each cell in the table to include a copy of the entire multiplication table for B. To avoid losing the information from the original table for A, retain its labels as well.

   In the middle of Figure 7.12, the large, blue numbers were retained from the C_4 table, and now they sit over red copies of the table for C_2.

3. Rename the elements in the table as pairs, using the A elements on the left of each pair and the B elements on the right.

   Each number in each pair in the $C_4 \times C_2$ table of this example kept its original color so that you can tell which table it came from.

This construction process, illustrated in Figure 7.12, can tell us a lot about the binary operation in the product group $C_4 \times C_2$. For starters, we can use its final table to do multiplication in $C_4 \times C_2$. For example, to find $(2, 1) \cdot (3, 1)$, we can look up the entry in

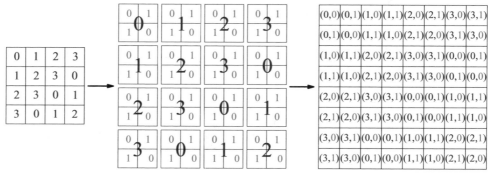

multiplication table for the group C_4 inflate each cell to also contain a copy of the multiplication table for C_2 join the little tables and element names to form the final direct product table

Figure 7.12. Constructing a multiplication table for $C_4 \times C_2$ from multiplication tables for C_4 and C_2, following the technique in Definition 7.3. Row and column headings could easily be added, but the process is more clearly illustrated without them.

7.1. The direct product

row $(2, 1)$ and column $(3, 1)$ of the table, finding the answer $(1, 0)$. But what we really want to know is *why* the table gives the answer that it does. The way the table was constructed answers that question.

The structure of the final table in Figure 7.12 was determined by the two overlaid tables in the middle of that figure. The pattern of the blue numbers in the direct product table was copied directly from the pattern of the large, blue numbers in the middle table, which in turn come from the original table for C_4. Thus the way the left (blue) elements in the pairs in the final table relate is determined by the original table; that is to say, it's determined only by their status in C_4, and in no way depends on the red elements with which they are paired. So the reason that the left component of the answer to $(2, 1) \cdot (3, 1)$ is 1 is that $2 + 3 = 1$ in C_4. The right elements are irrelevant when computing the left half of the answer. The left table in Figure 7.13 illustrates this by showing that any element whose left entry is 2 times any element whose left entry is 3 results in an element whose left entry is 1.

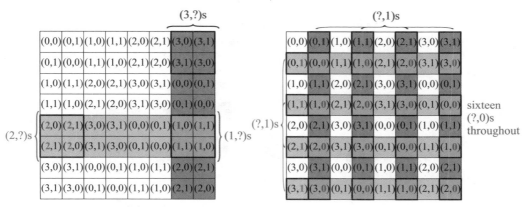

Figure 7.13. The multiplication table for $C_4 \times C_2$ constructed in Figure 7.12, used to show the independence of the factors in a direct product. The left table shows that any element $(2, ?)$ times any element $(3, ?)$ results in some element $(1, ?)$. The right table shows that any element $(?, 1)$ times any element $(?, 1)$ results in some element $(?, 0)$.

A similar fact holds for the red numbers in the final table of Figure 7.12. Their pattern is made from many copies of a C_2 table. The blue number under which a C_2 copy lay in the middle table in no way affected that table; it was the same everywhere. Thus the right component of the answer to $(2, 1) \cdot (3, 1)$ is 0 because $1 + 1 = 0$ in C_2; the 2 and 3 are irrelevant. This is illustrated on the right of Figure 7.13, which shows that any element whose right entry is 1 times any element whose right entry is 1 gives an element whose right entry is 0. The pattern depends upon the fact that all the C_2 copies are the same; when we do the $1 + 1$ computation in 16 identical copies of C_2, we get the same answer all 16 times.

So we find that to compute $(2, 1) \cdot (3, 1)$ in $C_4 \times C_2$ we only need to know how to compute $2 + 3$ in C_4 and $1 + 1$ in C_2. Each component in the pairs is treated completely independently, as summarized in the following equation.

$$(2, 1) \cdot (3, 1) = (2 + 3, 1 + 1) = (1, 0)$$

We have already seen the independence of the factors in a direct product show up in

Cayley diagrams; the principle above describes how that same independence appears in multiplication tables.

Now we are in a position to express that independence in algebraic terms. Take elements $a \in A$ and $b \in B$. If we use e to stand for the identity in each group, then the corresponding elements in the direct product group are (a, e) and (e, b). The following chain of equalities shows that (a, e) and (e, b) commute, based on the independence of factors discussed above.

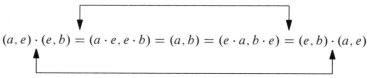

$$(a, e) \cdot (e, b) = (a \cdot e, e \cdot b) = (a, b) = (e \cdot a, b \cdot e) = (e, b) \cdot (a, e)$$

The equation as a whole says that (a, e) and (e, b) commute.

Thus the independence of factors in a direct product means that elements from one factor commute with elements from the other. Returning for a moment to considering Cayley diagrams, this means that arrows from one factor commute with arrows from the other. Figure 7.14 illustrates this idea using two different diagrams.

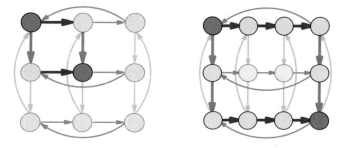

Figure 7.14. The left Cayley diagram illustrates the equation $a \cdot b = b \cdot a$ in $C_3 \times C_3$, and the right Cayley diagram illustrates the equation $a^3 \cdot b^2 = b^2 \cdot a^3$ in $C_4 \times C_3$. In each case, a is the red generator and b is the blue generator.

We now have a well-rounded view of the direct product, having seen it from both the actions-based and algebraic viewpoints. It is time to progress to another kind of product operation, also useful for breaking large groups down into smaller conceptual pieces. Although it is more intricate than the direct product, they have in common that if we understand the ingredients in the construction and the construction process, then we gain the ability to understand the resulting large group, even if it is too complex to diagram.

7.2 Semidirect products

The clean and simple direct product operation we just learned has two more complex cousins. I introduce the *semidirect product* here, but do not cover the more complex *knit product* in this book. Even the semidirect product depends on some concepts we will not learn until Chapter 8, so this section gives an initial taste of semidirect products, but the remainder must wait until the following chapter. As was the case with direct products,

7.2. Semidirect products

the reason for learning semidirect products is to have understandable ways to describe the structure of large groups more simply.

We've learned that direct products are made up of identical copies of one factor, with corresponding elements connected according to the pattern of the other factor. Semidirect products are very similar, in that they have two factors, and one determines the pattern for connecting corresponding elements in copies of the other. Where they differ is that those copies need not all be laid out in an identical way. Some twisting is permitted in the copies, and thus more intricacy is possible. To describe the kind of twists that semidirect products allow, I introduce the following term. It is not standard group theory terminology; in fact, we will replace it with a more technical term later. But for now, a visual term is appropriate.

Definition 7.4 (rewiring). I call one Cayley diagram a *rewiring* of another if all the following conditions are met.

(a) The two diagrams must have the same arrangement of nodes.

(b) The diagrams may have their arrows arranged differently.

(c) The algebraic relationships among group elements must be the same in both diagrams.

Condition (b) is not really a restriction, but appears on the list to highlight the major feature of rewirings, which is the differing patterns of arrows between the two diagrams. Another way to state condition (c) is that although the two Cayley diagrams are different, they correspond to the same multiplication table and thus describe the same group.

Figure 7.15 shows C_3 and a rewiring of it. Let's consider how it satisfies each of the three criteria of Definition 7.4. Condition (a) requires that the nodes be arranged the same in both diagrams, which is obvious from the figure. Condition (b) permits the arrows to connect the nodes differently in the two diagrams, which they do; one diagram cycles clockwise, the other counterclockwise. Verifying condition (c) means verifying that every equation that one diagram satisfies, the other does as well. (This is what "algebraic relationship" means here.) For example, the left diagram satisfies $a \cdot a = a^2$, because following two red a arrows from e reaches a^2. So we must verify that the right diagram also satisfies this equation. It does satisfy it, but for a different reason than the left diagram. In the right diagram, a is represented by two red arrows in succession, so $a \cdot a$ is four red arrows, which leads from e to a^2 only after going around the whole diagram once. We could also verify that both diagrams satisfy $a^3 = e$, and many other equations. To be thorough, we could convert each Cayley diagram into a multiplication table, and we would find the resulting multiplication tables to be the same. So the two diagrams describe the same group. Thus Figure 7.15 satisfies all three conditions.

Figure 7.15. The left diagram is a familiar Cayley diagram for C_3, and the right is a rewiring of it. Reversing either diagram's arrows yields the other diagram.

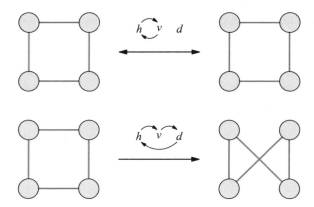

Figure 7.16. Two ways to rewire a Cayley diagram for V_4, described as permutations of horizontal (h), vertical (v), and diagonal (d). The permutations refer to diagonal (d) arrows even though the original diagram has only horizontal and vertical arrows, because they can be rearranged to diagonal positions.

Figure 7.16 shows a slightly larger example, two ways to rewire V_4, each described as a permutation of the diagrams' arrows. The top rewiring performed twice returns the diagram to its original state, because it is simply an exchange of horizontal arrows with vertical ones. The bottom rewiring, however, must be repeated three times to return to the original diagram, and it is therefore described by a permutation with order 3.

It may also be helpful to contrast these first two examples with two examples of what is *not* a rewiring, in Figure 7.17. Its top row shows a way to rearrange a Cayley diagram's arrows that does not produce another Cayley diagram. It is therefore not a rewiring because Definition 7.4 applies only to two Cayley diagrams. The bottom row of the same figure shows a transformation that does not keep the algebraic relationships among the group elements the same, even though it transforms one 4-cycle into another. For instance, $2 + 2 = 1$ in the right diagram, but to satisfy condition (c) of Definition 7.4 it ought to be $2 + 2 = 0$, as in the left diagram.

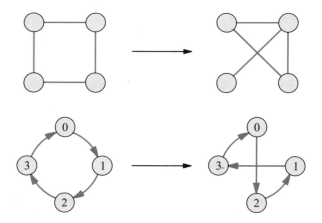

Figure 7.17. Two rearrangements of Cayley diagram arrows that are *not* rewirings.

7.2. Semidirect products

As you may have suspected from the fact that Figure 7.16 uses permutations to describe rewirings, we can put a collection of rewirings together to form a group. The simplest example is in Figure 7.15; not only does it show a way to rewire C_3, but if we view each Cayley diagram in Figure 7.15 as one single node, then the overall picture has the structure of the group C_2, whose only arrow is the one in the center of the diagram, labeled "reverse arrows." A more interesting example is the group of all rewirings of V_4, generated by the two rewirings from Figure 7.16. It has the structure of S_3, and is shown in Figure 7.18.

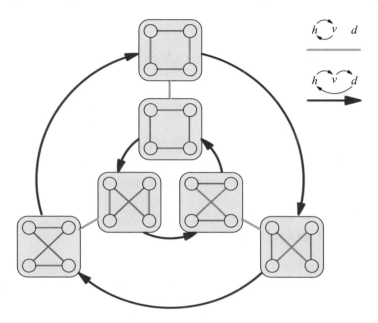

Figure 7.18. The group of all rewirings of V_4, generated by the two in Figure 7.16. The meanings of the arrows are given by the legend in the upper right.

Diagrams like those in Figures 7.15 and 7.18 are reminiscent of the direct product construction, because they treat Cayley diagrams as nodes to be connected. This is what Step 2 of Definition 7.1 asks us to do, as shown in the middle diagrams in Figures 7.2 and 7.4. The important difference between the direct product construction and groups of rewirings is that not all the Cayley diagrams in a rewiring group are identical. On the contrary, each is different! If we connect the rewirings in Figure 7.15 or 7.18 as Step 3 of Definition 7.1 taught us to do, the group that results is called a *semidirect product* group. Figure 7.19 illustrates doing such a connection to the diagram from Figure 7.15. We call the result a semidirect product of C_3 with C_2, because the individual groups in Figure 7.19 are rewirings of C_3, while its overall structure is that of C_2. A semidirect product of C_3 and C_2 is written $C_3 \rtimes C_2$.

This semidirect product construction can be stated more generally as follows: From a group G, draw a Cayley diagram of its rewiring group, and connect the rewirings as we connected copies in a direct product (in Step 3 of Definition 7.1). You could use this procedure to draw a Cayley diagram of the group $V_4 \rtimes S_3$ based on Figure 7.18. But semidirect product groups created using this procedure are just *some* of the semidirect

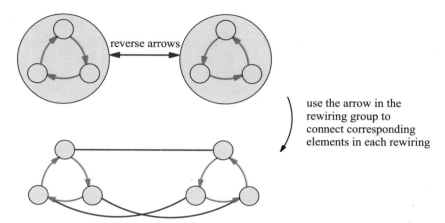

Figure 7.19. Applying Step 3 of Definition 7.1 to Figure 7.15 yields this semidirect product of C_3 and C_2, structurally identical to S_3.

product groups that exist. To be able to describe and build other semidirect product groups requires using homomorphisms to describe rewirings and combinations thereof. We will do so in Chapter 8, and will see how to construct various $A \rtimes B$ for any two groups A and B.

7.3 Normal subgroups and quotients

In Section 7.1.3 we saw that we could make some large groups more easily understandable if we described them as direct product groups. Understanding the factors and the direct product process gave us a simple way to picture an otherwise complex group. We need a similar technique for deconstructing semidirect products as well. It is therefore natural to wonder how to recognize a Cayley diagram of a product group when we see one. That is, if you didn't construct the Cayley diagram yourself, how could you tell that it represented $A \times B$ or $A \rtimes B$, for some groups A and B? And how could you tell what A and B are?

The most general way to deconstruct a large group into two factors is called taking a *quotient;* it reveals not only direct products but semidirect products as well. When a quotient exposes a direct or semidirect product, it gives us clear insight into the group's structure. (This will be even more true after we learn more about semidirect products in Chapter 8.) Quotients can even be used to organize groups that were not built by a direct or semidirect product operation, and the structure they reveal is often helpful for understanding the large group in those cases as well. Group products multiply groups, and group quotients divide them. We divide a group by one of its subgroups as follows. As with the other processes introduced in this chapter, I include in the definition an example that illustrates the process.

Definition 7.5 (quotient operation). To attempt to divide a group G by one of its subgroups H, follow these steps.

1. Organize a Cayley diagram of G by H (as we first did in Section 6.3).

    The top diagram in Figure 7.20 shows this for $G = C_6$ and $H = \langle 2 \rangle$. Notice that the cosets of H have been grouped together in rows, in preparation for the following steps.

7.3. Normal subgroups and quotients

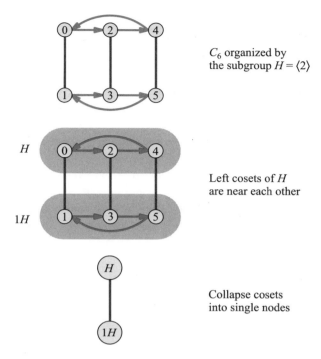

C_6 organized by the subgroup $H = \langle 2 \rangle$

Left cosets of H are near each other

Collapse cosets into single nodes

Figure 7.20. Application of the quotient process of Definition 7.5 to the group $G = C_6$ and subgroup $H = \langle 2 \rangle$, isomorphic to C_3. Notice that organizing the Cayley diagram of G by the subgroup H facilitates collapsing of the cosets of H.

2. Collapse each left coset of H into one large node.[2] Unite those arrows that now have the same start and end nodes. This forms a new diagram with fewer nodes and arrows.

   The middle and bottom diagrams in Figure 7.20 show this process in two steps.

3. If the new diagram is a Cayley diagram of a group, then you have obtained **the quotient group of G by H,** and the new diagram shows that group. If not, then G cannot be divided by H.

   The bottom diagram in Figure 7.20 is of the group C_2, so C_6 divided by C_3 is C_2.

We can already see one obvious way in which this quotient operation is like a reversal of both product operations we've seen. Both product processes created copies and arrows among them, while the quotient process collapses both copies and arrows. Another important detail to notice and understand in this example and in each example to come is that the *elements* in a quotient group are the *cosets* of the original subgroup H. The names of the nodes in the bottom diagram in Figure 7.20 make this clear.

But what was the original structure that this quotient operation deconstructed? Is C_6 a direct product, a semidirect product, or something else? We can tell that it is a direct product because the Cayley diagram for C_6 on the top of Figure 7.20 could have been created by the direct product construction: It is two copies of C_3 with corresponding elements

---

[2]I will use the informal word "collapse" from here on to refer to this clustering or collecting of several nodes together into one node. Probably the most natural English word for this type of uniting is "grouping," but obviously we're already using the word "group" to mean something else!

connected after the pattern of C_2. Step 1 of Definition 7.5 is essential because it reorganizes the Cayley diagram to bring out a structure that another layout of the same diagram may disguise. The group C_6 is therefore the group $C_3 \times C_2$, although we already knew this from the last chapter (and Figure 6.5 in particular). If you're wondering which other cyclic groups hide direct products inside them, we'll answer that question in Section 8.4!

Using the quotient process to deconstruct any direct product group reveals the two factors as follows. The left factor A is the subgroup by which Step 1 organizes the diagram, and the right factor B is the result of taking the quotient, the final diagram that the process creates.

A subgroup isomorphic to C_3 also appears in the group S_3, a group with a lot in common with C_6. Let's try to divide S_3 by C_3, and compare it to the previous example. Figure 7.21 shows the three steps of the quotient process applied to the group S_3. Although both quotients $\frac{C_6}{C_3}$ and $\frac{S_3}{C_3}$ give the result C_2, the groups are not the same. The quotient process has removed their differences.

I mentioned earlier that the quotient $\frac{C_6}{C_3}$ disassembled a direct product, so it is natural to ask whether the quotient $\frac{S_3}{C_3}$ also did. Notice that the blue arrows in S_3 cross in the top diagram of Figure 7.21; they do not connect elements of the C_3 subgroup with corresponding elements of its coset. Thus S_3 is *not* the direct product of C_3 with C_2. If we rearrange the nodes in the bottom row of S_3 to uncross the blue arrows, the result is shown

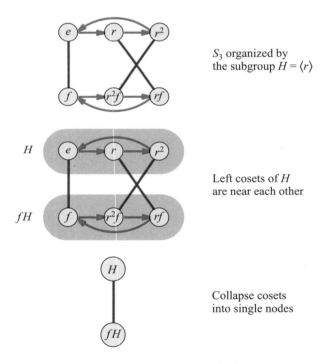

Figure 7.21. Application of the quotient process of Definition 7.5 to the group $G = S_3$ and subgroup $H = \langle r \rangle$, isomorphic to C_3. The result is the same as when we divided C_6 by a subgroup isomorphic to C_3, because the differences in the original groups have been removed by the quotient process.

7.3. Normal subgroups and quotients

in Figure 7.22. Although this means that corresponding elements are now connected, the bottom row is no longer an exact copy of the top row; it is a rewiring of it instead. Thus while S_3 is not a direct product, it is a semidirect product of C_3 with C_2. In fact, Figure 7.22 is just another layout of Figure 7.19.

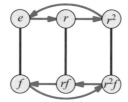

Figure 7.22. A rearrangement of the bottom row of S_3 from Figure 7.20 to uncross the vertical arrows.

We have therefore seen how quotient operations can reveal each of two different product structures. When a quotient operation yields identical copies with corresponding elements connected, it reveals a direct product structure. When it yields a group of rewirings with corresponding elements connected, it reveals a semidirect product structure. We will also see quotients that reveal neither of these patterns, but that nonetheless help us see structure in a large group (e.g., Exercise 7.18).

These first examples have been informative, but small. An instructive larger example is the group A_4, which we saw in Figure 5.27. That figure emphasizes the relationship between A_4 and the symmetries of the tetrahedron, but it doesn't tell us much about the group's internal structure. Finding a way to take a quotient of A_4 reveals a simpler structure therein. Let's take a quotient of A_4 by a subgroup isomorphic to V_4.

Following Step 1 of Definition 7.5, the top of Figure 7.23 shows A_4 organized by the subgroup $\langle x, z \rangle$. This reorganization required using arrows for the generators x and z, as well as a generator that connects cosets of $\langle x, z \rangle$, in this case a. Step 2 of Definition 7.5 asks us to collapse each of the three left cosets of $\langle x, z \rangle$ into a single node, uniting any arrows that are thereby rendered redundant. Doing so reveals a three-step cyclic structure in A_4, as shown on the bottom of Figure 7.23. Therefore $\frac{A_4}{V_4}$ is isomorphic to C_3.

Has this quotient reversed any kind of product? We can tell that it is not a direct product because the Cayley diagram on top of Figure 7.23 shows that the arrows connecting the cosets of H do not connect corresponding elements. For example, the top right node of H is not connected to the top right node of aH. Just as in the S_3 example above, this is not an issue simply of layout. Although we could rearrange the diagram's nodes to make the arrows connect corresponding elements of each coset, it would alter the layouts of those cosets, as in Figure 7.24. This trade-off is an unavoidable consequence of the tangled structure of A_4; you can either arrange the arrows the same or the cosets, but not both.

So A_4 is not $V_4 \times C_3$, but it may still be a semidirect product. Having rearranged the cosets of V_4 in Figure 7.24 so that the arrows connect corresponding elements, we notice that they form the outer ring of the semidirect product group $V_4 \rtimes S_3$ whose blueprint appears in Figure 7.18. Figure 7.25 shows the pattern. The group A_4 therefore forms one of that large group's subgroups, and when we learn the full generality of semidirect

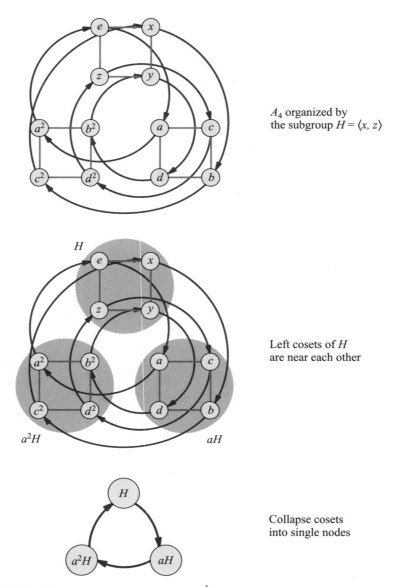

Figure 7.23. The quotient process computing $\frac{A_4}{V_4}$. All blue arrows from H lead to aH, and all blue arrows from aH lead to a^2H, and then back to H.

products in Chapter 8, we will learn how to express A_4 as a semidirect product of C_3 and V_4. For now, the quotient process has revealed to us that A_4 has the simple structure shown in Figure 7.25.

As the wording of Step 3 of Definition 7.5 suggests, in some cases an attempted quotient operation does not go smoothly, but instead results in an invalid Cayley diagram. It is therefore important for us to see an example of just such an invalid result, and A_4 can provide us with an example of this type as well. That example will help lead us to a theorem about when the quotient operation will succeed and when it will fail.

7.3. Normal subgroups and quotients

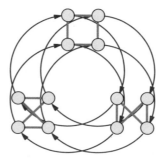

Figure 7.24. The cosets of V_4 in A_4 rewired to demonstrate its semidirect product structure.

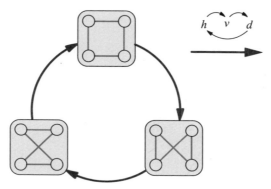

Figure 7.25. The pattern in A_4 as a subgroup of a rewiring group. To create A_4, connect the corresponding nodes in each rewiring with arrows in the three-cycle shown.

When we originally met A_4 in Figure 5.27, it was organized by the three-element subgroup $\langle a \rangle$. Figure 7.26 shows an attempt to divide A_4 by this subgroup, but the result is not a valid Cayley diagram. It is invalid for several reasons, but the most obvious one is that the only arrow type in the diagram (the blue) is ambiguous. For instance, from the top node, the blue arrows lead not to just one node, but to all three other nodes. Therefore this diagram does not depict a group. (Recall the visual criteria from Exercises 2.14 through 2.17, also mentioned in Section 6.1.)

So quotients succeed for some subgroups and not for others. In order to say we truly understand the quotient process, we need to know the reason for this. What is it about a group or a subgroup that determines whether a quotient is possible? As the title of this section suggests, the deciding factor is whether the subgroup is normal.

Theorem 7.6. *If $H < G$, then a quotient group $\frac{G}{H}$ can be constructed just when $H \triangleleft G$.*

Proof. The quotient process from Definition 7.5 succeeds just when the resulting diagram is a valid Cayley diagram. Most aspects of valid Cayley diagrams are guaranteed by the quotient process. For instance, because we begin with a diagram that has an arrow of every color exiting every node, our resulting diagram has this property as well. Because we begin with a regular diagram (as in Definition 6.1) and collapse identically structured sections distributed uniformly throughout the diagram, we end up with a regular diagram as well.

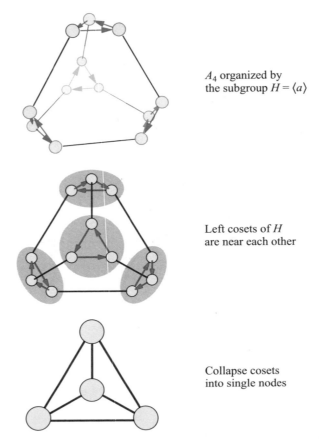

Figure 7.26. An organization of A_4 by a subgroup that is not normal, and therefore does not allow us to take a quotient.

One problem can arise. Recall the requirement that the final diagram in Figure 7.26 violates: only one arrow of a given color must enter or leave any given node. Because we collapse many nodes into one, dragging their various arrows along for the ride, the resulting diagram may not satisfy this requirement. So this theorem boils down to the following fact, for which I have yet to give evidence. Collapsing left cosets results in unambiguous arrows just when the subgroup is normal. I show that this is true by comparing the appearances of arrows that will become ambiguous after collapsing to the appearances of those that will not.

Collapsing cosets produces ambiguous arrows when some arrows of a given color go from nodes of one left coset to nodes of two or more different left cosets. The left illustration in Figure 7.27 shows this, and it is what caused problems in Figure 7.26. On the other hand, when all the like-colored arrows from any one coset go unanimously to another coset, no ambiguity arises, as the right illustration in Figure 7.27 exemplifies. This is what happened in the first three examples we saw of the quotient process, in Figures 7.20, 7.21, and 7.23.

The reason that unanimity of arrows occurs among left cosets just when the subgroup is normal lies in our original visualizations of left and right cosets from Chapter 6. Look

7.3. Normal subgroups and quotients

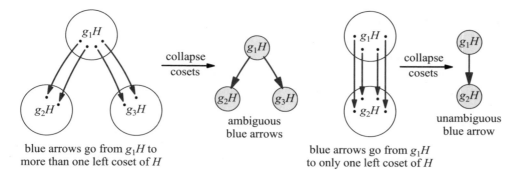

Figure 7.27. Arrows sending a left coset to several others cause ambiguity when the cosets are collapsed. Arrows unanimously sending a coset to one other coset do not create ambiguity after collapsing.

again at Figure 6.8 on page 104, which helped us picture a left coset gH as the copy of H to which the g arrow from e leads. That figure also depicts the right coset Hg as the nodes that g arrows can reach from H. These visualizations tell us that the defining characteristic of normal subgroups, $gH = Hg$, can be restated as follows.

> To whichever coset one g arrow leads from H (the left coset gH), all g arrows lead unanimously (because it is also the right coset Hg).

This is illustrated in Figure 7.28, which unites the two illustrations in Figure 6.8. Thus a subgroup is normal just when the arrows leaving the subgroup H are unanimous. Furthermore, because Cayley diagrams are regular, the pattern of arrows connecting H to each of its cosets will be repeated at every copy (left coset) of H throughout the whole diagram. So a subgroup is normal just when every color of arrow between cosets is unanimous, which is when the quotient operation succeeds. □

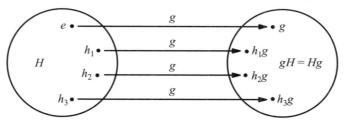

Figure 7.28. When H is normal, $gH = Hg$, and thus the illustrations in Figure 6.8 on page 104 lead us to picture g arrows as unanimously connecting H to gH.

This theorem shows us the very important role that normal subgroups play in group theory. Originally we described them as those subgroups whose left and right cosets match, which may not have seemed very significant at the time. But now we know its implications, that normal subgroups are those by which we can take a quotient, and thereby expose an overall structure that can help us better understand a large group. I therefore spend the final two sections of this chapter broadening our experience with normal subgroups and quotients in preparation for the more advanced material in Chapters 9 and 10.

7.4 Normalizers

Our return to studying normal subgroups begins with a simple question. Is there a way to measure how close a subgroup is to being normal? The answer to this question reveals several interesting patterns that can help our intuition about normal subgroups.

We have said that a subgroup H is normal when *every* element g in the group satisfies the requirement $gH = Hg$. So a simple way to measure how close a subgroup is to normal is to check how many of the $g \in G$ satisfy this requirement. As if G were a democracy, we give to each $g \in G$ a vote for whether H should be normal, and see how many vote in favor. Only those subgroups that earn a unanimous victory are given the coveted title "normal."

At a minimum, we know that every $g \in H$ must vote in favor, because for any $g \in H$, both gH and Hg are simply H. (Candidates usually get the votes in their hometown.) The other extreme is when H is normal, and all the $g \in G$ vote in favor. But there may be levels in between these two extremes as well, and in those cases we can use the number of elements that voted for H as a measure of how close it came to getting elected normal. The set of elements in G that vote in favor of H's normality are called the ***normalizer of H in G,*** usually written $N_G(H)$.

In this democracy, cosets of H always vote together. If some $g \in G$ satisfies $gH = Hg$, then every other element of gH does also. The reason comes from Observation 6.5, which told us that any $k \in gH$ could be used as the coset representative, so $gH = kH$ and $Hg = Hk$. Thus k votes for H ($kH = Hk$) just when g does ($gH = Hg$) because gH and kH are synonyms, as are Hg and Hk. This voting pattern narrows down the possibilities for what $N_G(H)$ might be, especially regarding its size. Because $N_G(H)$ is made up of whole cosets, its size will always be a multiple of $|H|$.

Furthermore, the deciding factor in how a left coset will vote is simply whether it is also a right coset, because gH votes for H just when $gH = Hg$. Thus the normalizer $N_G(H)$ is made up of all the left cosets that are also right cosets. This gives us a straightforward way to pick out $N_G(H)$ visually in a Cayley diagram. Because it is made up of those left cosets that are also right cosets, we can describe it more visually by saying that it is made up of those copies of H that are connected to H by unanimous arrows. Let's look at a few examples.

I start with a normal subgroup for simplicity. We saw in Figure 7.23 that $\langle x, z \rangle \triangleleft A_4$, and therefore each of its three left cosets is also a right coset; they are all connected to one another by unanimous arrows. Thus $N_{V_4}(A_4) = A_4$ because the subgroup is normal. The other extreme appears in Figure 7.26, showing $\langle a \rangle$, a non-normal subgroup of A_4. It has four left cosets (counting itself) but the arrows are spread so broadly that *no* left coset is also a right coset. Thus no element outside the subgroup $\langle a \rangle$ votes for it, making $\langle a \rangle$ as non-normal a subgroup as we'll ever find; $N_{A_4}(\langle a \rangle) = \langle a \rangle$.

Of course a more interesting example would be neither of these extremes, a subgroup that measures as being somewhat normal, but not all normal. Figure 7.29 shows D_6 organized by left cosets of $\langle f \rangle$. Consider the left coset $r\langle f \rangle$, labeled in the figure. There is no unanimous collection of arrows connecting $\langle f \rangle$ to $r\langle f \rangle$. The arrow from e to r makes $r\langle f \rangle$ a left coset, but the other r arrow leaving H disagrees, heading to $r^5\langle f \rangle$ instead. Verify for yourself that the same is true for each of the cosets $r^2\langle f \rangle$, $r^4\langle f \rangle$, and

7.4. Normalizers

$r^5\langle f\rangle$; none have a connection to $\langle f\rangle$ by a unanimous set of arrows. But the case of $r^3\langle f\rangle$ is different; the r^3 paths do indeed head from $\langle f\rangle$ to $r^3\langle f\rangle$ unanimously, although in this Cayley diagram one goes clockwise and the other counterclockwise. Similarly, r^3 paths returning from $r^3\langle f\rangle$ to $\langle f\rangle$ are also in agreement. Thus the normalizer $N_{D_6}(\langle f\rangle)$ does not contain only $\langle f\rangle$, but one of its cosets is as well, $r^3\langle f\rangle$. Though $\langle f\rangle$ does not get elected to be a normal subgroup, it does get some votes outside its hometown.

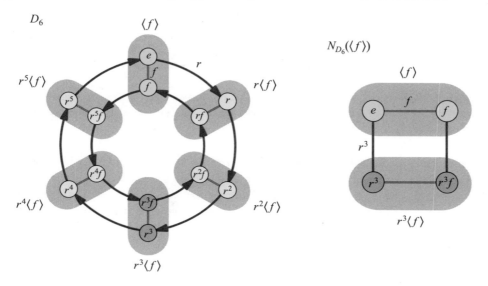

Figure 7.29. The subgroup $\langle f\rangle$ in D_6 is not normal, but neither is it its own normalizer. The normalizer $N_{D_6}(\langle f\rangle)$ is $\langle f, r^3\rangle$, a group isomorphic to V_4, as shown on the right.

The right side of Figure 7.29 shows $N_{D_6}(\langle f\rangle)$ lifted out of D_6. As you can tell, it is not just a set of nodes, but is in fact a group in its own right, and $\langle f\rangle$ is normal in it. These two facts are not coincidences. We may not have expected that the normalizer would be a group, but knowing that it is should make it obvious that the original subgroup must be normal inside its own normalizer. After all, the normalizer consists of just those elements that voted in favor of the subgroup being normal! So let us see why every normalizer is a group.

Theorem 7.7. *For any $H < G$, $N_G(H) < G$ as well.*

Proof. In the past, I've used the notation $\langle a, b, c\rangle$ to indicate a subgroup generated by the elements a, b, and c. Generating the subgroup means combining a, b, c, and their inverses in all possible ways. We know this will be a subgroup because this kind of exploration from generators is how we built groups in the first place, in Chapter 2.

Say the elements of $N_G(H)$ are $g_1, g_2, \ldots, g_n$, and consider the subgroup they generate, $\langle g_1, g_2, \ldots, g_n\rangle$. I prove that generating this subgroup does not actually reach any new elements, but rather $\langle g_1, g_2, \ldots, g_n\rangle$ is just $N_G(H)$. This proves that $N_G(H)$ is a subgroup of G; it is the subgroup generated by its own elements. I do this by first proving that multiplying two elements of $N_G(H)$ never reaches outside of $N_G(H)$, then proving that every inverse of an $N_G(H)$ element is also an $N_G(H)$ element.

Consider two elements g_1 and g_2 in $N_G(H)$. Each corresponds to some path in a Cayley diagram for G, and those paths must connect cosets of H unanimously because they are in $N_G(H)$. Figure 7.30 illustrates the straightforward conclusion that the combined path $g_1 g_2$ must also connect cosets of H unanimously. Thus $g_1 g_2$ is also in $N_G(H)$, and we see that combining elements in $N_G(H)$ stays within $N_G(H)$.

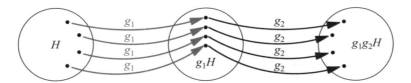

Figure 7.30. Following sequentially from H g_1 and g_2 arrows, two different sets of arrows that connect the cosets of H unanimously, shows that the product $g_1 g_2$ must also connect the cosets of H unanimously.

Inverses of elements in $N_G(H)$ are also in $N_G(H)$, as follows. Take any g in $N_G(H)$ and consider a particular unanimous set of g arrows from some left coset aH to another left coset bH. There are no other g arrows entering bH besides those from aH, because there is no place for them to go; there is already one from each aH element (matching the number of bH elements) and no two arrows can have the same destination. Thus g arrows into bH come only from aH, making g^{-1} arrows from bH lead unanimously to aH. This shows that the inverse of any g in $N_G(H)$ is also in $N_G(H)$.

So $N_G(H)$ is a subgroup because it is the subgroup generated by its own elements. □

This theorem shows that $N_G(H)$ is always a group, and I mentioned earlier that it is a group in which H is normal. It is called the normalizer because it answers the question of how much must be removed from G to make H normal, a variation of the question that opened this section.

A few important aspects of the relationship $H \triangleleft N_G(H) < G$ are highlighted in Figure 7.31. First, the size of $N_G(H)$ is some multiple of $|H|$, and the size of G is some multiple of $|N_G(H)|$, so that the three are only different when H is fairly small compared to G (at most one fourth its size). Second, the boundary of $N_G(H)$ falls on the boundaries of left cosets of H, cutting through none of them; it includes cosets whole or not at all, as we saw earlier. Third, it contains exactly those left cosets of H that are also right cosets. Although all of G is full of copies of H (its left cosets), the $N_G(H)$ boundary is as large as it can be while still keeping all of the relationships $H \triangleleft N_G(H) < G$ true. Normalizers will be a powerful tool in the advanced study of subgroups in Chapter 9.

7.5 Conjugacy

So far we've been using the most common definition of normal subgroup, that $H < G$ is normal if $gH = Hg$ for every $g \in G$. Another common (and equivalent) way to write the equation $gH = Hg$ is as $gHg^{-1} = H$. This alternate form was obtained from the original by multiplying both sides by g^{-1} on the right, then canceling, as follows.

7.5. Conjugacy

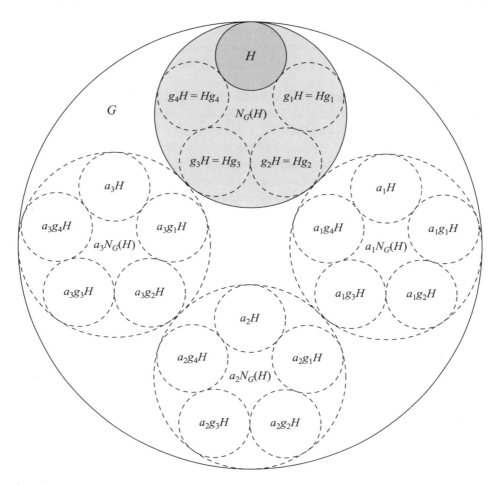

Figure 7.31. Abstract visualization of the relationships $H \triangleleft N_G(H) < G$. Cosets of H are labeled $g_1 H$, $g_2 H$, etc., while cosets of $N_G(H)$ are labeled $a_1 N_G(H)$, $a_2 N_G(H)$, etc.

Original equation:	$gH = Hg$
Multiply both sides by g^{-1}:	$gHg^{-1} = Hgg^{-1}$
Cancel gg^{-1} from the right:	$gHg^{-1} = H$

This equation expresses the same information as the first, and thus it is an acceptable alternative way to describe what it means for H to be normal.

Just as each side of the equation $gH = Hg$ is a set of group elements, so is each side of the equation $gHg^{-1} = H$. This new equation says that if you take all the elements $h \in H$ and multiply each on the left by g and on the right by g^{-1}, yielding ghg^{-1}, then the new collection of elements you obtain, written gHg^{-1}, is really just the subgroup H. Visually, that same equation says that all the g^{-1} arrows lead back from the left coset gH to the subgroup H.

The motivation for rewriting $gH = Hg$ as $gHg^{-1} = H$ is to bring out the connection between normal subgroups and an operation called *conjugation*. Taking h and surrounding it with g and its inverse (yielding ghg^{-1}) is called *conjugating h by g,* and the result is called *the conjugate of h by g,* or just a conjugate. Similarly, gHg^{-1} is also called the conjugate of H by g. Because normal subgroups can be defined using conjugation, the two concepts are tightly related. This section explores some of the consequences of their relationship.

The expression ghg^{-1} probably seems rather arbitrary, so I begin by explaining why conjugating one group element h by another element g often has a natural and useful meaning. Viewed as a sequence of actions, a conjugate ghg^{-1} is the action g, followed by the action h, followed by a reversal of the action g. Many everyday events fit this pattern. Let h stand for the simple action of opening a jar. When the lid is tight so that you're having trouble with h, you may heat the lid under hot running water (g), open the jar (h), and let the lid cool again (g^{-1}). The action h is temporarily moved into a different context by the action g, and the result is therefore different than if h had been done alone.

As another example, say that action h is a band playing their songs, and action g is the band driving to the auditorium where they will perform. This band does action h frequently when rehearsing, but on a concert night things are different. They conjugate their usual action h by g, thereby moving it into a new setting (and in this case a more exciting and lucrative one). They drive to the auditorium (g), perform their songs (h), and then drive back (g^{-1}). A good way to think of any conjugate action ghg^{-1} is that it is the action h moved into a new context by the action g. For this reason, an action and its conjugates often have a lot in common.

These first two examples of conjugacy are easy to understand, but are not examples from group theory. There are countless group theoretic examples to choose from; let's look at one from the group A_4, the symmetries of the tetrahedron. As before, I use the names a, b, c, and d in A_4 to represent the four 120-degree clockwise rotations about each of the four different vertices of the tetrahedron. In Figure 7.32 I've labeled the four vertices A, B, C, and D, corresponding to those four actions. Each of these four rotations is a conjugate of each of the others. This resonates with the non-group-theoretic examples above, because each of these 120-degree clockwise rotations is essentially the same as any other, except in a different context. Let's see exactly why they are conjugates.

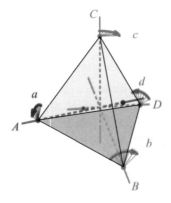

Figure 7.32. A tetrahedron with vertices A, B, C, and D. From each vertex a line through the opposite face is drawn, and the actions of rotating the entire object 120 degrees clockwise around each such line are labeled a, b, c, and d, respectively. (Here "clockwise" is interpreted from the point of view that looks toward the vertex from outside the tetrahedron.)

7.5. Conjugacy

I show how to conjugate a by another element of A_4 to obtain b, but the work would be very similar if I chose a different pair of rotations instead. (I ask you to try another in Exercise 7.37.) The action a is a clockwise rotation about the vertex A. We need an action that will reorient the tetrahedron so that the action a rotates instead about the vertex B. This can be done by any action in A_4 that repositions B to be the leftmost vertex (where A is in Figure 7.32), in order to "fool" action a into rotating the tetrahedron about the vertex B. Action c is just such an action.

Figure 7.33 shows what happens when we conjugate a by c. Each portion of the figure represents a different one of the three actions in the conjugate, and shows the orientation of the tetrahedron after that action is completed (assuming it began oriented as in Figure 7.32). The final result has the vertices A, C, and D rotated one step clockwise, as if all we had done was the one action b. Thus conjugating a by c does indeed result in b. We can express this algebraically by the equation $cac^{-1} = b$. Similar work could show that any two of the 120-degree clockwise rotations are conjugates.

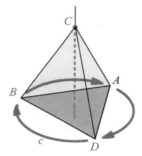 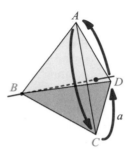 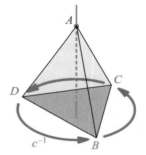

After doing c, we find that B has taken the place of A.

Therefore, the a action now rotates about the point B.

The c^{-1} action puts the vertex B back where it began.

Figure 7.33. Conjugating a by c results in b, written $cac^{-1} = b$. This figure assumes the tetrahedron began in the configuration shown in Figure 7.32. Thus you can verify the equation $cac^{-1} = b$ by picturing the effect of b on the tetrahedron in Figure 7.32; your result should be the result of the conjugation shown here.

Because conjugate elements often share many qualities, it can be informative to partition a group into sets of conjugates. There are three key facts that enable us to use a relationship like conjugacy to partition a group. You will be asked to prove each of these facts in Exercise 7.36, and therefore I will not prove them here. One of these facts is that whenever an element a is a conjugate of an element b, then b is also a conjugate of a. For this reason, the simpler (and appropriately ambiguous) phrase "a and b are conjugates" is common.

Table 7.1 partitions the elements of A_4 so that the elements in each row are all conjugate to one another, but not to any elements in any other row. Notice how each row has a simple, meaningful description that states the similarities the conjugates share. The four rows in Table 7.1 are called ***conjugacy classes.***

A common way to summarize the information in a table of conjugacy classes is in an equation saying that the sizes of the conjugacy classes add up to the size of the group. For example, from Table 7.1, with conjugacy classes of sizes 1, 3, 4, and 4, we would

Elements	Description
e	the identity element
x, y, z	180-degree flips (illustrated in Exercise 5.25)
a, b, c, d	120-degree clockwise rotations about vertices
a^2, b^2, c^2, d^2	120-degree counterclockwise rotations about vertices

Table 7.1. Conjugacy classes of A_4 described.

write
$$1 + 3 + 4 + 4 = 12.$$

This is called the **class equation** for A_4; each number on its left side represents a conjugacy class and the number on its right side is the group order. Class equations carry a surprising amount of information for how small and simple they are. The important use of class equations to which I now draw your attention brings us back to where our acquaintance with conjugacy started, its connection to normal subgroups.

I began the topic of conjugacy by pointing out that the defining characteristic of normal subgroups could be expressed as $gHg^{-1} = H$. Consequently, any element h in a normal subgroup H must have *all* its conjugates in H as well, because any conjugate ghg^{-1} is in gHg^{-1}, which for normal subgroups is H. Another way of saying this is that for every element in a normal subgroup, *its whole conjugacy class is also in the subgroup*. Therefore every normal subgroup is made up of *whole* conjugacy classes.

A class equation therefore enables us to dramatically narrow down a search for normal subgroups, as follows. From only the class equation for A_4, we can tell that the group has four conjugacy classes, with sizes 1, 3, 4, and 4. Any normal subgroup of A_4 will be one or more of these classes united together. That is, based on the previous paragraph, we can't put just half of a conjugacy class in a normal subgroup; we must put the whole thing. So to find all the normal subgroups of A_4, we experiment with different combinations of the numbers 1, 3, 4, and 4, trying to build subgroups out of conjugacy classes. While doing so, we have two restrictions. First, because we're trying to build subgroups, we must always include the conjugacy class containing the identity. Second, Lagrange's Theorem tells us that the size of the resulting subgroup must divide the size of the whole group, in this case 12. Once we've taken these two principles into account, only three possibilities emerge.

The whole group A_4, $\quad 1 + 3 + 4 + 4$
The identity and the flips, $\quad 1 + 3$
The trivial subgroup, $\quad 1$

No other combinations of 1 with 3, 4, and 4 add up to a factor of 12. The whole group and the trivial subgroup are normal subgroups of any group, and thus it is not a surprise to find them on this list. The interesting one is the middle one, standing for the first two rows of Table 7.1, $\{e, x, y, z\}$.

From this work alone, we cannot conclude that $\{e, x, y, z\}$ is a normal subgroup. We would need to look at a Cayley diagram or multiplication table to tell if $\{e, x, y, z\}$ is a subgroup of A_4, and not simply a subset. Of course, we have already done so in this chapter, even verifying that $\{e, x, y, z\}$ is normal by dividing A_4 by it in Figure 7.23. But in general, combining conjugacy classes does not always create subgroups. What it does is give us a much shortened list of things to check when seeking *normal* subgroups.

So class equations provide a shortcut when searching for normal subgroups. Rather than checking many subgroups for normality by attempting the quotient process in each case, we only have to check a few. (In this example, we only had to check one.) Although this connection of conjugacy with normal subgroups is useful, it is not the only value we will derive from the concept of conjugacy. Some of the exercises in Section 7.6.5 reveal new uses for conjugacy, and each of the remaining chapters in this book depends upon it as well.

7.6 Exercises

Recall the organization of the chapter stated at its outset: Each section from this chapter corresponds to a section in the exercises. For example, the exercises in Section 7.6.1 correspond to the material in Section 7.1. Those exercises are therefore only about direct products and do not cover additional material beyond Section 7.1. This allows you to try exercises periodically throughout your reading, to boost your command of the material.

7.6.1 Direct products

Exercise 7.1. How many elements are in each of the following groups?

(a) $C_2 \times C_6$

(b) $S_3 \times A_5$

(c) C_3^5, which means $C_3 \times C_3 \times C_3 \times C_3 \times C_3$

Exercise 7.2.

(a) Consider two Cayley diagrams, one for the group A with two arrow types (indicating two generators) and one for the group B with just one arrow type. How many arrow types will be in the Cayley diagram for $A \times B$, constructed by Definition 7.1?

(b) What is the answer if the diagram for A has n arrow types and the diagram for B has m?

Exercise 7.3. For each of the following statements, determine if it is true or false.

(a) If A and B are any two groups, then $|A \times B| = |A| \cdot |B|$.

(b) The group $C_3 \times C_4$ has the same elements as the group $C_4 \times C_3$.

(c) The group $A \times B$ is abelian, for any groups A and B.

(d) The group $C_2 \times C_2$ has the same structure as the group C_4.

(e) If A and B are any two groups, then $A \triangleleft A \times B$.

(f) The group D_n has the same structure as the group $C_2 \times C_n$.

Exercise 7.4.

(a) Create a Cayley diagram for $C_4 \times C_4$, which can be called C_4^2, "cee four squared."

(b) Create a Cayley diagram for $C_3 \times C_3 \times C_3$, which can be called C_3^3, "cee three cubed."

(c) Create a Cayley diagram for $C_2 \times C_2 \times C_2 \times C_2$, which can be called C_2^4, "cee two to the fourth."

(d) Are C_4^2 and C_2^4 the same?

Exercise 7.5.

(a) Describe the construction of $C_5 \times C_1$. To what is it isomorphic?

(b) Describe the construction of $C_1 \times C_5$. To what is it isomorphic?

(c) What can you say about $C_1 \times G$ and $G \times C_1$ in general?

Exercise 7.6. If $|A| = n$ and $|B| = m$, then what is $|A \times B|$?

Exercise 7.7. Although all parts of this question can be answered after only having read Section 7.1, parts (b) and (c) are easier if you have also read Section 7.3.

(a) Use direct product notation to describe the group depicted by the following Cayley diagram (shown from two different angles, to clarify its structure).

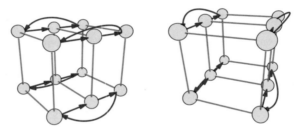

(b) The group C_{10} is a direct product. What are its factors?

(c) Is the group depicted by the following Cayley diagram a direct product group? Justify your answer.

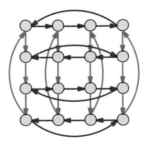

Exercise 7.8.

(a) If A and B are abelian, is $A \times B$?

7.6. Exercises

(b) Justify your answer to (a) visually. If you answered yes, give evidence by explaining why the direct product process for two abelian Cayley diagrams must produce an abelian Cayley diagram. If you answered no, give Cayley diagrams for abelian groups A and B and the corresponding non-abelian group $A \times B$.

(c) Justify your answer to (a) algebraically (either by reference to the groups' binary operations or their multiplication tables).

(d) If A is non-abelian, what can you conclude about $A \times B$?

(e) Justify your answer to (d) visually (by reference to Cayley diagrams).

(f) Justify your answer to (d) algebraically (either by reference to the groups' binary operations or their multiplication tables).

Exercise 7.9. A Cayley diagram for Q_4 appears in Exercise 4.4.

(a) Reorganize the diagram to show the subgroup $\langle i \rangle$ and its left cosets. Is $\langle i \rangle$ a normal subgroup of Q_4?

(b) Let's determine whether Q_4 is a direct product of $\langle i \rangle$ with some other subgroup $A < Q_4$. What size must A be?

(c) Based on part (b), what are the possibilities for A?

(d) Is Q_4 a direct product $\langle i \rangle \times A$ for some A? If so, what is A? If not, why not?

Exercise 7.10. Explain succinctly why A_4 is not a direct product $\langle x \rangle \times A$, $\langle y \rangle \times A$, or $\langle z \rangle \times A$ for any group A.

Exercise 7.11. Come up with a way to take any positive whole number n and create a group whose Cayley diagram requires at least n arrow types.

Exercise 7.12. Prove that $A \triangleleft A \times B$ and $B \triangleleft A \times B$. You may find the equations on page 128 useful for an algebraic argument, or Figure 7.14 for a visual one.

Exercise 7.13. Draw a representative portion of the infinite Cayley diagram for the group $\mathbb{Z}^2$.

7.6.2 Semidirect products

Exercise 7.14.

(a) Create and diagram the rewiring group for C_5.

(b) Create and diagram the rewiring group for C_7.

(c) What conjecture would you make about rewiring groups for C_p, when p is prime?

(d) What is the rewiring group of S_3?

Exercise 7.15.

(a) What is the semidirect product of C_4 with its rewiring group?

(b) What is the semidirect product of C_6 with its rewiring group?

(c) Do you suspect that the semidirect product of C_5 with its rewiring group will follow the pattern suggested by parts (a) and (b)? Why or why not?

(d) Draw a Cayley diagram of the semidirect product of C_5 with its rewiring group.

(e) Think about and then describe (without necessarily drawing it) the Cayley diagram for the semidirect product of C_7 with its rewiring group.

Exercise 7.16. What is the rewiring group of $\mathbb{Z}$? Compute the corresponding semidirect product group.

7.6.3 Quotients

Exercise 7.17. Consider the quotient taken in Figure 7.23.

(a) What is the subgroup by which the quotient is taken? Where in the figure can you see that subgroup?

(b) What is the order of that subgroup? How does the figure show that order?

(c) What is the index of that same subgroup? How does the figure show that index?

(d) Does A_4 have any subgroups of order 3? How does the figure show you such a subgroup, or show you that there are not any?

(e) Can A_4 be divided by any of its other subgroups?

Exercise 7.18. For each of the following H and G (with $H < G$), attempt the quotient process from Definition 7.5. If it succeeds, show a diagram like Figure 7.20 and state the name of the quotient group. If the quotient operation reveals a direct or semidirect product structure, say which it is and name the factors. If the quotient operation fails, show a diagram like Figure 7.26.

(a) $G = C_4$, $H = \langle 2 \rangle$

(b) $G = V_4$ with generators named a and b, $H = \langle a \rangle$

(c) $G = C_{10}$, $H = \langle 2 \rangle$

(d) $G = D_4$, $H = \langle r^2 \rangle$

(e) $G = D_4$, $H = \langle f \rangle$

(f) The group G shown in the Cayley diagram below, with H standing for the two-element subgroup generated by the green arrow

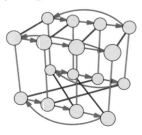

7.6. Exercises

(g) The group G shown in the same Cayley diagram (above), but this time with H standing for the two-element subgroup generated by the blue arrow

(h) The group G shown in the Cayley diagram below (sometimes called $G_{4,4}$), with H standing for the two-element subgroup generated by the red arrow

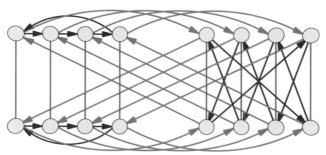

Exercise 7.19. In any group G, the relationships $G \triangleleft G$ and $\{e\} \triangleleft G$ are true (where e stands for the identity element).

(a) What is $\frac{G}{G}$?

(b) What is $\frac{G}{\{e\}}$?

Exercise 7.20.

(a) Develop a quotient procedure for multiplication tables. (Hint: You learned last chapter how to expose subgroups and their cosets.) Your procedure should succeed just when the subgroup by which you divide is normal. In that case, the result should be a multiplication table for the quotient group. Take care with how you name the elements in that table.

(b) Test your procedure on the following two multiplication tables. The left table is a multiplication table for S_3 organized by the normal subgroup $\langle r \rangle$. The right table is the same table, but organized by the non-normal subgroup $\langle f \rangle$.

	e	r	r^2	f	fr^2	fr
e	e	r	r^2	f	fr^2	fr
r	r	r^2	e	fr^2	fr	f
r^2	r^2	e	r	fr	f	fr^2
f	f	fr	fr^2	e	r^2	r
fr^2	fr^2	f	fr	r	e	r^2
fr	fr	fr^2	f	r^2	r	e

	e	f	r	fr^2	r^2	fr
e	e	f	r	fr^2	r^2	fr
f	f	e	fr	r^2	fr^2	r
r	r	fr^2	r^2	fr	e	f
fr^2	fr^2	r	f	e	fr	r^2
r^2	r^2	fr	e	f	r	fr^2
fr	fr	r^2	fr^2	r	f	e

(c) Choose one of the groups from Exercise 7.18, create a multiplication table for it, and apply your technique to one of its subgroups. Was the subgroup normal? If so, what was the quotient group?

(d) Write your technique down carefully and clearly, using some of your work from this exercise as examples to illustrate your explanation.

Exercise 7.21. Explain why every subgroup of an abelian group is normal.

Exercise 7.22. Is every quotient of an abelian group also abelian? Explain why. (Your answer to Exercise 7.20 may help.)

Exercise 7.23.

(a) Prove that if $H < G$ and $[G : H] = 2$, then $H \triangleleft G$.

(b) What will be $\frac{G}{H}$ in this case? Justify your answer.

(c) What relationship does $[G : H]$ have with $\left|\frac{G}{H}\right|$ in general?

Exercise 7.24. Recall the group $\mathbb{Q}$ (under addition) and the group $\mathbb{Q}^*$ (under multiplication) introduced in Exercise 4.33.

(a) Describe the quotient group $\frac{\mathbb{Q}}{\langle 1 \rangle}$.

(b) Describe the quotient group $\frac{\mathbb{Q}^*}{\langle -1 \rangle}$.

7.6.4 Normalizers

Exercise 7.25. Compute the normalizer of H in G for each of the following cases. None of these requires drawing; all are possible by thinking (and perhaps using your mind's eye).

(a) $G = C_9$, $H = \langle 3 \rangle$

(b) G is any group and $H = G$

(c) G is any group and $H = \{e\}$

(d) $G = D_n$ and $H = \langle r \rangle$

Exercise 7.26. Illustrate the normalizer of H in G in each of the following cases, as Figure 7.29 did for $\langle f \rangle < D_6$.

(a) $G = D_3$, $H = \langle f \rangle$

(b) $G = D_4$, $H = \langle f \rangle$

(c) $G = D_4$, $H = \langle f, r^2 \rangle$

(d) $G = D_5$, $H = \langle f \rangle$

Exercise 7.27. From what you observed as you did Exercise 7.26, can you say for which n the normalizer $N_{D_n}(\langle f \rangle)$ is not equal to $\langle f \rangle$?

7.6.5 Conjugacy

Exercise 7.28. What is the result of conjugating an element by itself?

Exercise 7.29. Show that all of an element's conjugates have the same order as the element.

Exercise 7.30. For each group below, an element has been singled out. Describe the conjugacy class of that element.

7.6. Exercises

(a) $r \in D_3$

(b) $r^k \in D_n$

(c) $m \in C_n$

(d) A 90-degree clockwise rotation about one face in the group of symmetries of the cube

(e) A 180-degree clockwise rotation about one face in the group of symmetries of the cube

(f) The permutation interchanging 1 and 2 in S_n (with $n \geq 2$)

(g) The following permutation in S_n (for $k \leq n$)

Exercise 7.31. Compute the class equations for the following groups.

(a) C_3

(b) V_4

(c) any abelian group of order n

(d) S_3

(e) Q_4 (as in Exercises 4.4 and 7.9)

(f) D_4

Exercise 7.32. Let c and t stand for the permutations shown below, members of S_n.

$$c = 1 \quad 2 \quad 3 \quad \cdots \quad n \qquad t = 1 \quad 2 \quad 3 \quad \cdots \quad n$$

Thus c stands for a cycle of n numbers in order, and t stands for the interchange of just the numbers 1 and 2, leaving the rest alone. This exercise determines what elements the subgroup $\langle c, t \rangle$ of S_n contains.

(a) What is the conjugate of t by c, written ctc^{-1}? What is the conjugate of t by c^k, for any k up to n?

(b) All the conjugates from part (a) are in $\langle c, t \rangle$. Describe that set of conjugate elements.

(c) What is the conjugate of t by the following permutation, which interchanges just the numbers 2 and 3, leaving the rest alone?

$$1 \quad 2 \quad 3 \quad 4 \quad \cdots \quad n$$

How could you use two of the elements in $\langle c, t \rangle$ to create a permutation that swaps any two numbers from 1 to n, leaving the rest alone?

(d) Describe the set of elements that part (c) shows to be members of $\langle c, t \rangle$.

(e) What permutation is obtained by doing t followed by the permutation shown in part (c)? How could you create any cyclic permutation using just elements of $\langle c, t \rangle$?

(f) All permutations can be broken into a sequence of non-overlapping cyclic permutations, as in the following example.

$$\begin{pmatrix} 1 & 2 & 3 & 4 & 5 \end{pmatrix} = \begin{pmatrix} 1 & 2 & 3 & 4 & 5 \end{pmatrix} \cdot \begin{pmatrix} 1 & 2 & 3 & 4 & 5 \end{pmatrix}$$

How does this help determine the subgroup $\langle c, t \rangle$ of S_n? What is that subgroup?

Exercise 7.33.

(a) Compute the class equation for the first few dihedral groups D_n with n odd, until you notice a pattern. State the pattern and give some justification for it.

(b) Compute the class equation for the first few dihedral groups D_n with n even, until you notice a pattern. State the pattern and give some justification for it.

(c) Class equations can be illustrated by coloring the elements of a group according to the sets in the conjugacy class partition, a different color for each set. Illustrate the patterns in each of parts (a) and (b) using colored Cayley diagrams.

(d) Cycle graphs display element order rather clearly. How is this relevant to conjugacy classes? Use *Group Explorer* to illustrate the patterns in each of parts (a) and (b) using cycle graphs.

Exercise 7.34. Which of the following equations could be class equations for a group? Find all the groups that have that class equation, if there are any. If there are not any, explain why.

(a) $1 + 2 = 3$

(b) $1 + 1 + 1 + 1 + 1 = 5$

(c) $1 + 2 + 3 = 6$

(d) $1 + 3 + 3 = 7$

(e) $1 + 3 + 4 = 8$

Exercise 7.35. How few elements might gHg^{-1} and H have in common? Find a minimal example and explain how you know it is minimal.

Exercise 7.36. An equivalence relation is one that can be used to partition a set; equivalence relations have three properties. This exercise asks you to prove that being a conjugate is an equivalence relation, using algebraic or visual evidence, whichever seems best to you.

(a) Show that being a conjugate is a *reflexive* relation: Any $g \in G$ is conjugate to itself.

(b) Show that being a conjugate is a *symmetric* relation: If g_1 is conjugate to g_2 (that is, there is some $h \in G$ such that $g_1 = hg_2h^{-1}$), then g_2 is also conjugate to g_1.

(c) Show that being a conjugate is a *transitive* relation: If g_1 is conjugate to g_2, which is conjugate to g_3, then g_1 is also conjugate to g_3.

7.6. Exercises

Exercise 7.37. Recall Figure 7.33, which showed that a and c are conjugates in A_4. Show that b and d are conjugates as well, by finding an element of A_4 by which to conjugate. This can be done algebraically (using a Cayley diagram or multiplication table for A_4 as reference) or visually (using Figure 7.32 or an actual tetrahedron for reference). Try to illustrate the conjugation like Figure 7.33 does.

8

The power of homomorphisms

Throughout this book I've said things like "this group has the same structure as that group" or "there is a copy of this group inside that group." The first time I did so (page 19, regarding the equivalence of Figures 2.7 and 2.8), I carefully explained how the two structures were the same. But since then, when stating that two structures are the same, I've depended on your ability to see how their patterns match, without my spelling out all the details. The purpose of this chapter is to create and study precise descriptions of how two structures correspond, because doing so will significantly advance our study of group theory.

To describe groups with the same structure, in Chapter 3 I introduced the technical term "isomorphic" without giving a formal definition. The related term "homomorphism" refers to a correspondence between two groups. The Greek roots "homo" and "morph" together mean "same shape." Isomorphisms are a specific kind of homomorphism, and this chapter explores both. Section 8.1 introduces homomorphisms through their relationship with two major topics we've already studied. The rest of the chapter tours four important ideas in group theory that homomorphisms help us better describe, understand, and visualize.

8.1 Embeddings and quotient maps

Homomorphisms are hiding in two important contexts we've already studied. It's time to bring them out of hiding and learn what they can teach us. First, we will find homomorphisms whenever one group is a subgroup of another. Second, we will find homomorphisms whenever we perform a quotient.

I begin with the case of subgroups. Consider the mathematical statement $C_3 < S_3$, which we first illustrated on the left of Figure 6.3 on page 100. The highlighting in that figure shows that S_3 contains a three-step cyclic subgroup, which is identical to C_3 in structure only. That is, none of the elements of C_3 (0, 1, or 2) are actually *in* S_3 at all. When we say that C_3 is a subgroup of S_3, we really mean that the structure of C_3 shows up, even if the elements are not 0, 1, and 2.

In Chapter 6, I expected you to visually recognize the structural correspondence in Figure 6.3 on your own; it is a rather obvious one. But this chapter studies such corre-

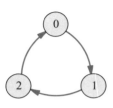 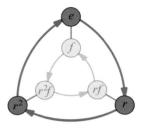

Figure 8.1. The diagrams for C_3 and S_3 side-by-side, to make it clear that the highlighted pattern in S_3 is the pattern of C_3, and to make it possible to describe the element-by-element correspondence

spondences, so let's begin by describing this one precisely. Figure 8.1 shows the Cayley diagrams of C_3 and S_3 side-by-side to facilitate the following explanation of the structural correspondence of C_3 to the orbit $\langle r \rangle$ in S_3.

The top element $0 \in C_3$ corresponds to the top element $e \in S_3$. If we proceed clockwise, following red arrows in each Cayley diagram, the correspondence continues. The first step along the red arrow path in C_3, to the element 1, corresponds to the first step along the red arrow path in S_3, to the element r. The path continues, making $2 \in C_3$ correspond to $r^2 \in S_3$, and then returning to both starting points simultaneously. The three-step walk through C_3 corresponds to the three-step walk through the r-orbit in S_3.

This explanation leaves no doubt about exactly what it means to say that C_3 is a subgroup of S_3, but it is far too wordy! It shouldn't take a whole paragraph to precisely describe such a simple correspondence. Homomorphisms are the mathematical tool for succinctly expressing precise structural correspondences. The homomorphism representing the explanation above is illustrated in Figure 8.2. It is depicted by dashed arrows connecting each C_3 element to its corresponding S_3 element, and it is also described in the legend on the right of the figure. The legend writes $0 \mapsto e$ to say that $0 \in C_3$ corresponds to $e \in S_3$, and so on.

Because homomorphisms describe how elements of one group correspond to elements of another, they are a kind of *function*. In mathematics, we often graph and study functions, which connect each value of an independent variable x to one and only one value of a dependent variable y. The x and y values are usually numbers, but not always; in group theory, our functions will often not be about numbers. Figure 8.2 specifies a function

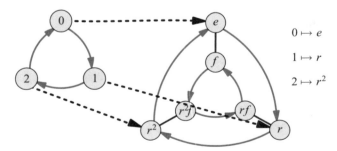

Figure 8.2. A diagram that spells out the correspondence between C_3 and the $\langle r \rangle$ orbit in S_3 by showing which element from S_3 corresponds to each element in C_3. The correspondence is shown using dashed lines and described using the legend on the right.

8.1. Embeddings and quotient maps

from the group C_3 (a group of numbers) to the group S_3 (not containing numbers). We say that this function *maps* elements of C_3 to elements of S_3.

Traditionally, Greek letters stand for maps between groups, so I'll use ϕ for the map in Figure 8.2. I can use the mathematical notation $\phi : C_3 \to S_3$ to say that ϕ maps C_3 to S_3. Just as standard function notation writes $f(2)$ to mean the number to which f maps 2, in group theory we can write $\phi(2)$. In the case of the homomorphism in Figure 8.2, $\phi(2) = r^2$. And just as it is standard to define a function using notation like $f(x) = x^2 + 1$, I can define ϕ succinctly with the expression $\phi(n) = r^n$. This definition works for any $n \in C_3$ (if we think of r^0 as e) and it highlights the connection of C_3 to the orbit of r.

The group from which a function originates is called its ***domain*** (in this case C_3), and the group into which the function maps is called its ***codomain*** (in this case S_3). The particular elements of the codomain that the function touches (in this case the orbit $\langle r \rangle$) are called the ***image*** of the function. The notation $Im(\phi)$ means the image of the homomorphism ϕ.

The original description I gave of ϕ a few paragraphs ago makes it clear that ϕ carefully mimics the structure of its domain C_3 in its codomain S_3. Group theory concerns itself most with functions that have this property, as described in the following definition.

Definition 8.1 (homomorphism). A ***homomorphism*** is a function between two groups that mimics the structure of its domain in its codomain. The following condition expresses this requirement; I state it in two equivalent ways, both of which are illustrated in Figure 8.3.

(1) In terms of Cayley diagrams: If an arrow b in the domain leads from the element a to the element c, then the $\phi(b)$ arrow in the codomain must lead from the element $\phi(a)$ to the element $\phi(c)$.

(2) In terms of multiplication tables: If the domain multiplication table says that $a \cdot b = c$, then the codomain multiplication table must say that $\phi(a) \cdot \phi(b) = \phi(c)$.

These statements must hold true for any a, b, and c in the domain. Take a moment to consider why each of these statements expresses the idea of mimicking the domain structure in the codomain.

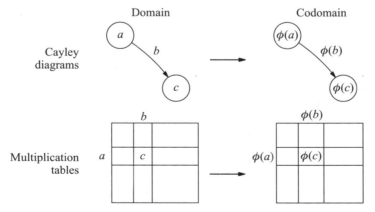

Figure 8.3. The criteria in Definition 8.1 illustrated. On the top left is a portion of the domain Cayley diagram, in which b connects a to c. Therefore somewhere in the codomain Cayley diagram, $\phi(b)$ connects $\phi(a)$ to $\phi(c)$ (top right). On the bottom left, the domain multiplication table includes $a \cdot b = c$, so the codomain multiplication table must include $\phi(a) \cdot \phi(b) = \phi(c)$ (bottom right).

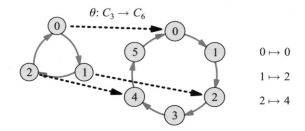

Figure 8.4. A diagram that spells out the correspondence between C_3 and the $\langle 2 \rangle$ orbit in C_6 by showing which element from C_6 corresponds to each element in C_3.

Another example can help us better appreciate this definition. Recall that C_3 is also a subgroup of C_6, as shown on the right of Figures 6.4 and 6.5 on page 100. We can express $C_3 < C_6$ using the homomorphism $\theta(n) = 2n$ from C_3 to C_6, as in Figure 8.4. The next few paragraphs use θ as an example to point out some important facts about homomorphisms in general.

Just as θ maps the node 1 to the node 2, it maps the 1-arrow in C_3 to the two-step path representing 2 in C_6. The 1 arrow traces the orbit $\{0, 1, 2\}$ in C_3, and the $\theta(1)$ path traces the corresponding orbit $\{\theta(0), \theta(1), \theta(2)\}$ in C_6, which is $\{0, 2, 4\}$. Thus homomorphisms not only map nodes of the domain's Cayley diagram to nodes of the codomain's, but also implicitly map *paths* in the domain's Cayley diagram to *paths* in the codomain's. Part (2) of Definition 8.1 requires that the path from a to b in the domain correspond to the path from $\theta(a)$ to $\theta(b)$ in the codomain. The Cayley diagram of C_6 in Figure 8.4 does not include arrows for $\theta(1)$, but they could be added. Or we can simply think of the $\theta(1)$ arrow as the path consisting of two red arrows in succession. So θ doubles both numbers and arrows. By viewing each pair of successive arrows in C_6 as one step that skips the odd-numbered nodes, we find the structure of C_3 in C_6.

Thus when using Cayley diagrams to visualize homomorphisms, we can think of the homomorphism on two levels, mapping nodes in the domain to nodes in the codomain and also mapping paths in the domain to paths in the codomain. Furthermore, if we know where a homomorphism maps the domain's arrows, this tells us enough to deduce where the homomorphism maps every element of the domain. The reason is that the arrows of a Cayley diagram represent the group's generators. Just as those elements generate the group, where they are mapped generates the homomorphism. For example, suppose there were another homomorphism $\theta' : C_3 \to C_6$ different from the θ in Figure 8.4. Given only that $\theta'(1) = 4$, it is possible to determine the rest of the homomorphism.

You could proceed using part (2) of Definition 8.1. For instance, it tells us that because $1 + 1 = 2$ in C_3, we must require $\theta'(1) + \theta'(1) = \theta'(2)$ in C_6. Therefore we can determine $\theta'(2)$ as follows.

$$\theta'(2) = \theta'(1) + \theta'(1) = 4 + 4 = 2$$

Similarly, from $1 + 2 = 0$ in C_3 we can determine that $\theta'(0)$ must be 0. So all of θ' is generated by the fact that $\theta'(1) = 4$. Try drawing this homomorphism and comparing it to the homomorphism θ shown in Figure 8.4.

Similar steps can be used to generate any homomorphism. Say $\phi : G \to H$ and assume that $G = \langle a, b \rangle$ and we are given the values of $\phi(a)$ and $\phi(b)$. We can determine the value

8.1. Embeddings and quotient maps

of $\phi(g)$ for any element $g \in G$ as follows. The element g can be written as some multiplied sequence of a's and b's, say $g = a \cdot b \cdot a \cdot a \cdot b$. Applying part (2) of Definition 8.1 several times yields

$$\phi(g) = \phi(a) \cdot \phi(b) \cdot \phi(a) \cdot \phi(a) \cdot \phi(b).$$

We can compute what $\phi(g)$ must be by multiplying $\phi(a) \cdot \phi(b) \cdot \phi(a) \cdot \phi(a) \cdot \phi(b)$ in the group H. In this way, any homomorphism is generated by what it does to a group's generators.

That G has two generators is irrelevant; similar reasoning works for three or more. Exercise 8.7 asks you to explain why this implies that any homomorphism must map the domain's identity to the codomain's identity. I will use this fact without having given evidence for it myself.

To understand Definition 8.1 better, it is helpful to contrast the example homomorphisms we've seen with examples of correspondences that are *not* homomorphisms. Figure 8.5 shows four examples of non-homomorphisms, with reasons for why each fails to satisfy Definition 8.1. Two are not homomorphisms because they are not even functions, and the other two are functions but do not mimic the structure of the domain or the codomain. Take a moment to understand each example in Figure 8.5.

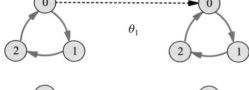

A function maps *each* element of the domain to some element of the codomain. But θ_1 ignores two elements of the domain, and thus is not a function.

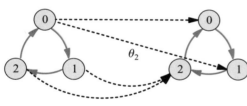

A function maps each element of the domain to *one* element of the codomain. But θ_2 maps 0 to two different elements of the domain, and thus is not a function.

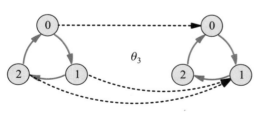

Although θ_3 is a function, it is not a homomorphism. In the domain, $1 + 1 = 2$, but in the codomain, $\theta_3(1) + \theta_3(1) \neq \theta_3(2)$.

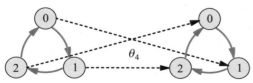

Although θ_4 is a function, it is not a homomorphism. Although the image is a 3-cycle like the domain, θ_4 does not map the domain identity element to the codomain identity element. For example, $\theta_4(0) + \theta_4(1) \neq \theta_4(1)$.

Figure 8.5. Four example maps that are not homomorphisms, with the corresponding reasons. For simplicity, I kept the domain and codomain C_3 the same in each case, but similar examples could be created for many domains and codomains.

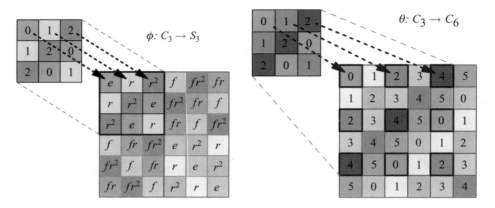

Figure 8.6. Multiplication tables representing the same homomorphisms from Figures 8.2 and 8.4.

Although I find it most helpful to visualize homomorphisms using Cayley diagrams, we could use multiplication tables or cycle graphs instead. For example, Figure 8.6 shows the same two homomorphisms as Figures 8.2 and 8.4, but using multiplication tables instead. Figure 8.7 shows them using cycle graphs.

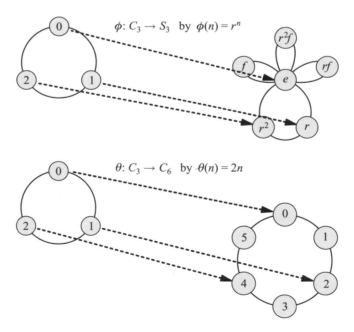

Figure 8.7. Cycle graphs representing the same homomorphisms from Figures 8.2 and 8.4. The homomorphism θ does not look much different than in Figure 8.4 because the cycle graphs of cyclic groups are almost the same as their Cayley diagrams.

8.1. Embeddings and quotient maps

8.1.1 Embeddings

I began Section 8.1 by saying that homomorphisms show up in two group theory situations we've seen before. The first is that homomorphisms help us get specific about how one group is a subgroup of another group, as in the examples we just saw. We call such homomorphisms *embeddings* because they show us how one group can be embedded in another.

Because any embedding finds a copy of the domain in the codomain, its image is therefore the same size as its domain. For this reason, embeddings have the interesting property that they never map two different domain elements to the same codomain element. If one did, its image would be smaller than its domain, and not a copy of it. This property is easy to verify in a figure, because it is easy to spot two homomorphism arrows that have the same destination.

An embedding whose image fills the whole codomain shows us that the domain and codomain actually have all the same structure. In such a case, we say that the function is not just a homomorphism, but an *isomorphism.* Figure 8.8 shows a simple example. Two groups are isomorphic if there is an isomorphism between them. This is the formal definition of the term I've been using informally since Chapter 3. Two isomorphic groups may name their elements differently and may look different based on the layouts of Cayley diagrams or multiplication tables, but the isomorphism between them guarantees that they have the same structure. When two groups G and H are isomorphic, we write $G \cong H$, read "G is isomorphic to H."

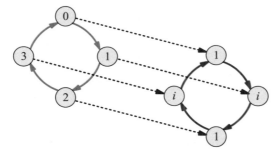

Figure 8.8. The domain of this isomorphism is C_4, and the codomain is a group of complex numbers $\{1, -1, i, -i\}$ under the operation of multiplication. The isomorphism points out that despite their different elements and operations, these groups have the same structure.

8.1.2 Quotient maps

Figure 8.9 shows two examples of homomorphisms that are not embeddings. They satisfy all the requirements from Definition 8.1 for being a homomorphism, but they map more than one domain element to the same codomain element. In doing so, their image becomes a smaller, simplified version of their domain. The diagrams in Figure 8.9 are intricate; take a moment to study them. Verify that they satisfy Definition 8.1, and try to notice patterns they have in common.

There are several useful patterns to notice that are not unique to these two examples, but hold for all non-embeddings. I make three observations below about non-embedding

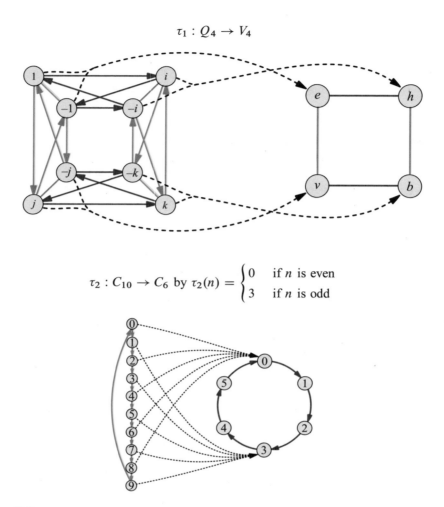

Figure 8.9. Two example homomorphisms that do not have the defining characteristic of embeddings. Each of these homomorphisms maps more than one domain element to the same codomain element.

homomorphisms and give reasons to support each. As you read Observations 8.2 through 8.4, verify that they are true about the two homomorphisms in Figure 8.9. This can help you gain some concrete intuition for both the observations and the reasoning behind them.

Observation 8.2. Every collection of domain elements that maps to the same codomain element has the same structure in the domain.

In other words, every non-embedding homomorphism follows a repeating pattern. For instance, in the homomorphism τ_1 in Figure 8.9, each element of the codomain is mapped to by a little copy of C_2 in the domain, two elements connected by a -1 arrow. In the homomorphism τ_2, each element in the image (0 and 3) is mapped to by five elements of the domain connected in a chain by pairs of red arrows. Let's prove that Observation 8.2 is true about any non-embedding.

8.1. Embeddings and quotient maps

Proof. Take two elements a and b in the image of a homomorphism ϕ. Let A stand for the set of domain elements that ϕ maps to a, and B the elements it maps to b. I will explain why A and B must have the same structure. It may help to visualize the following argument with a concrete example. For ϕ, think of τ_2 from Figure 8.9 and let $a = 0$ and $b = 3$ in its codomain, so that $A = \{0, 2, 4, 6, 8\}$ and $B = \{1, 3, 5, 7, 9\}$ in its domain.

The structure of A and B is comprised of the paths within them. Choose two starting points, a_1 in A and b_1 in B, as illustrated in Figure 8.10. I explain why every path from a_1 to another element in A also leads from b_1 to some other element in B.

Let p be a path from a_1 to some point a_2 in A. Part (1) of Definition 8.1 tells us that $\phi(p)$ leads from $\phi(a_1)$ to $\phi(a_2)$. But $\phi(a_1) = \phi(a_2) = a$, so $\phi(p)$ is a path leading from a to a. It is a path with zero steps (or you can think of it as a path that goes somewhere but then returns to end at its starting node). So where does the path p lead if we start from b_1? I'll call the point to which it leads b_2, and the important question is whether b_2 is in B. Well, $\phi(p)$ leads from $\phi(b_1)$ to $\phi(b_2)$, making them equal since $\phi(p)$ is an empty path. Thus $\phi(b_2) = \phi(b_1) = b$, which means that b_2 is in B, because B is the set of elements that ϕ maps to b.

This shows that my path from a_1 that stays within A has a corresponding path from b_1 that stays within B. Similar reasoning could show that any path from b_1 that stays within B has a corresponding path from a_1 that stays within A. Thus A and B have all structure in common. Because they stood for any two sets in the domain that map to single elements of the codomain (a and b), this shows that *all* such sets have the same structure. Observation 8.2 is true for any non-embedding. □

As Figure 8.10 suggests, this creates a partition of the domain into identical copies of a structure. This is reminiscent of when we partitioned a group by cosets of a subgroup in Chapter 6. In fact, that is exactly what is happening here as well.

Observation 8.3. The collections of domain elements mentioned in Observation 8.2 are actually a subgroup and its left cosets.

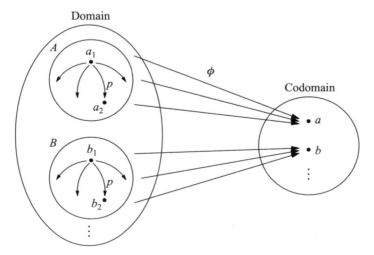

Figure 8.10. A non-embedding homomorphism ϕ and the sets of elements A and B that it maps to the elements a and b. This illustrates the reasoning following Observation 8.2.

From Chapter 6, we know that the left cosets of any subgroup are structural copies, and they partition the group. Thus if any one of the identical structures mentioned in Observation 8.2 is a subgroup, the others must be its left cosets. If one such structure were to be a subgroup, it would need to be the one containing the identity element, and therefore it would need to be the one that maps to the codomain identity. I give evidence for Observation 8.3 by explaining why the structure that maps to the codomain identity is indeed a subgroup of the domain.

I use the same type of argument that I did for Theorem 7.7. Let C stand for the collection of elements that map to the codomain identity. I show that the subgroup generated by C is just C. Obviously C generates at least its own elements, but I must show that it generates no more. For any two elements a and b in C, must their product ab remain in C?

The element ab is where the b arrow reaches from a, and thus the element $\phi(ab)$ is where the $\phi(b)$ arrow reaches from $\phi(a)$. But since b is in C, ϕ maps b to the identity and thus $\phi(b)$ is an empty path. So $\phi(a) = \phi(ab)$, and because ϕ maps a to the identity, it must map ab there as well. Thus ab is in C.

So the collection of elements that map to the codomain identity is a subgroup of the domain, and each of its left cosets maps to a different, single element of the codomain. The subgroup C is called the **kernel** of the homomorphism ϕ, and we write it $Ker(\phi)$.

Observation 8.4. The left cosets mentioned in Observation 8.3 are also right cosets, and thus the subgroup is normal.

In Chapter 7 we saw that normal subgroups are those with no ambiguous arrows among their cosets, as illustrated in Figure 7.27. If the arrows for any element g in the domain were ambiguous, connecting $Ker(\phi)$ to both $gKer(\phi)$ and $kKer(\phi)$ as in Figure 8.11, then part (1) of Definition 8.1 requires that $\phi(g)$ in the codomain connect e to both $\phi(g)$ and $\phi(k)$. Since we know that the codomain is a group, this cannot happen. The kernel must therefore be a normal subgroup.

Putting together Observations 8.2 through 8.4 we can see that the homomorphisms in Figure 8.9 are functioning much like the quotient process from Definition 7.5. The quotient process collapsed cosets of a normal subgroup into single nodes representing the coset; the way that a homomorphism maps each coset of its kernel to a single node is another way of doing the same collapsing. For example, the homomorphism τ_1 in Figure 8.9

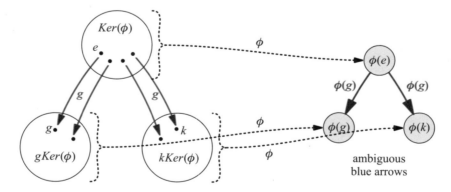

Figure 8.11. The pattern depicted here never happens because the ambiguity of blue arrows in the codomain is impossible; the codomain is a group.

8.2. The Fundamental Homomorphism Theorem

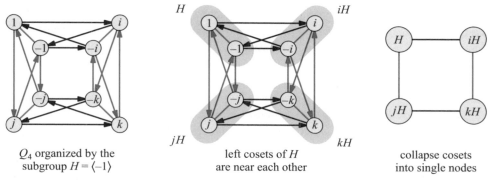

| Q_4 organized by the subgroup $H = \langle -1 \rangle$ | left cosets of H are near each other | collapse cosets into single nodes |

Figure 8.12. The quotient of Q_4 by C_2 corresponding to the homomorphism τ_1 in Figure 8.9. It follows the quotient procedure from Chapter 7, as depicted in Figure 7.20 (and others).

collapses copies of C_2 in Q_4 to single nodes of V_4. It corresponds to the quotient of Q_4 by the subgroup $\langle -1 \rangle$, isomorphic to C_2. That quotient process is shown in Figure 8.12.

This relationship of homomorphisms to quotients is the subject of Section 8.2, which begins with a famous theorem stating the connection formally. Because of this relationship, I call non-embedding homomorphisms **quotient maps.** This is one of the key facts about homomorphisms: they come in only two types, embeddings and quotient maps. Embeddings never collapse two domain elements into one codomain element, and quotient maps do so in the same way that a quotient process does. You have spent this section learning about embeddings first and quotient maps second, and there is no other kind of homomorphism.

8.2 The Fundamental Homomorphism Theorem

The following theorem has an important-sounding name for good reason; it sums up a lot of what we've already observed about homomorphisms, and more. I begin by simply stating it, then discuss its implications and its proof.

Theorem 8.5 (Fundamental Homomorphism Theorem). *If $\phi : G \to H$ is a homomorphism, then $Im(\phi) \cong \frac{G}{Ker(\phi)}$.*

As we move toward understanding why this theorem is true, let's first see some of its simplest implications. First, consider what the theorem says about embeddings. Any embedding ϕ must map only one element to the codomain identity, and so its kernel is just the trivial subgroup $\{e\}$. Theorem 8.5 therefore says that $Im(\phi)$ is isomorphic to $\frac{G}{\{e\}}$, which is isomorphic to G. Thus the theorem says that embeddings deserve their name, because their image is a copy of their domain.

When ϕ is not an embedding, somewhere it must collapse two domain elements to one codomain element. In fact, because quotient maps follow a repeating pattern, every coset of $Ker(\phi)$ will have at least two elements in it. Thus the quotient $\frac{G}{Ker(\phi)}$ collapses cosets with several elements into single elements, resulting in a smaller group (unlike the quotient $\frac{G}{\{e\}}$). So Theorem 8.5 tells us that quotient maps, too, deserve their name, because they perform a nontrivial quotient process. In fact, after we prove the theorem, we will use it to see exactly how quotient processes and quotient maps relate.

Proof. For any homomorphism $\phi : G \to H$, we know from Observations 8.2 to 8.4 that $Ker(\phi) \triangleleft G$, so we know that we can take a quotient $\frac{G}{Ker(\phi)}$ and obtain a group. Now I must explain why that group is isomorphic to $Im(\phi)$, a subgroup of H. We learned how to compute quotient groups like $\frac{G}{Ker(\phi)}$ by collapsing cosets of $Ker(\phi)$ in G, and Observations 8.2 to 8.4 already showed that homomorphisms collapse cosets of $Ker(\phi)$; each maps to a single node in $Im(\phi)$.

The question is whether ϕ also collapses arrows in the same way that the quotient by $Ker(\phi)$ does. Let's recall what the quotient process does to arrows between cosets. If b arrows lead from the coset $aKer(\phi)$ to the coset $cKer(\phi)$ in G, then when the quotient collapses cosets to nodes, the b arrows also collapse to one b arrow connecting the $aKer(\phi)$ node to the $cKer(\phi)$ node in the quotient group. That is, the quotient process requires the arrows to connect nodes in the same way after the quotient as before, but duplicate arrows become one. This is perfectly analogous to what homomorphisms do. Part (1) of Definition 8.1 requires that whenever an arrow b connects a to c in G, we must have $\phi(b)$ mapping $\phi(a)$ to $\phi(c)$ in H. Furthermore, because H is a group, there will be no duplicate arrows between nodes. Any homomorphism ϕ collapses both cosets and arrows just as does a quotient by $Ker(\phi)$, and the results therefore have the same structure.

If $\frac{G}{Ker(\phi)}$ and $Im(\phi)$ are the same structures, then there must be a suitable renaming, an isomorphism converting cosets of $Ker(\phi)$ to elements of $Im(\phi)$. Figure 8.13 diagrams this relationship and calls the isomorphism $i : \frac{G}{Ker(\phi)} \to Im(\phi)$. A specific example following the pattern of Figure 8.13 is shown in Figure 8.14. The homomorphism $\phi : A_4 \to C_3$ is the same as the quotient process q that divides A_4 by $Ker(\phi)$, followed by a renaming of cosets to numbers using the isomorphism i.

But what is the formula for i? As suggested by Figure 8.14, i must map a coset in $\frac{G}{Ker(\phi)}$ to where ϕ maps all that coset's elements. After all, we want q followed by i to agree with ϕ. Thus the isomorphism i should map any coset $gKer(\phi)$ to the element $\phi(g)$ in H. There is no ambiguity here, because regardless of which element is chosen to represent the coset, ϕ maps them all to the same destination. This isomorphism i is what makes $\frac{G}{Ker(\phi)}$ isomorphic to $Im(\phi)$. □

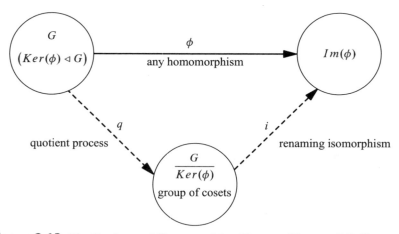

Figure 8.13. The Fundamental Homomorphism Theorem (Theorem 8.5) illustrated

8.3. Modular arithmetic

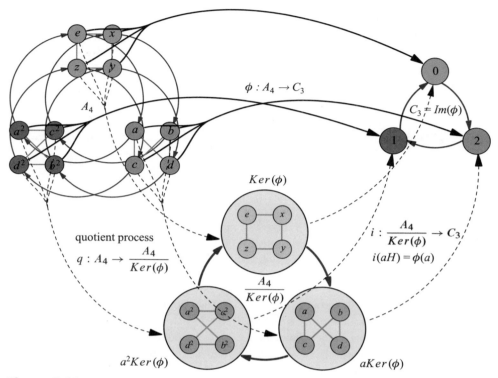

Figure 8.14. The Fundamental Homomorphism Theorem (Theorem 8.5) exemplified using the group A_4 and the quotient map ϕ whose kernel is the subgroup $\{e, x, y, z\}$ of A_4.

We can interpret Theorem 8.5 as saying that any homomorphism can be imitated by combining a quotient process with a suitable renaming isomorphism. What you may also suspect from these figures is that the quotient process is itself a homomorphism. It is a function from a group to a quotient group, and it respects the group operation. For this reason the quotient processes in Figures 8.13 and 8.14 are named q and diagrammed like the other homomorphisms in those figures. In fact, the term "quotient map" becomes rather ambiguous. I've been using it to mean a non-embedding homomorphism, which acts like a quotient process, as τ_1 in Figure 8.9 acts like the quotient process in Figure 8.12. But the quotient process is itself a homomorphism, and thus is also called a quotient map.

We now have a strong foundation for using homomorphisms in our study of group theory. The rest of this chapter explores a variety of uses for both homomorphisms and Theorem 8.5. There are many uses, but I choose four of the most significant and famous ones on which to spend our time, one topic each in Sections 8.3 through 8.6.

8.3 Modular arithmetic

In Section 5.1.2 we learned that the binary operation for the groups C_n is addition mod n. I introduced this concept by comparing C_{12} to a clock. A clock and a Cayley diagram for C_{12} are numbered almost exactly the same, but where the clock has 12, the Cayley diagram has 0 instead. This leads to a very useful way to think of the group C_{12}. It is the group obtained by taking the group $\mathbb{Z}$ of all integers and imposing the requirement that 12 and 0 should be the same. Theorem 8.5 can help us do this.

Begin with a Cayley diagram for $\mathbb{Z}$ (using one arrow for the generator 1), as shown in Figure 8.15. The recent discussion of Theorem 8.5 tells us that dividing $\mathbb{Z}$ by a normal subgroup containing 12 corresponds to a homomorphism that maps 12 to 0. Both represent the idea that 0 and 12 should become the same. Let's perform this quotient.

Figure 8.15. The infinite cyclic group $\mathbb{Z}$ of integers

To avoid removing too much, let's divide by the smallest normal subgroup of $\mathbb{Z}$ containing 12. Well, the smallest subgroup containing 12 in $\mathbb{Z}$ is of course $\langle 12 \rangle$, the contents of which are exactly the multiples of 12.

$$\langle 12 \rangle = \{\ldots, -36, -24, -12, 0, 12, 24, 36, \ldots\}$$

Because $\mathbb{Z}$ is abelian, all subgroups (including this one) are normal. Thus I now compute $\frac{\mathbb{Z}}{\langle 12 \rangle}$ using the visual quotient process from Chapter 7.

First I must organize a Cayley diagram of $\mathbb{Z}$ by $\langle 12 \rangle$ and its cosets. To do so, we must know what those cosets are. Because $\mathbb{Z}$ is a group whose binary operation is addition (rather than multiplication, as in many of the groups we've seen), we will indicate cosets of $\langle 12 \rangle$ using the notation $k + \langle 12 \rangle$ (rather than $k\langle 12 \rangle$). A few cosets are shown in Figure 8.16; notice that they are structural copies of $\langle 12 \rangle$, as expected. The coset $1 + \langle 12 \rangle$ contains all the numbers that are one more than a multiple of 12, that is, those numbers whose remainder is 1 when the number is divided by 12. The coset $2 + \langle 12 \rangle$ contains elements that give remainder 2 when divided by 12, and so on.

These cosets are called **congruence classes mod 12,** and we say two elements are **congruent mod 12** if they're in the same congruence class—they have the same remainder when divided by 12. When a and b are congruent mod 12, we write $a \equiv_{12} b$. Thus, for

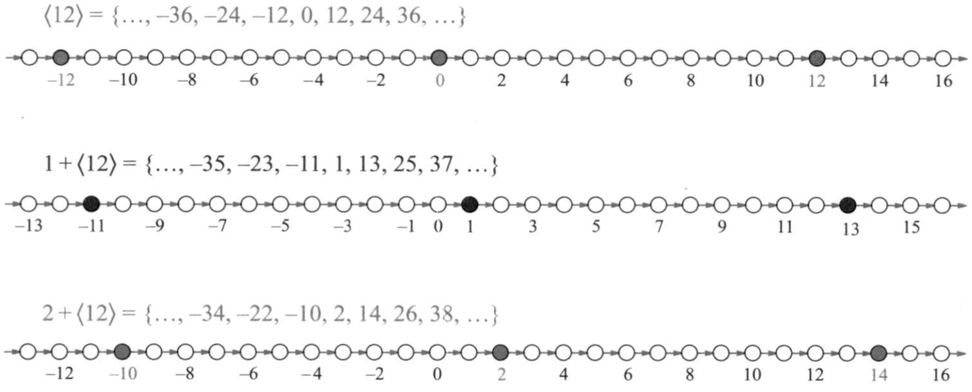

$\langle 12 \rangle = \{\ldots, -36, -24, -12, 0, 12, 24, 36, \ldots\}$

$1 + \langle 12 \rangle = \{\ldots, -35, -23, -11, 1, 13, 25, 37, \ldots\}$

$2 + \langle 12 \rangle = \{\ldots, -34, -22, -10, 2, 14, 26, 38, \ldots\}$

Figure 8.16. Segments of $\mathbb{Z}$ with a few cosets of $\langle 12 \rangle$ highlighted. Other cosets follow the pattern suggested here.

8.3. Modular arithmetic

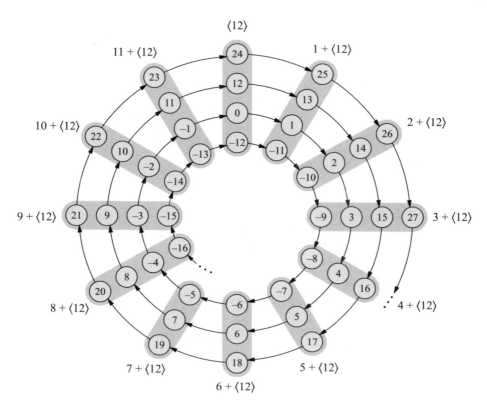

Figure 8.17. A Cayley diagram of the group $\mathbb{Z}$ organized by cosets of $\langle 12 \rangle$. The quotient group structure becomes clear; we can see that the cosets form C_{12}.

example, the following congruences are true.

$$6 \equiv_{12} 18 \qquad 1 \equiv_{12} 25 \qquad 12 \equiv_{12} 1200 \equiv_{12} 0$$

The group $\mathbb{Z}$ organized by these cosets forms an infinite spiral, as shown in Figure 8.17. The coset grouping in this layout shows that $\frac{\mathbb{Z}}{\langle 12 \rangle} \cong C_{12}$.

But there's more to learn from this example. Figure 8.18 diagrams this quotient after the pattern of Figures 8.13 and 8.14. The homomorphism $\phi : \mathbb{Z} \to C_{12}$, the combination of the quotient and the isomorphism $i : \frac{\mathbb{Z}}{\langle 12 \rangle} \to C_{12}$, is a natural and meaningful function.

For any integer k, $q(k)$ is the coset containing all integers with the same remainder as k when divided by 12. Then i maps this coset to the element it has in common with C_{12}. Because ϕ is q followed by i, we know it maps all elements in the coset $k + \langle 12 \rangle$ to the element $k \in C_{12}$. Of course, C_{12} contains just the possible remainders (0 through 11), so for any integer k, $\phi(k)$ is the remainder when k is divided by 12. The homomorphism ϕ is the natural arithmetic operation of finding a remainder.

This analysis of $\frac{\mathbb{Z}}{\langle 12 \rangle}$ could have been done for any $\frac{\mathbb{Z}}{\langle n \rangle}$, not just $n = 12$. Any C_n is isomorphic to $\frac{\mathbb{Z}}{\langle n \rangle}$, and the corresponding homomorphism ϕ computes remainders mod n. Two numbers are congruent mod n ($a \equiv_n b$) if and only if ϕ maps them to the same element of C_n.

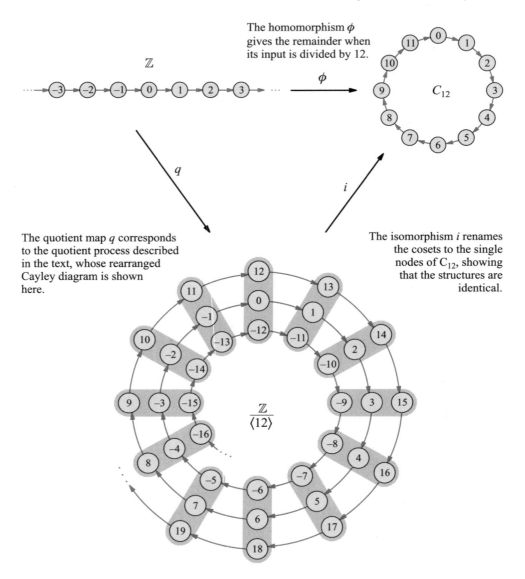

Figure 8.18. The combination of a quotient map and renaming isomorphism yields the map ϕ. For any integer k, $\phi(k)$ is the remainder when k is divided by 12.

8.4 Direct products and relatively prime numbers

Some cyclic groups can be disguised as direct products and some cannot. Earlier in this book, I disguised C_6 by using arrows for the generators 2 and 3, revealing its alternate identity, the direct product of C_2 and C_3. (See Figure 6.5 on page 101 and Section 7.3.) In contrast, it is not possible to disguise C_8 as anything but a cyclic group. You may add unneeded arrows, or leave off so many arrows that it's not a connected diagram anymore (and therefore does not clearly show C_8), but a valid Cayley diagram for C_8 will always include an order-8 arrow, and thus be recognizably cyclic. Why the difference? Why can

8.4. Direct products and relatively prime numbers

some C_n be disguised and others not? The answer can be expressed using isomorphisms, but it depends on the notion of *relatively prime numbers*.

What it means for two numbers a and b to be relatively prime can be defined in several ways. The definition most convenient here is based on the numbers into which both a and b divide evenly, their **common multiples**. Every two numbers a and b have their product ab as a common multiple. But they may also have a smaller common multiple; for example 6 and 15 both divide 30, which is smaller than $6 \cdot 15 = 90$.

Definition 8.6 (relatively prime[1]). Two integers a and b are **relatively prime** when their *least* common multiple is their product ab.

Let's see how this definition applies to Cayley diagrams of cyclic groups and their direct products. The numbers 3 and 4 are relatively prime, and Figure 8.19 shows a Cayley diagram for $C_3 \times C_4$ with the beginning of the orbit of $(1, 1)$ highlighted. What order does the element $(1, 1)$ have? Take a moment to determine the answer by tracing the rest of the orbit in the Cayley diagram.

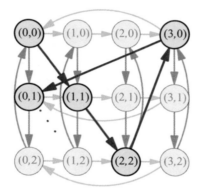

Figure 8.19. Tracing the orbit of $(1, 1)$ in the group $C_3 \times C_4$

As you may have noticed while tracing the orbit, each step of $(1, 1)$ moves to the right one column, eventually returning from the rightmost column to the leftmost on the fourth step, and again on the eighth step, and so on. It returns to the first column every fourth step because of course there are four columns in the diagram. The orbit of $(1, 1)$ ends when it comes back to $(0, 0)$, an element in the first column. Thus the number of steps in the orbit must be a multiple of 4. A similar fact is true about rows; each step of $(1, 1)$ moves down one row, returning to the top row every third step. Because $(0, 0)$ is also in the first row, the number of steps in the orbit of $(1, 1)$ must be a multiple of the number of rows in the diagram, 3. So the order of $(1, 1)$ is a multiple of both 3 and 4, a *common multiple*. Since the least common multiple of 3 and 4 is their product 12, the order of $(1, 1)$ must be at least 12. Thus its orbit covers all of $C_3 \times C_4$, making $C_3 \times C_4$ a cyclic

---

[1] An equivalent definition can be given in terms of common factors. Any two numbers a and b have 1 as a common factor, but they may also have a greater common factor. For example, 6 and 15 share the factor 3. You could define a and b to be relatively prime when their only common factor is 1 (making it also their *greatest* common factor). The reason this definition is equivalent to Definition 8.6 is a fact of number theory I will not prove here.

group generated by $(1, 1)$. The Cayley diagram in Figure 8.19 could be unwound to a single, 12-step cycle generated by $(1, 1)$.

The only assumption about 3 and 4 used in the above reasoning was that they are relatively prime. You could rewrite the previous paragraph about a Cayley diagram for $C_n \times C_m$ for any two relatively prime numbers n and m, and it would still be true. This proves one half of the following important theorem, which we will use to answer the question from the first paragraph of this section.

Theorem 8.7. $C_n \times C_m \cong C_{nm}$ *if and only if n and m are relatively prime.*

The phrase "if and only if" is a common mathematical phrase indicating that the two halves of the sentence always occur together. That is, if n and m are relatively prime, then the groups are isomorphic, *and* if the groups are isomorphic, then n and m must be relatively prime. I have already explained why one of these two statements is true. The paragraphs following Definition 8.6 demonstrate that if n and m are relatively prime, then $C_n \times C_m$ is isomorphic to the cyclic group C_{nm}.

Furthermore, they gave enough information to determine the isomorphism explicitly. The orbit of $(1, 1)$ in $C_n \times C_m$ is a cycle filling the whole group, and thus both $C_n \times C_m$ and C_{nm} are cycles containing exactly nm elements. Each element in the cycle C_{nm} corresponds to a step in the orbit of $(1, 1)$ in $C_n \times C_m$ by the isomorphism

$$\theta : C_{nm} \to C_n \times C_m \quad \text{by} \quad \theta(k) = \text{step } k \text{ in the orbit of } (1, 1).$$

I prove the other half of the theorem now.

Proof. If $C_n \times C_m$ is cyclic, it must be generated by one of its elements; let's call it (a, b) because we do not know specifically which $a \in C_n$ or $b \in C_m$ it involves. Because the orbit of (a, b) includes every element in $C_n \times C_m$, the orbit of a must therefore include every element in C_n and the orbit of b must include every element in C_m. (If $\langle a \rangle$ did not include some $c \in A$, how would $\langle (a, b) \rangle$ ever include an element starting with c?) Thus a generates C_n (and is of order n) and b generates C_m (and is of order m). Figure 8.20 is therefore an acceptable way to organize a Cayley diagram of $C_n \times C_m$, with generators

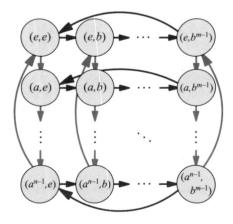

Figure 8.20. A Cayley diagram for the group $C_n \times C_m$, for some n and m

(a, e) and (e, b) as arrows. The orbit of (a, b) walks through this diagram much as the orbit of $(1, 1)$ walks through the diagram in Figure 8.19.

Let's see why the order of (a, b) must be the least common multiple of n and m. The orbit visits column 1 of the diagram after step m, step $2m$, step $3m$, and so on, all multiples of m, the number of columns in the diagram. Similarly, the orbit visits row 1 after steps that are multiples of n. Thus it visits the identity element, in both row 1 and column 1, after steps that are multiples of both n and m, any one of their common multiples. The order of (a, b) is defined as the *first* power that returns to the identity element. So in this case, the order of (a, b) will be the *least* common multiple of n and m. But we know that the orbit of (a, b) includes all nm elements of $C_n \times C_m$, and so the least common multiple of n and m is nm, and they are relatively prime. □

Theorem 8.7 answers the question of when it is possible to disguise a cyclic group as a direct product. If a cyclic group has an order that can be factored into two relatively prime numbers, as $6 = 2 \cdot 3$, then it can be diagrammed as a direct product, $C_6 \cong C_2 \times C_3$. By contrast, the number 8 factors only into $2 \cdot 4$, two numbers that are not relatively prime, and thus $C_8 \not\cong C_2 \times C_4$. Theorem 8.7 has other applications, some of which you will explore in this chapter's exercises. It can help determine the shapes of certain cycle graphs (Exercise 8.23) and determine efficient embeddings for C_n in S_m (Exercise 8.27). It also applies in the following section and the related exercises.

8.5 The Fundamental Theorem of Abelian Groups

Abelian groups have a simple definition: all pairs of elements commute ($ab = ba$). But the consequences of this one small requirement are so significant that a statement of its full implications directs the course of much of the study of group theory! This section tells us why.

We can get an idea of just how special abelian groups are by looking at Cayley diagrams for a representative sample of them. Figure 8.21 shows six abelian groups' Cayley diagrams. It includes two one-dimensional diagrams for one-generator abelian groups (cyclic ones), two two-dimensional diagrams for two-generator abelian groups, and two three-dimensional diagrams for three-generator abelian groups. The pattern typifying abelian groups (the leftmost pattern in Figure 5.8 on page 69) appears throughout each Cayley diagram in Figure 8.21. It makes all arrows parallel to arrows of the same color and perpendicular to arrows of other colors. (Those slightly curved arrows in some parts of Figure 8.21 could be flattened out, but I curved them to improve readability.)

As Figure 8.21 shows, abelian groups' Cayley diagrams are grids. Some are one-dimensional grids (just a line), some are two-dimensional grids (like graph paper), and some are three-dimensional grids (one or more boxes). But all have the parallel-and-perpendicular pattern just mentioned. And most importantly, that pattern is not only indicative of a group's being abelian, *but also of a direct product structure*. The Fundamental Theorem of Abelian Groups states this important connection precisely.

Theorem 8.8 (Fundamental Theorem of Abelian Groups). *Every finite abelian group A is isomorphic to a direct product of cyclic groups. That is, there is some list of integers*

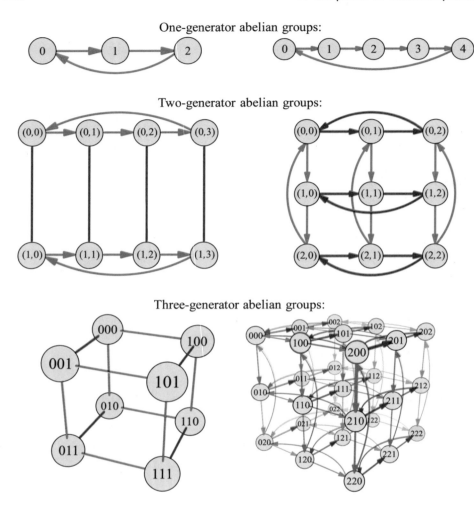

Figure 8.21. Cayley diagrams for six abelian groups, two one-generator groups, two two-generator groups, and two three-generator groups. The one-generator (cyclic) groups are arranged linearly to emphasize their one-dimensionality. Four- and higher-dimensional groups exist but are not shown here.

$n_1, n_2, \ldots, n_m$, such that

$$A \cong C_{n_1} \times C_{n_2} \times \cdots \times C_{n_m}.$$

Although we learned about abelian groups in Chapter 5 and direct products in Chapter 7, we did not encounter isomorphisms until this chapter, and thus were unable to state this theorem until now. I do not include the proof of Theorem 8.8 here, because it is lengthy. I guide interested readers through constructing a proof in the exercises in Section 8.7.10.

The important consequence of this theorem for the study of group theory is that abelian groups are very easy to understand! We understand cyclic groups and direct products very well, so abelian groups are built from two simple group theoretic concepts. Thus the grid-like Cayley diagrams in Figure 8.21 give an appropriately clean and organized impression; all the tangled intricacy possible in groups takes place in the non-abelian ones.

You could say that groups are like cities. A map of Long Island is very different from a map of Paris. It's much easier to give directions, to follow directions, and to picture your location when on Long Island, because all streets are straight, all turns are right angles, and all blocks are the same size. It's comprehensible in the same way abelian groups are, and so its map is a grid, just as is the Cayley diagram of an abelian group. Paris is a much harder city to navigate; the streets curve, connect at varying angles, and have varying lengths. It is much more difficult to get to know your way around, just as non-abelian groups require more study.

This is what I meant earlier by saying that the Fundamental Theorem of Abelian Groups impacts where we direct our efforts; because abelian groups are so well understood, the bulk of the study of group theory is aimed at non-abelian groups. The following section and Chapter 9 both contain material aimed primarily at better understanding non-abelian groups.

8.6 Semidirect products revisited

In Chapter 7, I introduced semidirect products by connecting rewirings of Cayley diagrams. I confessed that I was only introducing *some* semidirect products then, and promised to finish the job later. Now we can use the power of homomorphisms to do just that.

Definition 7.4 specified that a rewiring rearranges a Cayley diagram's arrows but not its nodes. For this reason, a rewiring that has both its nodes and its arrows labeled carries a complete record of how the diagram was rewired. Consider, for example, Figure 8.22, a rewiring of V_4. You can tell how the green and red arrows originally connected the diagram, because obviously the v arrow (green) must originally have connected e to v, and the h arrow (red) must originally have connected e to h. Or we could state this in terms of what the rewiring action did: It moved the h arrows to where v arrows were, and the v arrows to where d arrows would have been, had they been included in the original diagram.

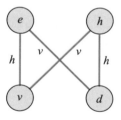

Figure 8.22. A rewiring of the group V_4, with both nodes and arrows labeled to make clear how the rewiring was done

Stating it this way reveals that the rewiring is a function; it maps each of the group's generators to one the group's elements. But it is not just any function; the definition of rewiring (Definition 7.4) requires that it result in no change to the group's multiplication table. In other words, the group structure after the rewiring must be no different than before. This is precisely what it takes to be an isomorphism! Every rewiring of a group is an isomorphism from the group to itself. Although rewirings only show where to map a Cayley diagram's arrows, those arrows generate the group, and as on page 161, knowing where to map the generators tells us how to generate the entire homomorphism (or in this case isomorphism).

Let's build an isomorphism from the rewiring in Figure 8.22. Let $\tau : V_4 \to V_4$ be generated by $\tau(h) = v$ and $\tau(v) = d$. We can generate the following full description of τ using the procedure exemplified on page 161.

$$\tau(e) = e \qquad \tau(h) = v \qquad \tau(v) = d \qquad \tau(d) = h$$

The standard mathematical term for an isomorphism from an object to itself is **automorphism.** The prefix "auto" means "self."

Chapter 7 told us that the set of rewirings (now called automorphisms) of a group G is also a group; we write it as $Aut(G)$. Look back at Figure 7.18, which showed the rewiring group $Aut(V_4)$. We used Cayley diagrams like that one to create semidirect product groups by connecting corresponding elements in the rewirings, as in Figure 7.19. The result was the semidirect product group $G \rtimes Aut(G)$. In general, however, there exist semidirect product groups $G \rtimes H$ for any group H at all. In Chapter 7 we did not have the tools to describe how such products can be constructed, but now we do.

The process is similar to that for the direct product, given in Definition 7.1. The direct product process fills nodes of one factor with copies of the other, then connects corresponding elements. The semidirect product process connects rewirings instead, so we need a way to fill each element of H with a rewiring of G. That is, we need a map of elements of H to elements of $Aut(G)$. An example homomorphism of this type is shown in Figure 8.23; it maps C_3 into $Aut(V_4)$, and its image is the outer ring of the automorphism group.

Any homomorphism whose codomain is an automorphism group, as in Figure 8.23, enables us to form a semidirect product group. The following definition explains the process completely. Compare it to the direct product process in Definition 7.1 and note how they are nearly the same. Like Definition 7.1, this definition includes an example, illustrated in Figure 8.24. Because different homomorphisms lead to different semidirect

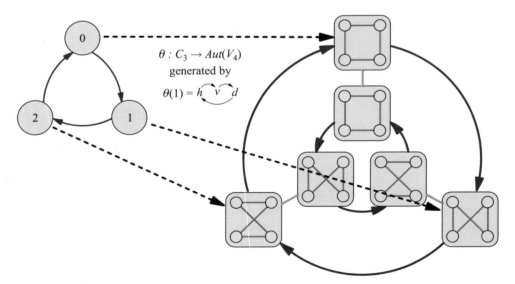

Figure 8.23. An example homomorphism into an automorphism group

8.7. Exercises

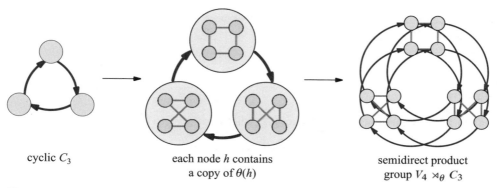

cyclic C_3 each node h contains a copy of $\theta(h)$ semidirect product group $V_4 \rtimes_\theta C_3$

Figure 8.24. An example for creating the semidirect product $V_4 \rtimes_\theta C_3$, with θ defined in Figure 8.23. The process exemplified here is described in Definition 8.9; it parallels the direct product construction illustrated in Figures 7.2 and 7.4.

products, we distinguish them using subscripts: $G \rtimes_\theta H$ means the semidirect product created using the homomorphism θ.

Definition 8.9 (technique for constructing semidirect products using Cayley diagrams). To create a Cayley diagram of $G \rtimes_\theta H$ from a Cayley diagram of G and a homomorphism $\theta : H \to Aut(G)$, proceed as follows.

1. Begin with the Cayley diagram for H.

   In the example in Figure 8.24, the group $H = C_3$ appears on the left.

2. Inflate each node h in the Cayley diagram of H and place in it a diagram of the rewiring $\theta(h)$.

   In the middle of Figure 8.24, each of the three nodes $h \in C_3$ grew larger to contain a copy of $\theta(h)$, using the θ from Figure 8.23.

3. Remove the (inflated) nodes of H while using the arrows of H to connect corresponding nodes from each rewiring, just as in the construction of a direct product group.

   In the example in Figure 8.24, each rewiring has four nodes. Thus the top left nodes in all the rewirings correspond to one another, and we connect them as a little copy of C_3. We do the same with the top right nodes, then bottom left, then bottom right, resulting in a complete Cayley diagram for $V_4 \rtimes_\theta C_3$.

As was promised in Chapter 7, we now see exactly how A_4 is a semidirect product of V_4 and C_3; it is the semidirect product in Figure 8.24. We could therefore label the elements in $V_4 \rtimes_\theta C_3$ with the names from A_4. But in any semidirect product, we also have the option to name the nodes as pairs, just as in the case of a direct product. Give the name (g, h) to the element that was g in the rewiring $\theta(h)$.

8.7 Exercises

This chapter opens new avenues of group theory by allowing us to use homomorphisms to analyze the relationships between groups. Homomorphisms enable us to state significant group theory facts, such as Theorems 8.5, 8.7 and 8.8, and to understand constructions

8.7.1 Basics

Exercise 8.1.

(a) In the homomorphism ϕ in Figure 8.2, what is $\phi(2)$?

(b) In the homomorphism θ in Figure 8.4, what is $\theta(1)$?

(c) In the isomorphism in Figure 8.8, the equation $1 + 2 = 3$ in the domain corresponds to what equation in the codomain?

(d) In the homomorphism τ_1 in Figure 8.9, what elements map to b?

(e) In the homomorphism τ_2 in Figure 8.9, what elements map to 0?

Exercise 8.2. For each statement below, determine whether it is true or false.

(a) For any groups H and G, there is some homomorphism from H to G.

(b) For any groups H and G, there is some embedding of H into G.

(c) Every homomorphism is either an embedding or a quotient map.

(d) Embeddings are those homomorphisms whose kernel is empty.

(e) When $A \cong B$, there is some isomorphism $i : A \to B$, and therefore there is also an isomorphism $j : B \to A$.

8.7.2 Homomorphisms

Exercise 8.3. If $\phi : G \to H$ maps every element of G to the identity element of H, is ϕ a homomorphism?

Exercise 8.4. For each part below, list all homomorphisms (both embeddings and quotient maps) with the given domain and codomain. Does each collection of homomorphisms form a group, as collections of automorphisms do?

(a) Domain C_3 and codomain C_2

(b) Domain C_2 and codomain C_3

(c) Domain and codomain both C_4

(d) Domain C_2 and codomain V_4

(e) Domain and codomain both V_4

Exercise 8.5. Consider the function $\phi : \mathbb{Z} \to \mathbb{Z}$ by $\phi(n) = 2n$. Justify your answer to each of the following questions about ϕ.

(a) Is it a homomorphism? If so, is it an embedding or a quotient map?

(b) Would ϕ be a homomorphism if it were to use a different coefficient than 2? If so, what numbers could be used in place of 2?

(c) What are $Ker(\phi)$ and $Im(\phi)$?

Exercise 8.6. Assume there is a homomorphism $\phi : G \to H$. Justify your answers to each of the following questions.

(a) If there is a subgroup $K < G$, will the set of elements in H to which ϕ maps elements of K also be a subgroup?

(b) If there is a normal subgroup $K \triangleleft G$, will the set of elements in H to which ϕ maps elements of K also be a normal subgroup?

(c) If there is a subgroup $K < H$, will the set of elements in G that ϕ maps to elements of K also be a subgroup?

(d) If there is a normal subgroup $K \triangleleft H$, will the set of elements in G that ϕ maps to elements of K also be a normal subgroup?

Exercise 8.7. Use the concept of generating a homomorphism (page 161) to explain why any homomorphism must map the domain's identity element to the codomain's identity element.

8.7.3 Embeddings

Exercise 8.8. Is it possible to embed C_n in $\mathbb{Z}$ with a homomorphism? Explain your answer.

Exercise 8.9.

(a) How many embeddings of C_4 are there into itself?

(b) How many automorphisms are there of C_4?

(c) Is an embedding of any group into itself always an automorphism?

Exercise 8.10. For each part below, describe all the embeddings with the given domain and codomain. Choose one from each part (if available) to diagram.

(a) Domain C_2 and codomain V_4

(b) Domain C_2 and codomain C_3

(c) Domain C_2 and codomain C_4

(d) Domain C_3 and codomain S_3

(e) Domain C_n and codomain $\mathbb{Z}$

(f) Domain and codomain both $\mathbb{Z}$

8.7.4 Quotient maps

Exercise 8.11.

(a) Diagram the quotient $\frac{\mathbb{Z}}{\langle 3 \rangle}$ similar to the diagram of $\frac{\mathbb{Z}}{\langle 12 \rangle}$ in Figure 8.17.

(b) What is the corresponding quotient map from $\mathbb{Z}$ to C_3?

(c) Can you devise a way to diagram that quotient using a multiplication table instead?

Exercise 8.12. For parts (a) through (c), a group G is given together with a normal subgroup H. Illustrate not only the quotient map $q : G \to \frac{G}{H}$, but also illustrate the embedding $\phi : H \to G$, chained together so that $Im(\phi) = Ker(q)$. Here is an example for $H = C_2$ and $G = C_6$. Elements of H (as well as elements to which they map) are highlighted.

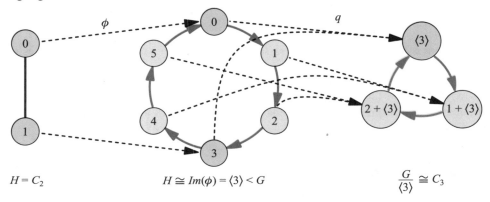

$H = C_2$ \qquad $H \cong Im(\phi) = \langle 3 \rangle < G$ \qquad $\frac{G}{\langle 3 \rangle} \cong C_3$

(a) $H = C_3$, $G = C_6$

(b) $H = C_3$, $G = S_3$

(c) $H = V_4$, $G = A_4$

Now answer each of the following questions about each of your answers to parts (a) through (c).

(d) What map θ into H would satisfy the equation $Im(\theta) = Ker(\phi)$? Choose one with the smallest possible domain.

(e) What map θ' from $\frac{G}{H}$ would satisfy the equation $Im(q) = Ker(\theta')$? Choose one with the smallest possible codomain.

(f) Add the two maps θ and θ' to your illustration.

The new chain of four homomorphisms is called a *short exact sequence*. It is one way to use homomorphisms to illustrate quotients, and it shows a connection between embeddings and quotient maps.

Exercise 8.13. For any group G and any number n we can create a homomorphism that raises every element to the nth power, $\phi : G \to G$ by $\phi(g) = g^n$. (In an additive group like $\mathbb{Z}$, we would write ng instead of g^n. Thus this ϕ is like the function in Exercise 8.5, but it works for any group G.)

8.7. Exercises

(a) What is the kernel of this homomorphism?

(b) When we compute $\frac{G}{Ker(\phi)}$, do we get a subgroup of G?

(c) Is $\frac{G}{H}$ always isomorphic to a subgroup of G (for any G and $H \triangleleft G$)?

Exercise 8.14. For any group G consider the homomorphism $\theta : G \to G$ by $\theta(g) = g^{-1}$. What are its image and kernel? What more can you say about it?

8.7.5 Abelianization

Section 8.4 showed how to divide $\mathbb{Z}$ by $\langle n \rangle$ to make all multiples of n zero. We now wish to divide a non-abelian group by its "non-abelian parts," to make them zero, leaving only abelian parts in the resulting quotient. This process is called *abelianization*, and is investigated in the exercises in this section.

Exercise 8.15. Figure 5.8 on page 69 shows the pattern in Cayley diagrams distinguishing abelian and non-abelian groups, the visualization of the equation $ab = ba$.

(a) Use algebra to show that the equation $aba^{-1}b^{-1} = e$ is equivalent to the original.

(b) Use algebra to show that it is also equivalent to the equation $ab(ba)^{-1} = e$.

(c) Create an illustration of what $aba^{-1}b^{-1} \neq e$ looks like in a Cayley diagram.

Based on Exercise 8.15, a group G containing an element $ab(ba)^{-1}$ that is not the identity e cannot be abelian. Such elements are called *commutators*. We wish to form the *commutator subgroup,* generated by the set of all commutators. Then we will divide G by it to eliminate all the elements that keep G from being abelian, and an abelian group will result.

Exercise 8.16. Explain why the commutator subgroup must be a normal subgroup.

Exercise 8.17. The abelianization of a group G is the quotient of G by its commutator subgroup.

(a) Compute the abelianization of S_3.

(b) Compute the abelianization of A_4.

(c) Compute the abelianization of D_5. What does it have in common with the abelianization of D_3 from part (a)?

(d) The group D_2 is isomorphic to V_4, which is abelian. What is its abelianization?

(e) Compute the abelianization of the groups D_4 and D_6.

(f) What general conclusion do you draw about the abelianizations of dihedral groups?

Exercise 8.18. Use the abelianizations in Exercise 8.17 to help you determine whether an abelianization of a group is the same thing as its largest abelian subgroup.

8.7.6 Modular arithmetic

Exercise 8.19. Why do Figures 8.16 and 8.17 write cosets of $\langle 12 \rangle$ using the notation $k + \langle 12 \rangle$ instead of $k \langle 12 \rangle$?

Exercise 8.20. For each number given below, find the smallest nonnegative integer to which it is congruent mod 12.

(a) 15

(b) 30

(c) 529

(d) −9

(e) −182

Exercise 8.21. If $a \equiv_{12} b$, what can you say about $a - b$?
 Hint: Use Figure 8.16 to help you visualize the situation.

Exercise 8.22. For each of the following statements, determine whether it is true or false.

1. If $a \equiv_6 b$ then $a \equiv_{12} b$.
2. If $a \equiv_6 b$ then $a \equiv_3 b$.
3. If $a \equiv_6 b$ then $a \equiv_5 b$.
4. If $a \equiv_{12} b$ then $a \equiv_2 b$.

8.7.7 Relatively prime numbers

Exercise 8.23. Let p be prime and consider the group $C_p \times C_p$.

(a) Let (a, b) be any non-identity element in the group. What is its order? How do you know?

(b) If (a, b) and (c, d) are both elements of $C_p \times C_p$ and neither one is in the orbit of the other, then do their orbits overlap at all?

(c) How many different orbits are there in $C_p \times C_p$?

(d) What does a cycle graph of $C_p \times C_p$ look like?

Exercise 8.24. Use Theorem 8.7 to prove that if n and m are relatively prime, then there must be a multiple of n that is just one greater than a multiple of m (that is, $am = bm + 1$).

Exercise 8.25. Section 8.4 showed that C_n can be disguised as a direct product if and only if n can be factored into two relatively prime numbers. Many numbers n have this property, but none of them are prime.

(a) Make a list of the first 10 numbers besides primes which cannot be factored into two relatively prime numbers.

(b) What do these numbers have in common?

8.7. Exercises

Exercise 8.26. Apply Theorem 8.7 to the following questions.

(a) Write C_{100} as a cross product of two cyclic groups.

(b) Write C_{308} as a cross product of three cyclic groups.

(c) Is there more than one way to answer either of parts (a) or (b)?

(d) If n is a positive whole number with prime factorization $p_1^{e_1} \times p_2^{e_2} \times \cdots \times p_n^{e_n}$ (for primes p_i and exponents e_i), then write C_n as a cross product of n cyclic groups.

(e) Is there more than one way to answer part (d)?

Exercise 8.27. Exercise 5.44 asked you to find, for various cyclic groups C_n, the smallest S_m into which C_n can be embedded. How can Theorem 8.7 be used to confirm your earlier results, and suggest the general pattern for any n and m?

8.7.8 Semidirect products

Exercise 8.28. The semidirect product in Figure 8.24 uses an embedding homomorphism; let's try a semidirect product using a quotient map. Consider the homomorphism $\theta' : C_4 \to Aut(C_4)$ shown below. It can be used to create a semidirect product group $C_4 \rtimes_{\theta'} C_4$ in which each rewiring of C_4 appears twice. Construct a Cayley diagram for that semidirect product group.

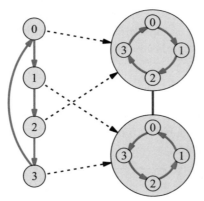

Exercise 8.29. Recall that Exercise 7.14 had you diagram several rewiring groups, which we now call automorphism groups, including $Aut(C_5)$, $Aut(C_7)$, and $Aut(S_3)$.

(a) Find an embedding $\theta : C_2 \to Aut(C_5)$ and diagram the semidirect product $C_2 \rtimes_\theta C_5$. What is a more common name for this group?

(b) Repeat part (a) for C_7. Make a general conjecture from these two semidirect products.

(c) How many embeddings are there of C_3 into $Aut(S_3)$? Create a diagram of the semidirect product group $C_3 \rtimes_\theta S_3$ for one such embedding θ.

(d) If n and m are positive whole numbers and n is even, consider $\theta : C_n \to Aut(C_m)$ defined as follows. Even numbers map to the automorphism that changes nothing (all elements and arrows correspond to themselves, the non-rewiring). Odd numbers map to the automorphism that reverses the C_m arrows.

Draw one such $C_n \rtimes_\theta C_m$, and describe them in general. Where have we seen one before?

Exercise 8.30. Specify θ for the semidirect product of C_4 with C_3 shown below.

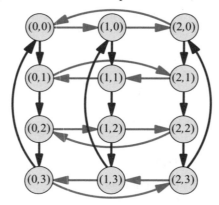

Exercise 8.31. What homomorphisms are there from C_3 to $Aut(V_4)$ besides the one in Figure 8.23? What semidirect products do they generate?

Exercise 8.32. An inner automorphism $\theta : G \to G$ is one that conjugates every element of G by some particular element of G chosen in advance. That is, from any element $g \in G$, we can create an inner automorphism θ defined by $\theta(x) = gxg^{-1}$. Obviously there are different inner automorphisms for different g, though sometimes several different g's result in the same θ.

(a) If G is abelian, what inner automorphisms does it have?

(b) Fill in the following table for S_3 so that the entry in row a and column b contains not ab (as multiplication tables do) but rather bab^{-1}, the conjugate of a by b. The result is a conjugation table.

	e	r	r^2	f	rf	r^2f
e						
r						
r^2						
f						
rf						
r^2f						

(c) What significance do the rows in the above table have?

(d) What significance do the columns in the above table have?

(e) What are all the inner automorphisms of S_3?

Exercise 8.33. Let $\theta : H \to Aut(G)$ be the map that sends every $h \in H$ to the identity element of $Aut(G)$, the non-rewiring (as in Exercise 8.3). What is $G \rtimes_\theta H$?

Exercise 8.34. Definition 8.9 required the function θ from H to $Aut(G)$ to be a homomorphism. This turns out to be necessary; not just any function from H to $Aut(G)$ will work. Find two groups G and H and a function $f : H \to Aut(G)$ that is not a homomorphism, and apply Definition 8.9 to them. Why is the resulting diagram not a Cayley diagram? What necessary property of Cayley diagrams does it fail to possess?

Exercise 8.35. Definition 8.9 defines the semidirect product process for Cayley diagrams. Come up with the semidirect product process for multiplication tables.

8.7.9 Isomorphisms

Exercise 8.36. Prove that $A \times B \cong B \times A$. Give the formula for the isomorphism.

Exercise 8.37. Which of the following equations is true for any G and H? If the equation is true, describe the isomorphism map. If the equation is false, find a particular G and H that make it false and explain why.

(a) $\frac{G \times H}{H} \cong G$

(b) $\frac{G \times H}{G} \cong H$

(c) $\frac{G \rtimes_\theta H}{H} \cong G$

(d) $\frac{G \rtimes_\theta H}{G} \cong H$

Exercise 8.38. Which of the following equations is true for any G and H? If the statement is true, describe the isomorphism map. If the statement is false, find a particular G and H that make it false and explain why.

(a) If $H \triangleleft G$ then $\frac{G}{H} \times H \cong G$.

(b) If $H \triangleleft G$ then $\frac{G}{H} \rtimes_\theta H \cong G$ for any $\theta : \frac{G}{H} \to Aut(H)$.

(c) If $H \triangleleft G$ then $\frac{G}{H} \rtimes_\theta H \cong G$ for some $\theta : \frac{G}{H} \to Aut(H)$.

Exercise 8.39.

(a) Explain why Q_4 is not isomorphic to any member of any of the families of groups we met in Chapter 5.

(b) Explain why the $C_4 \rtimes_\theta C_3$ from Exercise 8.30 is not isomorphic to any member of any of the families of groups we met in Chapter 5.

Exercise 8.40. Recall the group $\mathbb{Q}$ (under addition) and the group $\mathbb{Q}^*$ (under multiplication) introduced in Exercise 4.33. Show that $\mathbb{Q} \times C_2 \cong \mathbb{Q}^*$ by specifying the isomorphism, and explaining why the function you give is indeed an isomorphism.

Exercise 8.41. The group U_n contains the numbers between 1 and n that are relatively prime to n, with the operation of multiplication mod n. So, for example, $U_8 = \{1, 3, 5, 7\}$, and has the following multiplication table.

	1	3	5	7
1	1	3	5	7
3	3	1	7	5
5	5	7	1	3
7	7	5	3	1

Notice that $1 \cdot 3 = 3$ as you would expect, but for instance $5 \cdot 7 \neq 35$, because we work mod 8, and the remainder of $35 \div 8$ is 3.

(a) To what more familiar group is U_8 isomorphic?

(b) What are the orders of the groups U_n for $n \leq 10$?

(c) What is the relationship between U_5 and U_{10}?

(d) Examine U_p for the first few primes p. What conjecture do you make about U_p for any prime?

(e) All the groups U_n belong in which of the families of groups we met in Chapter 5?

The family of groups U_n has several interesting properties. For instance, every finite abelian group is isomorphic to a subgroup of some U_n. More information on U-groups can be found in [18].

Exercise 8.42. This exercise assumes knowledge of matrix multiplication. If that topic is new to you or if you would like a refresher, refer to the hint for this problem in the Appendix.

For each part below, consider the group generated by the two matrices shown, using matrix multiplication as the binary operator. To what common group is it isomorphic? What is the isomorphism?

(a)
$$\begin{bmatrix} 0 & -1 \\ -1 & 0 \end{bmatrix} \quad \begin{bmatrix} 0 & 1 \\ 1 & 0 \end{bmatrix}$$

(b) The constant i stands for the complex number $\sqrt{-1}$.
$$\begin{bmatrix} 0 & -1 \\ -1 & 0 \end{bmatrix} \quad \begin{bmatrix} 0 & i \\ i & 0 \end{bmatrix}$$

(c)
$$\begin{bmatrix} 0 & 1 & 0 \\ 1 & 0 & 0 \\ 0 & 0 & 1 \end{bmatrix} \quad \begin{bmatrix} 1 & 0 & 0 \\ 0 & 0 & 1 \\ 0 & 1 & 0 \end{bmatrix}$$

8.7. Exercises

Exercise 8.43. If a group G has two subgroups H and K, we write HK to mean the set of elements obtained by multiplying any $h \in H$ by any $k \in K$, as in hk. In other words, combine all the elements of all the left cosets hK for any $h \in H$; this is the same as combining all the right cosets Hk for all the $k \in K$.

This problem deals with the special case when H and K are both normal subgroups. Consider the homomorphism $\theta : H \times K \to G$ by $\theta(h, k) = hk$, which takes pairs of elements from $H \times K$ and multiplies them in G. Notice that $Im(\theta) = HK$.

(a) If H and K intersect only at the identity element, explain why θ is an isomorphism (and thus $H \times K \cong HK$).

(b) Is the reverse also true? That is, if $H \times K \cong HK$, must H and K only overlap at e?

8.7.10 Finite Abelian Groups

Exercise 8.44. For each order given below, list all abelian groups of that order. Use the Fundamental Theorem of Abelian Groups (Theorem 8.8) to obtain your answers. (If you need some help getting started, refer to the Appendix, where the solutions to parts (a) and (d) appear.)

(a) 4

(b) 8

(c) 10

(d) 30

(e) 81

(f) 200

In the following chapter, we will learn to list *all* groups of a given order (if the order is relatively small).

The remaining exercises in this section ask you to complete a proof of the Fundamental Theorem of Finite Abelian Groups, following an outline developed elsewhere [14, 20]. Some hints are available in the Appendix.

Any composite number can be factored, and then its factors can be factored, and so on, until only prime numbers are left. This is called a ***prime factorization.*** Here is an example.

$$600 = 30 \cdot 20 = 2 \cdot 15 \cdot 2 \cdot 10 = 2 \cdot 3 \cdot 5 \cdot 2 \cdot 2 \cdot 5$$

If we write the primes in increasing order and combine duplicates using exponents, we can write $600 = 2^3 \cdot 3 \cdot 5^2$.

The proof of Theorem 8.8 begins by breaking up an abelian group in a similar way. For example, if A is an abelian group of order 600, we can find groups G_2, G_3, and G_5 so that $A \cong G_2 \times G_3 \times G_5$ and $|G_2| = 2^3$, $|G_3| = 3$, and $|G_5| = 5^2$. The following exercise asks you to prove that we can separate any abelian group into a direct product of factors whose orders are the powers of primes in the prime factorization of the group's order.

Exercise 8.45. Let A be any abelian group, and let p be a prime number that divides the order of A. Make the following definitions.

$G_p =$ the set of all $g \in A$ whose orders are powers of p (including $p^0 = 1$)

$S =$ the set of all $g \in A$ whose orders do not have p as a factor

(a) Show that G_p and S are both subgroups of A.

(b) What elements do G_p and S have in common? Justify your answer.

Exercise 8.43 defined multiplication of two subgroups. Let's show that $G_p S$ is actually the whole group A; take any $g \in A$ and we will show that it is in $G_p S$.

Let p^k be the highest power of p that divides $|g|$. Then $|g| = p^k m$ for some number m into which p does not divide, making p^k and m relatively prime. From Exercise 8.24, we take the equation $ap^k = bm + 1$, and we can then write the equation $g^{ap^k} = g^{bm+1}$.

(c) How does this equation help show that $g \in G_p S$?

(d) Therefore what is the relationship between A, G_p, and S? Why?

(e) Explain how repeating steps (a) through (d) for other values of p achieves the desired result. (You may assume that $G_p \neq \{e\}$; this is guaranteed to be true by Theorem 9.6 of Chapter 9.)

The previous exercise showed that any abelian group A can be factored into groups whose orders are powers of primes. Thus if we can prove Theorem 8.8 just about abelian groups whose orders are powers of primes, we can then apply it to the factors of A and the entire result is proven. Exercise 8.47 accomplishes this, and the argument it uses depends upon Exercise 8.46.

Exercise 8.46. This exercise shows that the following fact holds for all abelian groups G whose order is a power of a prime p: If G has only one subgroup of order p, then G is cyclic. (We already know that if G is cyclic, then it has only one subgroup of order p; that was Exercise 6.8. Now we're working in the other direction.)

Assume we have an abelian group G whose order is a power of p, and H is its only subgroup of order p. If $|G| = p$, then we know that G is cyclic from Exercise 6.12. Since that is the easy case, assume $|G| > p$.

(a) Let $\phi : G \to G$ be the homomorphism that raises all elements to the p power, as in Exercise 8.13. Show that if $Im(\phi)$ is a cyclic group, then G must be a cyclic group.

Even if the group $Im(\phi) < G$ is not a cyclic group, it is still an abelian group whose order is a power of p, like G. Therefore set up a sequence of homomorphisms as follows.

Let $\phi_1 : G \to G$ be the homomorphism ϕ from part (a), $\phi_1(g) = g^p$.

Let $\phi_2 : Im(\phi_1) \to Im(\phi_1)$ also be defined by $\phi_2(g) = g^p$.

Let $\phi_3 : Im(\phi_2) \to Im(\phi_2)$ also be defined by $\phi_3(g) = g^p$, and so on for ϕ_4, etc.

(b) For which of the images $Im(\phi_n)$ is $H < Im(\phi_n)$?

(c) How many subgroups can $Im(\phi_n)$ have of order p?

(d) Illustrate the chain of connected homomorphisms ϕ_1, ϕ_2, ϕ_3, and so on. How does the chain end, and what are the last few groups before the end?

8.7. Exercises

(e) Put together the facts from parts (a) through (d) to prove the goal stated at the beginning of this exercise.

Exercise 8.47. This exercise shows that you can factor an abelian group G whose order is a power of a prime p into a direct product $C \times H$ for some cyclic subgroup C of G and another subgroup H of G. The process for finding C and factoring it out can then be reapplied to H to factor out another cycle, and so on, until G is shown to be a product of cycles.

Take any element in G with the highest order (or tied for it) and let C be the cyclic subgroup generated by that element. If $C = G$, then G is cyclic, and so we don't have to do any work to factor it into cycles. Thus the interesting case is when G is not cyclic.

(a) When G is not cyclic, what does Exercise 8.46 tell us?

(b) Because C is cyclic, how many subgroups of order p does it have?

(c) Use the previous two parts to explain why there is a subgroup $K < G$ of order p that is not a subgroup of C. How many elements do C and K have in common?

(d) What does Exercise 8.43 say about C and K? (Keep in mind that both are normal in G because it is abelian.)

We can take a quotient $\frac{G}{K}$ because G is abelian; let $q : G \to \frac{G}{K}$ be the quotient map.

(e) To what does q map the subgroup $CK < G$? To what is that subgroup isomorphic?

(f) Explain why $\frac{G}{K}$ contains a cycle of the same size as C.

As in the previous exercise, let's build a chain of homomorphisms; I'll use G_1, C_1, K_1, and q_1 to refer to G, C, K, and q, as the start of the sequence. Let G_2 stand for $\frac{G_1}{K_1}$ and let C_2 stand for the cycle of the same size as C_1 in G_2. Exercise 7.22 and Exercise 7.23 part (c) guarantee that $\frac{G_1}{K_1}$ is an abelian group whose order is a power of p. We can therefore find a subgroup $K_2 < G_2$ that is not a subgroup of C_2, and take a quotient $\frac{G_2}{K_2}$ with the map q_2, just as in part (c).

This process could continue with $G_3 = \frac{G_2}{K_2}$ and so on. Yet of course the orders of these groups are decreasing, $|G_1| > |G_2| > |G_3|$.

(g) Where will this process lead? That is, when does it stop, and what is the final G_n?

(h) As in Exercise 8.46, work backwards from the last group G_n to show that at every step, G_i is isomorphic to a product $C_i \times H_i$ for some subgroup H_i of G_i.

Hint: $H_n = K_n$. What are H_{n-1}, H_{n-2}, and so on?

Exercise 8.48. Summarize the argument in the past three exercises. That is, explain (without the details) how to factor any abelian group A into a product of cyclic groups, and why the combination of Exercises 8.45 through 8.47 proves that it is possible.

Exercise 8.49. The exercises in this section have proven something stronger than Theorem 8.8, which put no requirements on what the subscripts n_1 through n_m had to be like. Yet our proof guarantees something special about those subscripts. What can you say about n_1 through n_m, based on Exercise 8.45?

Exercise 8.50. If an infinite abelian group G is generated by $g_1, \ldots, g_n$, only some of which have finite order, then how might we write G as a cross product of cyclic groups?

(The proof of this infinite version of Theorem 8.8 is lengthier than the proof built from Exercises 8.45 through 8.47. I do not ask you to prove your answer to this question, but rather make a reasonable conjecture about an infinite version of the theorem.)

9

Sylow theory

This chapter is about one question: What groups are there? In Chapter 5, we met a variety of groups, to gain a breadth of exposure to the subject. But it was just a sample, not a comprehensive list. We have since seen groups outside the five families of that chapter. Seeing a list of all groups would surely benefit our knowledge of (and intuition for) group theory. This chapter begins to make such a list.

Of course, there are infinitely many groups, so we cannot list them all! But we can start with the smallest ones and grow our list to include larger ones. For starters, we can ask what groups there are of order 1. (There's just one, C_1.) What groups are there of order 2? If we answer that question, and then move on to order 3, then 4, and so on, we build up a reference library of all groups up to a certain size. To answer such questions, we need a method for finding all the groups of a given order.

Exercise 8.44 is a simplified version of this problem, asking you to find all abelian groups of a given order. The Fundamental Theorem of Abelian Groups (Theorem 8.8) is a powerful tool for answering that exercise. But the complexity of group theory lies in the non-abelian groups, so to find *all* groups of a given order, we need stronger tools. Those tools are the Sylow Theorems, and although they do not give us a foolproof method for finding all groups of a given order, they guide our search by telling us what structures such groups must contain. From just a group's order, they can answer questions like these.

1. How big are its subgroups?
2. How are those subgroups related?
3. How many subgroups are there?
4. Are any of them normal?

The Sylow Theorems do not always fully answer each of these questions, but their power comes from the fact that they can give answers based only on a group's order. This chapter proves the three Sylow Theorems and uses them to find all groups of order 10 or less and some orders beyond 10. In this way the chapter begins to answer the question "What groups are there?"

The Sylow Theorems also answer another question, one that's been lingering since Section 6.5, about Lagrange's Theorem. That theorem tells us that if G has a subgroup of order n, then n must be a factor of $|G|$. I asked whether this statement would still be true if we turned it around: If n is a factor of $|G|$, then must G have a subgroup of order n? This is called the *converse* of Lagrange's Theorem, and we saw that it is not true; just because Lagrange's Theorem permits a subgroup of a certain size does not guarantee that such a subgroup will exist. Sylow Theory digs deeper, finding which sizes of subgroups are guaranteed to exist.

The common thread among this chapter's theorems is that they determine group structure based on group order. I illustrate this common thread by applying each theorem to an unknown group A of order 200. I make no assumptions about A except that it has order 200; it could be an abelian group such as C_{200}, a non-abelian group such as D_{100}, or a group we've never heard of. To illustrate how little we've assumed about A, Figure 9.1 contains a diagram with almost no content. The group A contains an identity e, but the scattered question marks indicate that we don't know anything about the other 199 elements of A. (Yet.)

In fact, there are several different groups of order 200, and we don't even know which one A is. The theorems in this chapter tell us what structures all groups of order 200 have in common, and we can conclude that A contains those structures. As we meet each new theorem in the chapter, I use it to update the illustration in Figure 9.1, removing question marks and replacing them with new information. To preview how Figure 9.1 will evolve, you can look ahead to Figures 9.6, 9.13, 9.18, and 9.20.

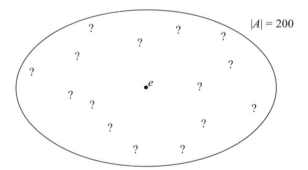

Figure 9.1. If we assume only that a group A has order 200, we do not yet know anything about its internal structure. This chapter teaches several theorems that will reveal much of that structure.

9.1 Group actions

I prepare us for the theorems of this chapter by expanding our understanding of groups as collections of actions. In Chapter 3 we saw many applications of groups as collections of actions, beginning with the rectangle puzzle. We studied groups that permute dancers, coins, and other objects. In Chapter 10 we will learn how studying groups that rearrange polynomial roots made mathematical history.

It is no coincidence that applications of group theory often mention permutations (rearrangements). We proved at the end of Chapter 4 a tight connection between the two

9.1. Group actions

concepts: Every group is isomorphic to a subgroup of a group of permutations. Using the language of homomorphisms, we say that every group can be embedded into some S_n.

Because of this connection, given any group of actions, we should be able to find a set of things it permutes. Sometimes such a set is obvious, as in the example of the dancers; their dancing rearranged them, and it was therefore the dancers themselves being permuted. In other examples, such as the rectangle puzzle, what's being permuted is less obvious. In fact, in the rectangle puzzle, we have two different ways to look at it, both of which are valid. The symmetry-measuring technique in Definition 3.1 suggests viewing the actions as permuting the numbered corners of the rectangle, which they do. However, I prefer an alternative that connects more closely to Cayley diagrams, like the Cayley diagram of the rectangle puzzle in Figure 2.7 on page 18. If we view the actions as permuting the configurations of the rectangle, then each arrow type in the Cayley diagram in Figure 2.7 clearly shows how its group element permutes the configurations.

Our knowledge of homomorphisms enables us to turn this principle into a mathematical definition. In the following definition, I write $Perm(S)$ to indicate the group of all possible permutations of the elements of a set S. (So the group structure of $Perm(S)$ is the same as S_n, for $n = |S|$.)

Definition 9.1 (group acting on a set). A group G **acts on** a set S if there is a homomorphism $\phi : G \to Perm(S)$.

Think of ϕ as a way to interpret every $g \in G$ as doing something to the elements of S (rearranging them). That is, $\phi(g)$ is the *action* that g takes on the elements of S. The definition requires that ϕ be a homomorphism (not just a function) because we want to interpret a combination of group elements as a combination of actions. That is, the interpretation of ab should be two actions in succession, the action for a and then the action for b. This requirement can be expressed as the equation $\phi(ab) = \phi(a)\phi(b)$, which is the defining characteristic of homomorphisms.

Let's see some example group actions, beginning with the familiar rectangle puzzle. If S is the set of four configurations from Figure 2.7, then the arrows show how the actions permute those configurations, and we can formalize that action to fit Definition 9.1 as follows. The group $V_4 = \langle h, v \rangle$ acts on S by the interpretation homomorphism $\phi : V_4 \to Perm(S)$ with the following definition.

$\phi(h)$ = the permutation swapping each configuration with its horizontal flip

$\phi(v)$ = the permutation swapping each configuration with its vertical flip

It is just as easy to rewrite any Cayley diagram of a group of symmetries to show that it fits Definition 9.1. Thus we can see the technique of Definition 3.1 as leading (indirectly) to a group G and an interpretation homomorphism $\phi : G \to Perm(S)$, with S being the set of configurations of the object. But the technique does not give just any G and ϕ; its symmetry-measuring ability came from the following two desirable traits.

1. The image of ϕ is exactly those permutations corresponding to actions possible by hand (guaranteed by part (2) of Definition 3.1).[1]

---

[1] Imposing different restrictions in part (2) of Definition 3.1 would create group actions whose homomorphisms have different images. For example, some definitions of symmetry include any manipulation that alters neither distances nor angles, including some manipulations not possible by hand. (One such manipulation is turning the object into its mirror reflection.)

2. The homomorphism ϕ is an embedding, so that G is isomorphic to that image. (Because we create ϕ from the Cayley diagram drawn in part (3) of Definition 3.1.)

Thus the technique from Definition 3.1 constructs a very specific kind of group action, one that corresponds to the object's three-dimensional symmetries. Of course, other groups may act on the object's configurations in ways that have nothing to do with symmetry. Thus our first example of group actions, while familiar, was a special case. To get a broader idea of what group actions can be like, we need to see some different examples.

Let S still stand for the configurations of the rectangle, and define $\theta : V_4 \to Perm(S)$ by

$\theta(h) = $ the permutation swapping each configuration with its horizontal flip,

and $\theta(v) = $ the identity permutation.

This homomorphism violates both conditions given above that ϕ satisfies, because it was not designed to measure or reflect the rectangle's symmetry. It will therefore result in a different kind of group action than ϕ did. Figure 9.2 diagrams that group action.

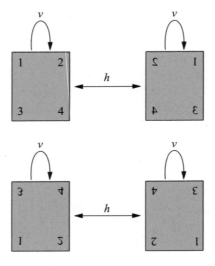

Figure 9.2. The group V_4 acts on the configurations of the rectangle puzzle by the homomorphism θ defined in the text. That homomorphism assigns v to the identity permutation (no action), so the v arrows in this diagram do not alter the rectangle at all.

It is not a Cayley diagram for two reasons. First, there are not enough arrows to connect all the configurations to one another. Second, although a group must have only one identity element, the non-identity element v is drawn as if it were the same as the identity element e. I call diagrams of group actions (like Figure 9.2) **action diagrams.** Some action diagrams are Cayley diagrams (Figure 2.7) and some are not (Figure 9.2). Like Cayley diagrams, action diagrams include arrows only for generators.

Because action diagrams are not Cayley diagrams, they have some new structures that we have not seen before. The following two definitions give names to two such structures that are important for the rest of this chapter.

9.1. Group actions

Definition 9.2 (orbit). When a group G acts on a set S, the **orbit** of any s in S is the set of elements of S that G arrows can reach from s. I write it as $Orb(s)$.

For example, in Figure 9.2, the orbit of the original rectangle configuration (top left) is the pair of configurations in the top row of the figure. The only other orbit in the figure is the pair of configurations in the bottom row; each orbit has just two elements. In contrast, in Figure 2.7, the orbit of any one configuration is the entire set of four configurations.

Obviously two different orbits are totally unconnected by an action diagram's arrows. (If they were connected, they would be just one orbit, not two.) This makes orbits easy to spot in an action diagram, and means that the orbits partition S. Orbits arise when there aren't enough arrows to connect the whole diagram, as in Figure 9.2. The redundant arrows in that same figure (the v arrows, which act like e arrows) give rise to the following complementary notion.

Definition 9.3 (stabilizer, stable). The **stabilizer** of an element s in S is the set of group elements g that don't move s. I write it as $Stab(s)$. A configuration s in S is called *stable* if no actions move s (that is, $Stab(s)$ is all of G).

For example, the stabilizer of any configuration in Figure 9.2 is $\{e, v\}$. Neither e nor v arrows lead from any configuration to a *different* configuration. But in Figure 2.7, the stabilizer of any configuration of the rectangle is just $\{e\}$. Notice that while orbits are collections of elements of S, stabilizers are collections of elements of G. Exercise 9.17 will ask you to explain why any stabilizer is a group (and therefore a subgroup of G).

We have not yet seen an example with any stable elements; the following example fills that gap. In this example, S_3 acts on its own subgroups, which is good preparation for the rest of the chapter, because Sylow theory is built on using groups to act on group-theoretic objects like subgroups and cosets. Let S stand for all the subgroups of S_3.

$$S = \left\{ \{e\}, \langle r \rangle, \langle f \rangle, \langle rf \rangle, \langle r^2 f \rangle \right\}$$

The homomorphism $\tau : S_3 \to Perm(S)$ will have S_3 act on S as follows.

$\tau(g) = $ the permutation that moves each subgroup H to the subgroup gHg^{-1}

For such an action, we say that S_3 acts on S "by conjugation." Figure 9.3 shows the corresponding action diagram and each permutation in the image of the homomorphism.[2] I adopt the convention of putting stable elements on the left of an action diagram and larger orbits on the right. This action diagram is not a Cayley diagram, but for a different reason than Figure 9.2; it is not regular, and therefore lacks the uniform symmetry of a Cayley diagram.

Now that we have seen examples of both orbits and stabilizers, let's see how they relate. The larger $Orb(s)$ is, the more group elements move s around, and the smaller

---

[2] Beware of a common misinterpretation of Figure 9.3: If, from a subgroup H of g, we follow two arrows in succession, say r then f, we have conjugated first by r and then by f.

$$H \xrightarrow{\text{first conjugate by } r} rHr^{-1} \xrightarrow{\text{then conjugate by } f} frHr^{-1}f^{-1}$$

The result of conjugating by r and then f is *not* the same as conjugating by the element rf, as we might expect. Rather, it is the same as conjugating by fr. This reversal is an unavoidable consequence of the definition of conjugacy.

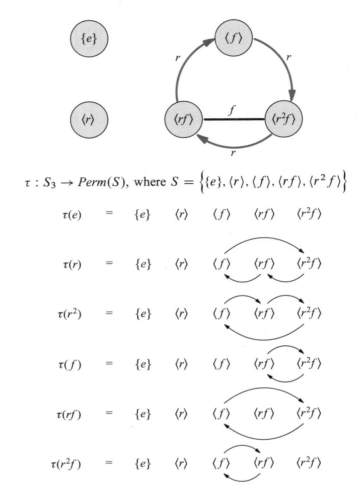

Figure 9.3. The group S_3 acts on its own subgroups by conjugation. Each element in S_3 maps to a permutation of S, as shown. Note that none of the permutations move $\{e\}$ or $\langle r \rangle$, and those two elements are stable.

$Stab(s)$ must be. Conversely, the larger $Stab(s)$ is, the fewer elements there are that move s, and so the smaller $Orb(s)$ must be. Although this give-and-take relationship makes sense as just stated, it is not very specific; the following theorem makes it precise.

Theorem 9.4 (Orbit-Stabilizer Theorem). *The size of an element's orbit times the size of its stabilizer is the size of the group.*

$$|Orb(s)| \cdot |Stab(s)| = |G|$$

This indeed makes the statements from the previous paragraph precise; for a given orbit size, the Orbit-Stabilizer Theorem immediately determines the size of the same element's stabilizer (and vice versa). Although $Orb(s)$ is not a group, I still use the notation $|Orb(s)|$ to mean the number of elements in it. I will continue to use this notation for the size of any set. Before reading the following proof, consider using Figure 9.3 to find $Orb(s)$

and $Stab(s)$ for each $s \in S$. It should help your intuition for the theorem, and make the following proof more concrete.

Proof. Because $Stab(s)$ is a subgroup of G, the definition of subgroup index (Definition 6.9) tells us that

$$\underbrace{|Stab(S)|}_{\text{size of subgroup}} \cdot \underbrace{[G : Stab(s)]}_{\text{number of cosets}} = |G|.$$

I prove that the number of elements in $Orb(s)$ equals the number of cosets of $Stab(s)$, so that I can substitute $|Orb(s)|$ for $[G : Stab(s)]$ in the above equation to prove the theorem. I show that each left coset of $Stab(s)$ pairs up with some element of $Orb(s)$ in a one-to-one relationship, and thus the two sets must be the same size.

First, let's see that the elements of any left coset $gStab(s)$ all move s to the same place, so that no coset corresponds to more than one element of $Orb(s)$. Consider two elements of $gStab(s)$, say gh_1 and gh_2. If ϕ is the interpretation homomorphism, then

$$\phi(gh_1) = \phi(g)\phi(h_1) \qquad \text{and} \qquad \phi(gh_2) = \phi(g)\phi(h_2).$$

But h_1 and h_2 are in $Stab(s)$, so neither $\phi(h_1)$ nor $\phi(h_2)$ moves s. Thus we can rewrite the above equations as

$$\phi(gh_1) = \phi(g) \qquad \text{and} \qquad \phi(gh_2) = \phi(g).$$

Both gh_1 and gh_2 have the same effect on s, as would any other element of $gStab(s)$.

Next, let's see that elements of different left cosets correspond to different operations on s, so that no element of $Orb(s)$ corresponds to more than one coset of $Stab(s)$. Consider two different cosets $gStab(s)$ and $kStab(s)$; I show that they affect s differently. In a Cayley diagram of G, any arrow h from $gStab(s)$ to $kStab(s)$ corresponds to the action that changes the results of the $gStab(s)$ action into the results of the $kStab(s)$ action. Thus if $gStab(s)$ and $kStab(s)$ have the same effect on s, then h should do nothing to s at all. But the h arrow begins in one copy of $Stab(s)$ and ends in another, and thus h is not in the subgroup $Stab(s)$. It *does* move s, and so $gStab(s)$ and $kStab(s)$ have different effects on s.

The previous two paragraphs establish a one-to-one relationship between the left cosets of $Stab(s)$ and the elements of $Orb(s)$. Thus the two sets have the same size, and we can substitute $|Orb(s)|$ for $[G : Stab(s)]$ in the first equation in this proof so that the desired relationship holds.

$$|Orb(s)| \cdot |Stab(s)| = |G|$$

□

9.2 Approaching Sylow: Cauchy's Theorem

Theorem 9.4 is a key insight into group actions that will be used to prove many of this chapter's theorems. The first, Cauchy's theorem, gives us a small, partial converse to Lagrange's Theorem, by guaranteeing the existence of subgroups of prime order. In Section 9.4, the Sylow Theorems will build on Cauchy's Theorem and extend it. Cauchy's Theorem uses the following short theorem, built on the Orbit-Stabilizer Theorem.

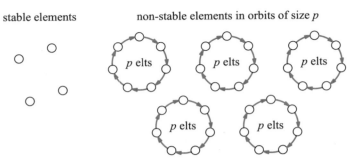

The number of non-stable elements is a multiple of p.

Figure 9.4. A group of prime order p acts on the elements of S. Therefore elements of S are either stable or part of an orbit of size p.

Theorem 9.5. *If a group G of prime order p acts on a set S, then the order of S and the number of stable elements in S are congruent mod p.*

Recall that what it means for two elements to be congruent mod p is that the distance between them on the number line is a multiple of p, so that they are in the same congruence class. We first pictured congruence classes this way in Figure 8.16.

By now it should come as no surprise that I find this theorem far easier to understand and to prove if I draw a picture of it. The following short proof is based on Figure 9.4, which shows what action diagrams for groups of prime order look like.

Proof. The Orbit-Stabilizer Theorem tells us that the size of each $Orb(s)$ is a factor of $|G|$, so when $|G|$ is a prime p, all orbits have size 1 or p. The stable elements are in orbits of size 1 (each by itself) and the rest of S is partitioned into orbits of size p. The number of non-stable elements in S is therefore a multiple of p, as in Figure 9.4.

Consider the following straightforward equation.

$$\text{number of stable elements in } S + \underbrace{\text{number of non-stable elements in } S}_{\text{This is a multiple of } p.} = \text{size of } S$$

Another way to read this equation is that the distance (on the number line) between the number of stable elements in S and the size of S is a multiple of p, and so they must be in the same congruence class mod p. □

Theorem 9.6 (Cauchy's Theorem). *If p is a prime number that divides $|G|$, then G has an element g of order p, and therefore a subgroup $\langle g \rangle$ of order p.*

Proof. Because p is prime, if I find some $g \neq e$ satisfying $g^p = e$, then g must have order p. Exercise 9.15 asks you to explain why this is so, but for now I use the fact without justification.

In most groups there are lots of ways to multiply p group elements together and result in the identity element e, but usually only a small minority of them are powers, like the g^p we're looking for. The equation $e^p = e$ is part of that minority and is true in any group, but we seek a g other than the identity.

9.2. Approaching Sylow: Cauchy's Theorem

$e \cdot e \cdot e = e$	$r \cdot e \cdot r^2 = e$	$r^2 \cdot e \cdot r = e$	$f \cdot e \cdot f = e$	$fr \cdot e \cdot fr = e$	$fr^2 \cdot e \cdot fr^2 = e$
$e \cdot r \cdot r^2 = e$	$r \cdot r \cdot r = e$	$r^2 \cdot r \cdot e = e$	$f \cdot r \cdot fr = e$	$fr \cdot r \cdot fr^2 = e$	$fr^2 \cdot r \cdot f = e$
$e \cdot r^2 \cdot r = e$	$r \cdot r^2 \cdot e = e$	$r^2 \cdot r^2 \cdot r^2 = e$	$f \cdot r^2 \cdot fr^2 = e$	$fr \cdot r^2 \cdot f = e$	$fr^2 \cdot r^2 \cdot fr = e$
$e \cdot f \cdot f = e$	$r \cdot f \cdot fr^2 = e$	$r^2 \cdot f \cdot fr = e$	$f \cdot f \cdot e = e$	$fr \cdot f \cdot r = e$	$fr^2 \cdot f \cdot r^2 = e$
$e \cdot fr \cdot fr = e$	$r \cdot fr \cdot f = e$	$r^2 \cdot fr \cdot fr^2 = e$	$f \cdot fr \cdot r^2 = e$	$fr \cdot fr \cdot e = e$	$fr^2 \cdot fr \cdot r = e$
$e \cdot fr^2 \cdot fr^2 = e$	$r \cdot fr^2 \cdot fr = e$	$r^2 \cdot fr^2 \cdot f = e$	$f \cdot fr^2 \cdot r = e$	$fr \cdot fr^2 \cdot r^2 = e$	$fr^2 \cdot fr^2 \cdot e = e$

Table 9.1. All equations $a \cdot b \cdot c = e$ for any $a, b, c \in S_3$

For example, consider the case when $G = S_3$ and $p = 3$. Table 9.1 lists all products of three S_3 elements that yield e. Direct your attention to the left-hand sides of these equations, the 36 expressions in the set

$$\{e \cdot e \cdot e, \ e \cdot r \cdot r^2, \ e \cdot r^2 \cdot r, \ \ldots, \ fr^2 \cdot fr^2 \cdot e\}.$$

In this example and for any G and p in general, I call this set of expressions S, and the set of powers in it (like e^p) I call P.

This proof focuses on S and P, and so it will be helpful to know how many elements are in those sets. Let's deal with S first. In the example of $G = S_3$ and $p = 3$, there are obviously 36 equations in Table 9.1. Let's see why that total comes to 36, and discern the general pattern from there. To create one of the equations in Table 9.1, you can choose any of the six S_3 elements to be the first in the equation and any of those same six elements to follow it, but then the final element of the equation's left side is constrained to be the one element that makes that left side equal e. If we call the first two elements a and b, then the third must be nothing other than $(ab)^{-1}$. In Table 9.1, these two choices (of a and b) correspond to each of the six columns and six rows, respectively, giving 36 equations in all.

If p were larger, creating an element of S could be done by a similar (but longer) sequence of choices. For $p = 5$, you could choose elements a, b, c, and d, but then the fifth element must be $(abcd)^{-1}$ to make the equation $a \cdot b \cdot c \cdot d \cdot (abcd)^{-1} = e$ true. For any size p, choose the first $p - 1$ elements from G any way at all, but then the last element must be the inverse of their product. Thus creating an element of S involves $|G|$ choices for each of the first $p - 1$ elements but no choice at all for the last element, giving the following equation for the size of S.

$$|S| = \underbrace{|G| \times |G| \times \cdots \times |G|}_{p-1 \text{ times}} = |G|^{p-1}$$

Out of this total of $|G|^{p-1}$ expressions, the ones I care about are those in P, those elements g satisfying the equation $g^p = e$. In the example of $G = S_3$ and $p = 3$, Table 9.1 shows that P contains three elements. But how many does it contain in general, for any G and p? It is impossible to get a precise answer from how little we know of G, but I do not need a precise answer. I need to prove only that $|P| > 1$, so that there is some $g \in P$ besides e. Strategic use of Theorem 9.5 can tell us something about $|P|$. That

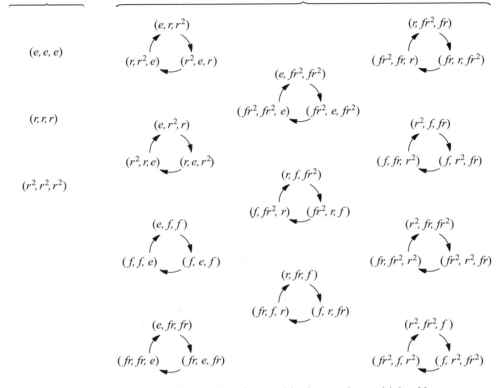

Figure 9.5. The group C_3 acts on the set of triples of elements of S_3 that multiply to e, as in the proof of Cauchy's Theorem (Theorem 9.6). Arrows represent the action $\phi(1)$, which cycles each triple to the right one step.

theorem deals with group actions and their stable elements, so the strategy is to devise a group action whose stable elements are the members of P.

Let the cyclic group C_p act on S with the interpretation homomorphism $\phi : C_p \to Perm(S)$ defined by

$\phi(n)$ = the permutation that cycles the elements in an expression to the right n steps.

For example, the action $\phi(1)$ sends the expression $e \cdot r \cdot r^2$ to the expression $r^2 \cdot e \cdot r$. The action $\phi(2)$ would cycle the expression $fr \cdot r^2 \cdot f$ to the right two steps, giving $r^2 \cdot f \cdot fr$. Figure 9.5 shows the action diagram in the case where $G = S_3$ and $p = 3$. In Exercise 9.16, I ask you to prove that each $\phi(n)$ will be a permutation of S, as I have claimed above.

I set up this action intending that its stable elements should be just the members of P. We can see that this is the case, because those expressions that $\phi(1)$ does not change are those in which each element is the same as the next, making them all equal. Thus the stable elements are just those in P. And because C_p is a group of prime order, we can apply Theorem 9.5 to give us some information about $|P|$. Theorem 9.5 gives us the

9.2. Approaching Sylow: Cauchy's Theorem

equation $|S| \equiv_p |P|$. But $|S| = |G|^{p-1}$, and $|G|$ is a multiple of p, making $|S|$ and $|P|$ multiples of p as well. Although we do not know exactly what $|P|$ is, we now know that it is in the set $\{0, p, 2p, 3p, \ldots\}$.

But it cannot be zero, because e^p is in P. So there must be at least p elements in P, which means at least 2 (the smallest prime). Thus there is at least one other equation $g^p = e$ besides $e^p = e$, which gives us the element g of order p that we have been seeking. □

After every important theorem in this chapter, we will take the theorem for a test drive to see what it can do. I have selected two applications of Cauchy's Theorem to show us its power. First, recall the unknown group A of order 200 from Figure 9.1; Cauchy's Theorem gives us some new information to add to the diagram. The number 200 factors into $2^3 \cdot 5^2$, so the only primes dividing 200 are 2 and 5. Cauchy's Theorem therefore says that A contains elements a and b of orders 2 and 5 respectively, as shown in Figure 9.6.

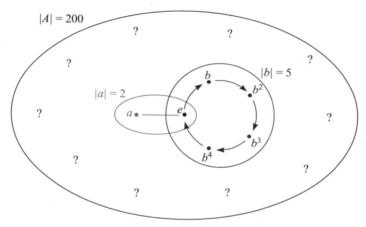

Figure 9.6. An unknown group A of order 200, with subgroups marked whose existence Theorem 9.6 guarantees

The elements a and b (and thus the subgroups $\langle a \rangle$ and $\langle b \rangle$) exist regardless of what group A is; Cauchy's Theorem requires only that 2 and 5 divide the group's order. I still include many question marks in the picture, because we only know about two small subgroups of this very large group! We do not yet know whether there are more subgroups of the same sizes as $\langle a \rangle$ and $\langle b \rangle$, or whether there are any larger subgroups. But Cauchy's Theorem did tell us some information, and in fact the kind of information it gives us can be even more significant when we apply the theorem to smaller groups. The following section does so, and is our second example application of Cauchy's Theorem.

9.2.1 Classification of groups of order 6

If you did Exercise 4.21 part (c), then you know that there are only two different groups of order 4. That exercise depends on earlier ones in which you must painstakingly create all possible four-by-four multiplication tables. In those exercises, the elements are named 0, 1, 2, and 3 because the names are unimportant; it is the structure of the group we care about. The result of these exercises is that any group of order 4 is isomorphic to either V_4 or C_4 by a suitable renaming isomorphism.

We say that those exercises "classified groups of order 4 up to isomorphism." Such work answers the question, "What groups are there with four elements?" The exhaustive, multiplication-table approach to classification becomes unworkably tedious for even slightly larger groups. The results of this chapter, starting with Cauchy's Theorem, give us a better way to find out how many groups there are of a given order, and what those groups are. For small group orders like 6, Cauchy's Theorem is sufficient, as this section shows. For larger orders we will need stronger tools, which is where the Sylow Theorems come in.

We already know that there are at least two groups of order 6, C_6 and S_3 (also known as D_3); we will see that there are actually none but these two. I begin by using Cauchy's Theorem to deduce some essential facts about any group of order 6. Then to find all groups of order 6, we start exploring using Cayley diagrams that fit those facts. Thus Cauchy's Theorem gives us a head start on our search, and keeps us from searching down some dead-end streets.

The head start Cauchy's Theorem gives us is that in any group of order 6, there must be an element of order 2 and an element of order 3, because those are the two primes that divide 6. Call those elements a and b, respectively. Therefore we can start drawing a Cayley diagram for any group of order 6 by drawing a subgroup $\langle a \rangle$ of order 2, like the top diagram in Figure 9.7. From the identity element in the Cayley diagram, one can also follow b arrows, which must create an orbit of size 3. Such an orbit contains only elements of order 3, and thus it does not contain a (which has order 2). So we can safely extend our diagram to the one on the lower left of Figure 9.7.

The subgroup $\langle b \rangle$ must have a left coset $a\langle b \rangle$, a copy of $\langle b \rangle$ that does not touch $\langle b \rangle$ and contains a. That coset therefore contains two new group elements, as shown in the lower right image in Figure 9.7. See how much our diagram has grown just from the seeds Cauchy's Theorem provided! *Any* Cayley diagram for a group of order 6 must contain the pattern on the lower right of Figure 9.7. Our search for groups of order 6 is therefore quickly drawing to an end; we have a nearly complete Cayley diagram already.

One choice remains: How will the a arrows connect the lower four nodes in this partial diagram? Or more simply, where does an a arrow lead from b? There are only

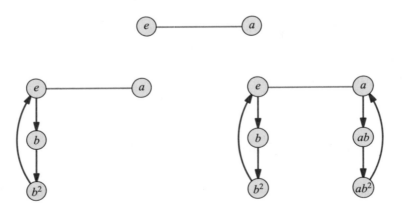

Figure 9.7. Three steps in the creation of any Cayley diagram for a group of order 6

9.3. p-groups

three other nodes that need an a arrow entering them, b^2, ab, and ab^2. Each of these choices can be expressed as an equation involving a and b. For example, an a arrow connecting b to b^2 implies the equation $ba = b^2$. Cancelling b from the left of each side yields $a = b$, which we know is not true, so the a arrow from b does not lead to b^2. I consider the remaining two options separately in the following two paragraphs.

If the a arrow from b leads to ab, then $ba = ab$. This equation determines all remaining connections in the diagram. The visual reason for this is that Cayley diagrams must be regular, and so the path ab must equal the path ba throughout the diagram. We could state the same idea algebraically, by multiplying the equation $ba = ab$ by various elements on the left, to see where a arrows lead from other nodes. For example, multiplying by b yields $b^2 a = bab$, telling us where the a arrow leads from b^2. Either the visual or algebraic view can be used to complete the diagram, resulting in the diagram on the left of Figure 9.8. The group is $C_2 \times C_3$, or simply C_6.

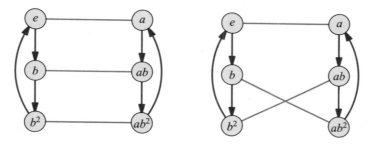

Figure 9.8. Two isomorphism classes of groups of order 6

If the a arrow from b leads to ab^2, the equation $ba = ab^2$ results. As in the previous case, either visual or algebraic means could use this equation to determine all other connections in the diagram. The Cayley diagram on the right of Figure 9.8 results, showing the group S_3.

We have therefore classified groups of order 6. There are only two of them up to isomorphism; we could also say that there are two "isomorphism classes" of groups of order 6. This classification tells us not to waste our time imagining any groups of order 6 besides these two. Trying to contemplate a group of order 6 all of whose non-identity elements have order 3 is nonsensical, as is imagining a group with class equation $1 + 1 + 2 + 2 = 6$. Our classification shows that the laws of group theory forbid such objects, and permit only the two in Figure 9.8.

9.3 p-groups

Cauchy's Theorem is a partial converse to Lagrange's Theorem, "partial" because it guarantees the existence of a subgroup only if its order is prime. As we just saw, in small groups this can be very helpful because the subgroups of prime order make up a large portion of the group. But in large groups (like our order-200 example), knowing whether there are larger subgroups would be more helpful. The Sylow Theorems include a more general converse to Lagrange's Theorem, using p-groups.

Definition 9.7 (*p*-group, *p*-subgroup). A ***p-group*** is a group whose order is a power of a prime p. A p-group that is a subgroup of G is called, for short, a ***p-subgroup*** of G.

For example, D_4 is a 2-group, because its order is 8, a power of the prime 2. It contains a subgroup isomorphic to C_4, which is therefore a 2-subgroup of D_4 because its order is 2^2. Similarly, C_1 and C_{13} are both 13-groups, because their orders are powers of the prime 13 (specifically, 13^0 and 13^1). We finish preparing for the Sylow Theorems by applying what we know of group actions to p-groups in the following two short results.

Theorem 9.8. *If a p-group G acts on a set S, then the order of S and the number of stable elements in S are congruent mod p.*

This theorem is nearly identical to Theorem 9.5. The only change is that the assumption $|G| = p$ has become $|G| = p^n$, making it more general.

Proof. As in the proof of Theorem 9.5, the size of each orbit must divide the order of G. In this case, only powers of p divide $|G|$, so the orbits are therefore of various sizes including 1, p, p^2, p^3, up to at most p^n. So simply modify the illustration for Theorem 9.5 (in Figure 9.4) so that the orbits are any of these various sizes, all powers of p. The rest of the proof is then the same as for Theorem 9.5. □

The following theorem is a bit dense with notation, but the subsequent paragraphs clarify what it's saying.

Theorem 9.9. *If H is a p-subgroup of G, then $[N_G(H) : H] \equiv_p [G : H]$.*

Recall that $[G : H]$ is the number of left cosets of H in G. Thus the theorem says that whether we count only those cosets in the normalizer $N_G(H)$ or all the cosets in G, the result is the same mod p. Another way to state this is that the difference (the number of cosets *outside* $N_G(H)$) is a multiple of p.

This theorem is here because the First Sylow Theorem depends on it. But secondarily, its proof reiterates a valuable proof strategy from Cauchy's Theorem that we will encounter several times in this chapter. One way to learn how many elements are in a set is to devise an action of a p-group whose stable elements are just those we wish to count. Then Theorem 9.5 or Theorem 9.8 can count those elements mod p. In the proof of Cauchy's Theorem the stable elements were the set P of powers g^p; a similar strategy is used here.

Proof. Let S be the left cosets of H in G and consider the group H acting on S by the interpretation homomorphism $\phi : G \to Perm(S)$ defined by

$$\phi(h) = \text{the permutation that sends a coset } gH \text{ to the coset } hgH.$$

Exercise 9.19 asks you to prove that each such $\phi(h)$ is indeed a permutation. In such a situation, we say that G acts on S "by left multiplication." This action was chosen because of the strategy mentioned above; I can prove that its stable elements are just those left cosets in the normalizer, as shown in Figure 9.9.

Recall that the left cosets of H in $N_G(H)$ are those left cosets that are also right cosets. Figure 9.10 illustrates and explains why they are also those cosets stabilized by the above group action. The explanations in that figure are an important part of this proof. Since the stable elements are those cosets in $N_G(H)$, the number of them is $[N_G(H) : H]$.

9.3. p-groups

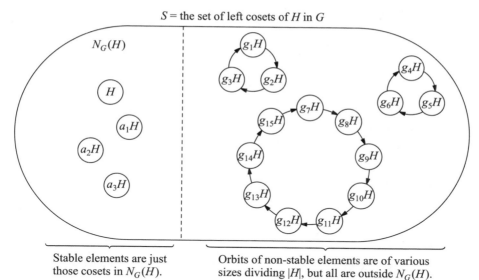

Figure 9.9. When H acts on its left cosets by left multiplication (as in the proof of Theorem 9.8) the stable elements are exactly those elements in its normalizer $N_G(H)$.

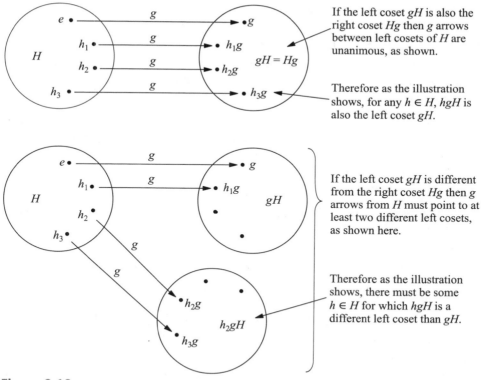

Figure 9.10. Those cosets gH stable under left multiplication by H are just those that are both left and right cosets, $gH = Hg$.

The final step of the strategy is to apply Theorem 9.8, which tells us that the number of stable elements must be congruent mod p to $|S|$, which in this case is the number of left cosets of H, $[G:H]$. Writing this as an equation yields the statement of the theorem,

$$[N_G(H):H] \equiv_p [G:H].$$

□

If you wish to pause and gain a better understanding of this theorem, I encourage you to try Exercise 9.10 now, which asks you to illustrate the group action used in the proof, for a specific group G and subgroup H.

9.4 Sylow Theorems

The Sylow Theorems give us the following three kinds of information about the p-subgroups of any group. I state them imprecisely here, just to give some intuition for what is to come.

Existence: In every group, p-subgroups of all possible sizes are guaranteed to exist.

Relationship: A group's p-subgroups have ties to one another through conjugacy.

Number: There are restrictions on how many p-subgroups a group can have.

This section proves the three Sylow Theorems and shows some of their applications, including extending our knowledge of the order-200 group A.

9.4.1 The First Sylow Theorem: Existence of p-subgroups

Theorem 9.10 (First Sylow Theorem). *If G is a group and p^n is the highest power of p dividing $|G|$, then there are subgroups of G of every order $1, p, p^2, p^3$, up to p^n. Also, every p-subgroup with fewer than p^n elements is inside one of the larger p-subgroups.*

The First Sylow Theorem generalizes Cauchy's Theorem in several ways, summarized in Table 9.2. Its proof deals with both statements in the theorem at once by using Cauchy's Theorem to expand smaller p-subgroups to create larger ones. The First Sylow Theorem also tells us a bit about the relationship among p-subgroups, but we will learn more about that relationship from the Second Sylow Theorem.

Proof. It is easy to find a p-subgroup of order 1 (which is p^0) because it is obviously $\{e\}$. We also know that there is a p-subgroup of order p (which is p^1) from Cauchy's

Cauchy's Theorem (Theorem 9.6)	First Sylow Theorem (Theorem 9.10)
If p divides $|G|$,	If p^i divides $|G|$,
then there is a subgroup of order p.	then there is a subgroup of order p^i.
It is cyclic and has no subgroups.	Each has subgroups of orders $1, p, p^2$, up to p^i.
There is also an element of order p.	There is not necessarily an element of order p^i.

Table 9.2. A comparison of the subgroup existence statements in Theorem 9.6 to those in Theorem 9.10. Most of Theorem 9.10's statements are stronger.

9.4. Sylow Theorems

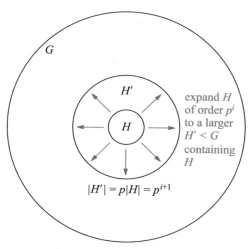

Figure 9.11. Theorem 9.10 gives a procedure for taking a subgroup of order p^i and finding a larger subgroup whose order is p^{i+1} (as long as p^{i+1} divides $|G|$).

Theorem (as long as $|G| > 1$). The main job of this proof is showing the existence of the larger subgroups, by explaining how to take any $H < G$ of any order $p^i < p^n$ and expand it to create a new subgroup $H' < G$ that contains H and is p times as large, as shown in Figure 9.11. We can then repeatedly expand the smallest p-subgroups, creating larger ones up to size p^n.

We can find the H' we seek inside the normalizer $N_G(H)$ by relying on the fact that $H \triangleleft N_G(H)$. Figure 9.12 illustrates the groups, subgroups, and homomorphism that come into play in the rest of this proof; refer to it to help visualize the following argument.

Create the quotient group $\frac{N_G(H)}{H}$ and call the quotient map q. The size of the quotient group is the number of cosets of H in its normalizer, $[N_G(H) : H]$, which Theorem 9.9

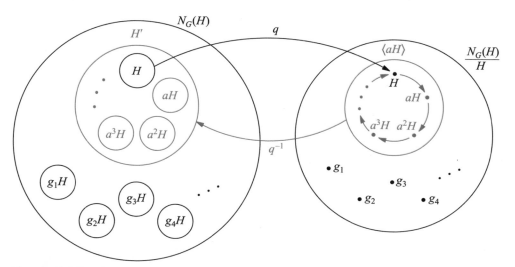

Figure 9.12. The quotient map and its inverse used in Theorem 9.10 to create a subgroup H' whose order is p times that of H

says must be congruent to $[G:H]$ mod p. So what do we know about $[G:H]$? We're given that the order of G is some multiple of p^n, say $p^n m$. So the number of cosets of H is

$$[G:H] = \frac{|G|}{|H|} = \frac{p^n m}{p^i} = p^{n-i} m.$$

Because $p^i < p^n$, we know that $p^{n-i} > 1$ and so p divides $p^{n-i} m$. Therefore $[G:H]$ and $[N_G(H):H]$ are both multiples of p.

The order of $\frac{N_G(H)}{H}$ is obviously not 0, so it must be a positive multiple of p. This lets us use Cauchy's theorem to find an element of order p in the quotient group; call that element aH. The cyclic subgroup $\langle aH \rangle$ will be very useful to us.

The collection of elements that q maps to $\langle aH \rangle$ obviously contains H, but as Figure 9.12 suggests it is also a subgroup of $N_G(H)$, by Exercise 8.6 part (c). It is the subgroup H' that we seek; it contains H and has size p^{i+1} for the following reason. There are p elements in $\langle aH \rangle$, and therefore p cosets of H in H'. Since H contains p^i elements, each of its cosets does as well, and H' contains p of them, for a total of p^{i+1} elements.

The preceding paragraphs give a way to expand any H of order $p^i < p^n$ into a larger H' of order p^{i+1}. Beginning with $H = \{e\}$, we can repeatedly expand it to create H', H'', and so on of orders p, p^2, up to p^n. □

The expansion technique in this proof is an example of conjugacy. It applies q to H, applies Cauchy's theorem in the quotient group to turn one element into p elements, and applies q^{-1} to bring those p elements back into G as p cosets forming a subgroup H'. Even though q^{-1} isn't really a function, this is a useful way to summarize the argument.

Let's see how much more power this theorem has than Cauchy's Theorem. First, let's return to the group A from Figure 9.6, of which we assumed only that its order is 200. Cauchy's Theorem told us that there must therefore be two subgroups of orders 2 and 5, but the First Sylow Theorem tells us much more. Those small, prime-order subgroups are inside groups of orders 4 and 25 respectively, and the subgroup of order 4 is in one of order 8. These larger, nested subgroups are shown in Figure 9.13.

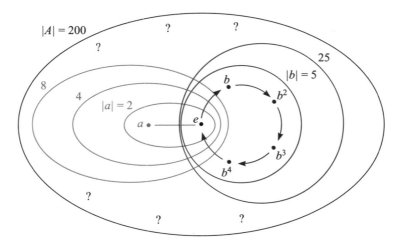

Figure 9.13. The arbitrary group A of order 200, some of whose details we discerned in Figure 9.6, here augmented with new information based on Theorem 9.10.

9.4. Sylow Theorems

We also know that there are no subgroups of order $2^4 = 16$, nor $5^3 = 125$, because those numbers do not divide 200. But we do not yet know how many subgroups there are of orders 2, 4, 8, 5, and 25. We know there is at least one of each of these sizes, but we do not yet know whether there are more. Hence, some questions marks remain in Figure 9.6.

9.4.2 Classification of groups of order 8

Cauchy's Theorem helped us classify groups of order 6, because the subgroups it gave us were a significant portion of the whole group. Because larger numbers may not factor so helpfully, the First Sylow Theorem becomes valuable in classifying groups of those orders.

Table 9.3 lists every group of order 1 through 7, and the reason justifying each of the seven classifications. Four of the seven classifications are based on Exercise 6.12, regarding groups of prime order necessarily being cyclic. Let's extend that table to groups of order 8. From Exercise 8.44 part (b) you know that there are three abelian groups of order 8, $C_2 \times C_2 \times C_2$, $C_2 \times C_4$, and C_8. The First Sylow Theorem helps simplify the search for non-abelian groups of order 8, and together these results add another row to Table 9.3.

Order	Groups	Reason
1	only one group, $\{e\}$	The one element must be the identity.
2	only one group, C_2	See Exercise 6.12.
3	only one group, C_3	See Exercise 6.12.
4	two groups, C_4 and V_4	See Exercise 4.21 part (c).
5	only one group, C_5	See Exercise 6.12.
6	two groups, C_6 and S_3	See Section 9.2.1.
7	only one group, C_7	See Exercise 6.12.

Table 9.3. Classification of groups of orders 1 through 7.

The First Sylow Theorem says that in any group of order 8, there will be at least one subgroup of order 4 (something Cauchy's Theorem could not tell us). From Table 9.3, we know that such subgroups must be isomorphic to either V_4 or C_4. If all of them were isomorphic to V_4, then the group would have no elements of order 4, and therefore only elements of order 2. From Exercise 5.38, we know that such a group must be abelian, and so we've already counted it ($C_2 \times C_2 \times C_2$). Therefore any non-abelian group of order 8 has a copy of C_4 in it.

Let a stand for the generator of this copy of C_4, so that we can speak of the subgroup $\langle a \rangle$ and its one left coset, which together form all eight elements of the group. Call the coset $b \langle a \rangle$, for some element $b \notin \langle a \rangle$. Just as in the classification of groups of order 6, our task is to determine the possibilities for how a and b relate, and so determine the structures possible in non-abelian groups of order 8. We do not yet know the order of b, though we know it must be 2 or 4. I consider each of these two possibilities. Notice how

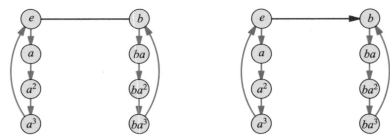

Figure 9.14. The First Sylow Theorem leads to these two starting points for seeking non-abelian groups of order 8. They differ only in that the b arrow in the left diagram has no arrowhead, because it is of order 2.

much the First Sylow Theorem has helped us; we know that non-abelian groups of order 8 are built from one of two partial Cayley diagrams, those shown in Figure 9.14.

First assume b has order 2, as diagrammed on the left of Figure 9.14. We must decide where b arrows send the element a. Any such decision can be expressed as an equation relating a and b, which must then hold throughout the entire diagram. I use a process of elimination to determine which of the alternatives lead to groups.

> The arrow for b cannot send a to the top row of the diagram because there are already b arrows touching both such elements.
>
> The b arrow cannot send a to the left column, because $b \notin \langle a \rangle$.
>
> If the b arrow were to send a to ba, we would have the equation $ab = ba$, resulting in an abelian group. We have already listed all abelian groups and are now seeking only the non-abelian ones.
>
> If the b arrow were to connect a to ba^2, we would have the equation $ab = ba^2$. By regularity, the paths ab and ba^2 should lead to the same place no matter where in the diagram we start. Starting from a, this requires drawing a b arrow from a^2 to b. But b already has an incoming b arrow, so we have hit a dead end; we cannot have $ab = ba^2$.

The only remaining choice is to connect a to ba^3, and all other b connections are then determined by the equation $ab = ba^3$. This creates the Cayley diagram on the left of Figure 9.15, of the group D_4.

We also must investigate the possibilities when b has order 4, as diagrammed on the right of Figure 9.14. We can determine the relationship between a and b by asking

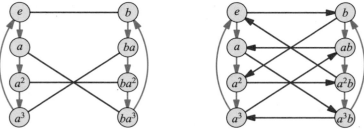

Figure 9.15. The only two non-abelian groups of order 8, as determined in Section 9.4.2. The left diagram shows D_4 and the right shows Q_4.

9.4. Sylow Theorems

where the b cycle continues from b. I again apply a process of elimination, this time to determine which element is b^2.

It is not any of the elements in the right column, because $b \notin \langle a \rangle$, and so $b^2 \notin b\langle a \rangle$.

It is not e because that would make $|b| = 2$, a case we've already considered.

It is not a because if $b^2 = a$, then b has order 8 (twice the order of a), and yields the group C_8, a case that's already done.

It is not a^3 because that, too, would make $|b| = 8$.

Thus our only choice is $b^2 = a^2$, an equation that we can use to determine the entire b orbit, as shown in Figure 9.16. Regularity requires that we duplicate this pattern of b arrows when connecting the coset $a\langle b \rangle$. The resulting group is Q_4, as on the right of Figure 9.15.

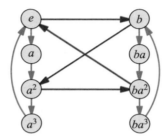

Figure 9.16. An intermediate step in classifying the last non-abelian group of order 8.

Thus there are two non-abelian groups of order 8, D_4 and Q_4. Together with the abelian groups $C_2 \times C_2 \times C_2$, $C_4 \times C_2$, and C_8, the total number of groups of order 8 is five (up to isomorphism). The First Sylow Theorem helped us in this exploration by giving us the two partial Cayley diagrams in Figure 9.14 as starting points. As in the classification of groups of order 6, much fruitless searching was thereby eliminated.

9.4.3 The Second Sylow Theorem: Relationship among p-subgroups

The First Sylow Theorem guarantees the existence of subgroups of certain sizes, and what it told us of the relationship among such groups was that all smaller p-subgroups are inside larger ones. The Second Sylow Theorem shows us how the largest such p-subgroups relate through conjugacy.

Definition 9.11 (Sylow p-subgroup). We call H a *Sylow p-subgroup* of G if it is a p-subgroup whose order is the highest power of p that divides $|G|$. In other words, H is a p-subgroup of G that's either the largest one, or tied for it.

Theorem 9.12 (Second Sylow Theorem). *Any two Sylow p-subgroups are conjugates.*

Before proving this theorem, let me mention some of its important consequences that may not at first be obvious. Conjugating by any group element creates an isomorphism from the group to itself called an inner automorphism (Exercise 8.32). Thus when two subgroups are conjugates (say $H = gKg^{-1}$), there is an inner automorphism mapping one to the other ($\phi(x) = gxg^{-1}$). Therefore conjugate subgroups are isomorphic.

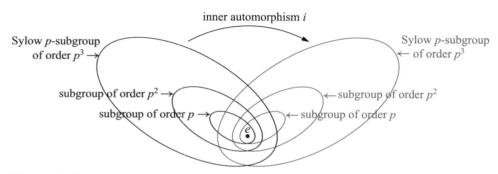

Figure 9.17. The inner automorphism i maps one Sylow p-subgroup (of order p^3) to another, which therefore has identical internal structure.

The Second Sylow Theorem tells us that all of a group's largest p-subgroups are one another's conjugates, and so they are all isomorphic to one another. Now recall the nesting relationship among p-subgroups given by the First Sylow Theorem, so that every p-subgroup is inside a Sylow p-subgroup. Conjugating any Sylow p-subgroup by any group element results in a (possibly different) Sylow p-subgroup, with identical internal structure, as shown in Figure 9.17. Therefore any smaller p-subgroup must have a copy of itself (one of its conjugates) in every Sylow p-subgroup.

This improves our picture of the group A of order 200, but in a subtle way. We still do not know whether there are any p-subgroups other than those we saw earlier in Figure 9.13, but *if there are,* then we know that they must have the same internal structure as the ones in that figure, and in fact must be their conjugates. Figure 9.18 shows a Sylow 2-subgroup different from the first, yet identical in structure. I draw it in dashed lines because its existence is not guaranteed by any of the theorems we have seen. Although

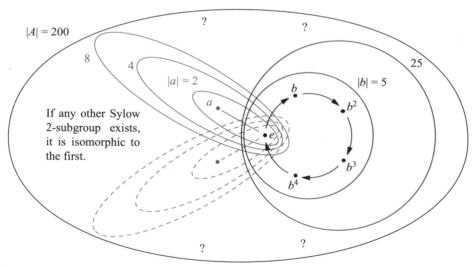

Figure 9.18. If the group A from Figure 9.13 has more than one Sylow 2-subgroup (for example), the Second Sylow Theorem tells us that such additional subgroups will be identical in structure to the first (and conjugate to it).

9.4. Sylow Theorems

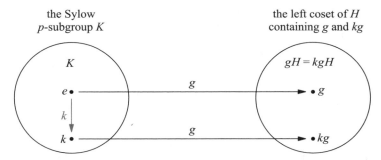

Figure 9.19. A stable element gH of the group action in the proof of the Second Sylow Theorem is one for which, for any $k \in K$, $kgH = gH$.

it appears to intersect the original Sylow 2-subgroup only at the identity, that is just one possibility. For instance, if $\langle a \rangle$ were a normal subgroup, then it would be conjugate only to itself, so every Sylow 2-subgroup would contain all of $\langle a \rangle$.

Proof. I use again the strategy described after the statement of Theorem 9.9. Let S be the left cosets of some Sylow p-subgroup $H < G$ and have another Sylow p-subgroup K act on S by left multiplication (as in the proof of Theorem 9.9).

A stable element of this action is a left coset gH for which $kgH = gH$ for every $k \in K$, as in Figure 9.19. The path from g to kg in that figure is $g^{-1}kg$, which is a member of H because it starts and ends inside the same copy of H. When this is true for any $k \in K$, we find that all of $g^{-1}Kg$ is in H, and therefore $g^{-1}Kg = H$, since they are the same size. Thus if there is even one stable element gH, then H and K are conjugates and the theorem is true.

Theorem 9.8 tells us that the number of stable elements must be congruent mod p to $|S|$, which is $[G : H]$. Because H is a Sylow p-subgroup, its order is the highest power of p dividing $|G|$. Thus $[G : H] = \frac{|G|}{|H|}$ is not a multiple of p, and thus it is not zero. So the number of stable elements must also be nonzero, making K and H conjugates. $\square$

Because a conjugate of a subgroup is a subgroup of the same size, any conjugate of a Sylow p-subgroup will be another Sylow p-subgroup. Thus the Sylow p-subgroups are not just *some* of one another's conjugates, but are *all* of one another's conjugates.

This is noteworthy when there is only one Sylow p-subgroup. Normal subgroups are those that are unmoved by conjugation (Section 7.5). Thus a Sylow p-subgroup is normal just when it has no conjugates, which is when it is the only Sylow p-subgroup. Therefore being able to deduce the number of Sylow p-subgroups of a group helps us label some of them as normal. The Third Sylow Theorem often helps us do just that. For this reason, the Second Sylow Theorem is most useful when combined with the third.

9.4.4 The Third Sylow Theorem: Number of p-subgroups

Theorem 9.13 (Third Sylow Theorem). *The number n of Sylow p-subgroups of G obeys the following two restrictions.*

$$n \text{ divides } |G| \qquad n \equiv_p 1$$

This theorem helps us narrow down, just based on a group's order, the possible number of Sylow p-subgroups that the group can have. It does not always narrow down the possibilities to only one number, but it is often a big improvement over no information at all. After proving this theorem, I conclude the chapter with two of its applications.

Proof. The first of the two restrictions is the easier to prove. Let H be one of the n Sylow p-subgroups of G. We know that the set of Sylow p-subgroups is the set of conjugates of H. Another way to say this is that if G acts on its subgroups by conjugation, then the set of Sylow p-subgroups is $Orb(H)$. The size n of that orbit must therefore divide $|G|$ by Theorem 9.4; this is the first restriction on n.

To see that n is also congruent to 1 mod p, I apply for the last time the proof strategy used several times already this chapter (from the proof of Theorem 9.9). Let S be the set of n Sylow p-subgroups, and consider just the subgroup H acting on S by conjugation. If I can show that the number of stable elements is congruent to 1 mod p, then n must also be (by Theorem 9.8). I show that there is actually just one stable element, which obviously must be H.

For a Sylow p-subgroup K to be a stable element means that $hKh^{-1} = K$ for every h in H. That is, H is in the normalizer of K. But H and K are also Sylow p-subgroups of the group $N_G(K)$, and therefore conjugates in that group. Yet since K is normal in $N_G(K)$, its only conjugate there is itself, meaning that K and H must be the same.

Thus the number of stable elements is 1, and thus $n = |S| \equiv_p 1$, the second restriction stated in the theorem. $\square$

Before seeing an extended application of this theorem, let's see how it impacts our running example of the group A of order 200. The Third Sylow Theorem requires that the number of Sylow 5-subgroups be a factor of 200 that's congruent to 1 mod 5. The factors of 200 are

$$1, 2, 4, 5, 8, 10, 20, 25, 40, 50, 100, \text{ and } 200,$$

and you can verify that the only one congruent to 1 mod 5 is the number 1 itself. Thus there is only one Sylow 5-subgroup, and it must therefore be normal. However, the Third Sylow Theorem does not narrow down the possible number of Sylow 2-subgroups as much. That number must be a factor of 200 that's congruent to 1 mod 2, which leaves three options, 1, 5, and 25. So the Third Sylow Theorem does not give us full information about the number of Sylow 2-subgroups. Perhaps some additional reasoning would narrow the number down further, or perhaps not.

The final modification of our running illustration of A is shown in Figure 9.20. It describes the Sylow 5-subgroup as both unique and normal, but leaves some question marks near the Sylow 2-subgroup, because its number of conjugates remains undetermined.

This example of the group A has shown us how much information we can glean just from a group's order. Smaller groups are usually easier to dissect with Sylow Theory than larger groups, but not always; what matters more is how the group order factors. For example, classifying groups of order 8 is harder than classifying groups of order 15, which is our final application of Sylow Theory. The Third Sylow Theorem gives us key information for shortening the job.

9.4.5 Classifying groups of order 15

Cauchy's Theorem tells us that in any group of order 15, there must be an element a of order 3 and an element b of order 5. Since no higher powers of 3 or 5 divide 15,

9.5. Exercises

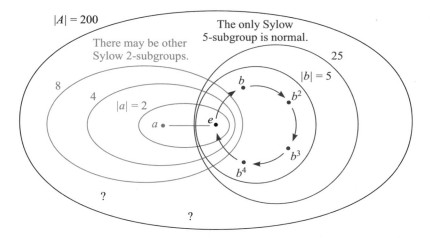

Figure 9.20. The group A illustrated throughout this chapter must have only one subgroup of order 25, which is therefore normal, based on Theorem 9.13. However, that theorem does not immediately tell us how many subgroups there are of order 8.

the subgroups $\langle a \rangle$ and $\langle b \rangle$ are Sylow 3- and 5-subgroups respectively. The Third Sylow Theorem tells us that the number of conjugates of $\langle a \rangle$ must be a factor of 15 that's congruent to 1 mod 3; the only such number is 1, making $\langle a \rangle$ normal. Similar reasoning shows that $\langle b \rangle$ must also be unique and normal.

Because $\langle a \rangle$ contains only elements of order 1 or 3 and $\langle b \rangle$ only elements of order 1 or 5, these subgroups intersect only at the identity. Thus the fact proven in Exercise 8.43 applies, telling us that the group must be isomorphic to $\langle a \rangle \times \langle b \rangle$, or $C_3 \times C_5$, which is isomorphic to C_{15} by Theorem 8.7. Thus there is only one group of order 15 up to isomorphism, the cyclic one.

We have classified all groups of order 8 or less, as well as those of order 15. In the exercises, you are asked to extend this work to order 10, and also to try classifying all groups of a few larger orders (12, 14, and 21). *Group Explorer* contains a visual library of all groups of order 40 or less, and the computational software package GAP has a database of all groups that goes beyond order 1000 [3].

Classifying groups of a given order is the most common use of Sylow Theory. Other applications arise in the culmination of this book, Chapter 10, which discusses the historic work of a young Frenchman named Évariste Galois. See Exercise 9.22 for a preview.

9.5 Exercises

9.5.1 Basics

Exercise 9.1. Name two questions that the Sylow Theorems answered about the group A of order 200. Name two questions they did not answer about it.

Exercise 9.2. If S has six elements, how many does $Perm(S)$ have?

Exercise 9.3. If G acts on S and $s \in S$, consider the two sets $Orb(s)$ and $Stab(s)$.

(a) Is either of them a group?

(b) How do the two sets relate?

(c) What is the smallest size that $Orb(s)$ can have?

(d) What is the largest size that $Orb(s)$ can have?

(e) Are all sizes in between possible?

Exercise 9.4. If $|G| = 28$, what sizes of subgroups does Cauchy's Theorem guarantee exist in G? How does the First Sylow Theorem improve on that guarantee?

Exercise 9.5. How many groups are there of order less than or equal to 8?

9.5.2 Group actions and action diagrams

Exercise 9.6. What do the h arrows in Figure 9.2 signify? What do the v arrows in that figure signify?

Exercise 9.7. Consider the action of C_3 on the set $\{A,B,C,D\}$ by the interpretation homomorphism $\phi : C_3 \to Perm(\{A,B,C,D\})$ generated by the following equation.

$$\phi(1) = A \quad \overset{\frown}{B \quad C \quad D}$$

(a) Draw the corresponding action diagram.

(b) What are the stable elements?

(c) What are the orbits?

Exercise 9.8. For each part below, create an action diagram satisfying the given criteria.

(a) condition (1) of the two on pages 195–196, but not condition (2)

(b) condition (2) of the two on pages 195–196, but not condition (1)

(c) neither of the two conditions on pages 195–196

(Although Figure 9.2 answers this part, create a different action diagram than that one.)

Exercise 9.9. If C_5 acts on the letters $\{A,B,C,D\}$, what will the action diagram be? Why?

Exercise 9.10. Consider the group action defined at the start of the proof of Theorem 9.9.

(a) Draw the action diagram for the specific example of $G = D_5$ and $H = \langle f \rangle$.

(b) Verify that for each $s \in S$, your diagram satisfies the Orbit-Stabilizer Theorem.

Exercise 9.11.

(a) Consider the group action defined at the start of the proof of Theorem 9.12. Draw the action diagram for the specific example of $G = Q_4$, $H = \langle j \rangle$, and $K = \langle k \rangle$.

(b) Are there any stable elements?

(c) Is either H or K normal? Why or why not?

9.5.3 Justification

Exercise 9.12. Give a counterexample to the general converse of Lagrange's Theorem. That is, give a group G and a number n dividing $|G|$ such that G has no subgroups of order n.

9.5. Exercises

Exercise 9.13. I claimed that the Cayley diagrams shown at the end of Section 9.2.1 represent the groups C_6 and S_3. Demonstrate clearly why this is so.

Exercise 9.14. The statement of Theorem 9.5 would not be true without the assumption that $|G|$ is prime. Show this by finding a particular G, S, and ϕ for which the statement would fail.

Exercise 9.15.

(a) If $g^p = e$ and p is prime, why must $|g|$ be 1 or p?

(b) In general, if $g^n = e$, must $|g|$ divide n?

Exercise 9.16. Demonstrate that, for the ϕ defined in the proof of Theorem 9.6, each $\phi(n)$ is a permutation of S, by showing both of the following statements to be true.

(a) Each $\phi(n)$ sends any element $s \in S$ to another element in S.

(b) No $\phi(n)$ sends two different elements of S to the same destination.

Exercise 9.17. Assume G acts on S.

(a) Why is $Stab(s) < G$ for any $s \in S$?

(b) In Figure 9.2, $Stab(s) = Ker(\theta)$. Are stabilizers always the kernel of the interpretation homomorphism?

Exercise 9.18. Consider the following alternate definition of p-group: A p-group is a group whose elements all have orders that are powers of the same prime p.

(a) If Definition 9.7 is true of a group, explain why this alternate version is also true of the group.

(b) If Definition 9.7 is not true of a group, explain why this alternate version is not true either.

Exercise 9.19. Let S be the set of left cosets of a subgroup H in a group G, as in the proof of Theorem 9.9. If G acts on S by left multiplication (like the ϕ in that proof), then explain why every $\phi(g)$ is a permutation. That is, how do we know it doesn't send two different elements of S to the same destination?

Exercise 9.20. How do the Sylow Theorems make doing Exercise 8.45 easier?

Exercise 9.21. The converse of Lagrange's theorem is true for abelian groups. Any n dividing $|G|$ has a subgroup $H < G$ of order n. Explain why.

9.5.4 Sylow p-subgroups

Exercise 9.22. In Chapter 10 the *simple group* A_5 will play an important role; simple groups are those that have no normal subgroups. For each n given below, use Sylow Theory to explain why there can be no simple groups of order n.

(a) 33

(b) 84

(c) 12

(d) $p^n m$ for any prime p and any positive integers n and m, with $m < p$

Exercise 9.23.

(a) Find all Sylow 3-subgroups in S_4.

(b) Find all Sylow 3-subgroups in S_5.

Exercise 9.24. What are the Sylow 2-subgroups of D_n when n is odd?

9.5.5 Classifying groups of a given order

Exercise 9.25. Classify all groups of order 9. Did you use any facts from the Sylow Theorems?

Exercise 9.26. Classify all groups of the following two orders. The same techniques are useful in both, because they are both of the form $2p$ for some prime p. Make a conjecture about what non-abelian groups exist of any order $2p$ when p is prime.

(a) 10

(b) 14

Exercise 9.27. Classify all groups of order 21.

Hint: For non-abelian groups, Cauchy's Theorem tells us that there must be an element of order 3 and an element of order 7. So start with a diagram like the one in Figure 9.21. A decision of where to draw an a arrow from b implies an equation like $ba = a^m b$, for some m. Use algebra (and the equation $b^3 = e$) to find helpful restrictions on m.

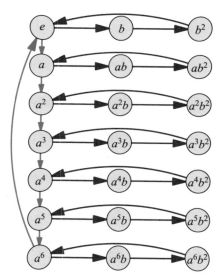

Figure 9.21. Exploring non-abelian groups of order 21 starts with this partial Cayley diagram.

Exercise 9.28. (Challenging) Classify all groups of order 12.

10

Galois theory

This book ends where group theory began. In the nineteenth century, two very young mathematicians, Neils Abel and Évariste Galois, answered a question about which the mathematical world had been curious for centuries. It is a question about the central activity of algebra, solving equations. Galois coined a new term to describe the mathematical objects that played a central role in his work on this question; he called them groups, and group theory was born.

But groups are only part of the solution; the work of Abel and Galois also involves algebraic structures called *fields*. I touch on field theory in this chapter, but many of its facts I state without proof; this book is about groups. Part of what makes the material in this chapter great is that Galois theory is built on a beautiful interplay between groups and fields, which we will see in Section 10.6. I hope that this final chapter also serves as a "teaser" for field theory. If field theory piques your interest (or to see the proofs I omit), I encourage you to study it further; refer to the list of good books I recommend at the end of this chapter.

Let's begin with the big question that Abel and Galois answered. It's a question from elementary algebra, about solving the simplest kinds of equations we can write. The next two sections give all the background, state the big question, and prepare us to appreciate its solution.

10.1 The big question

One of our first exposures to mathematics as children is counting. The set of counting numbers $\{1, 2, 3, \ldots\}$ is often called $\mathbb{N}$, the natural numbers.[1] Later we learn the basic operations of arithmetic, starting with addition and subtraction, and then moving up to multiplication and division. I call these four operations "arithmetic."

As children learn arithmetic, they also learn new kinds of numbers. Although adding two natural numbers gives another natural number (nothing new), subtracting two natural

---

[1] Some prefer to include zero in $\mathbb{N}$, others leave it out. For this chapter, I use $\mathbb{N} = \{1, 2, 3, \ldots\}$.

numbers like $7-7$ or $3-6$ requires a knowledge of zero and the negative numbers. Adding and subtracting natural numbers broadens children's horizons to a set of numbers we know well, the integers, $\mathbb{Z}$. Soon, however, we move beyond $\mathbb{Z}$ as well. Although multiplication does not reach beyond $\mathbb{Z}$, division can. For example, $6 \div 2$ is an integer, but $6 \div 5$ is not. This prompts the introduction of fractions like $\frac{6}{5}$, leading to the set of all rational numbers, $\mathbb{Q}$. This brings us to the first of five key points I emphasize to organize the background material of this chapter.

Key Point 10.1. To do arithmetic, we need at least the rational numbers.

Figure 10.1 shows the growing number system, from $\mathbb{N}$ to $\mathbb{Z}$ to $\mathbb{Q}$, required by the operations of arithmetic. Children who know only how to count live in $\mathbb{N}$. When they learn to add and subtract, they move up to $\mathbb{Z}$; when they learn to multiply and divide, they move up to $\mathbb{Q}$. I use diagrams like Figure 10.1 to explain relationships among number systems throughout this chapter.

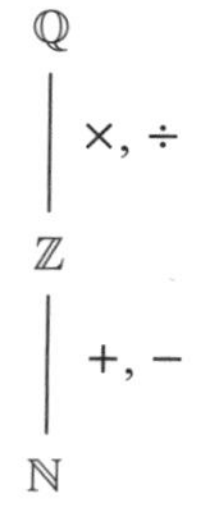

$$\begin{array}{c} \mathbb{Q} \\ | \ \times, \div \\ \mathbb{Z} \\ | \ +, - \\ \mathbb{N} \end{array}$$

Figure 10.1. Three number systems. The smallest one, the natural numbers, appears at the bottom. Expanding it (upwards) with addition and subtraction gives the integers, and expanding that set with multiplication and division gives the rational numbers.

Our knowledge of groups and their operations allows us to explain the expansion of $\mathbb{N}$ to $\mathbb{Z}$ in terms of group theory. The set $\mathbb{N}$ is not a group under addition, because there is no identity (zero) and there are no inverses (negative numbers). To find them, we must expand to $\mathbb{Z}$, which is a group under addition. In that group, negative numbers are the inverses of positive ones, and so subtraction is not really a new operation, but a particular use of addition; $a - b$ can be thought of as a added to the inverse of b, as in $a + (-b)$.

Similarly, expanding from $\mathbb{Z}$ to $\mathbb{Q}$ can be described as generating a group under multiplication, with one exception. As in Exercise 4.33, zero has no multiplicative inverse, so we remove it and use the remaining integers as generators. Their multiplicative inverses are fractions, with which we generate all the rational numbers. Exercise 4.33 used the notation $\mathbb{Q}^*$ to mean $\mathbb{Q}$ without zero, a group under multiplication. Thus on $\mathbb{Q}$ we have two groups, addition on all of $\mathbb{Q}$ and multiplication on most of it, $\mathbb{Q}^*$. This prompts us to define a new type of algebraic structure.

Definition 10.1 (field). A set S with addition and multiplication operations is a *field* if the following three conditions hold.

1. S and its addition operation form an abelian group.

2. Removing the identity element from that group (the zero) leaves a set that, under the multiplication operation, also forms an abelian group.

10.1. The big question

3. Addition and multiplication relate through the distributive law, $a(b + c) = ab + ac$.

The rational numbers $\mathbb{Q}$, the real numbers $\mathbb{R}$, and the complex numbers $\mathbb{C}$ are all fields. (See Exercise 10.8.) There are finite fields as well, which have Cayley-diagram-like visualizations that you can investigate in Exercise 10.29. But finite fields do not enter into the work of Abel and Galois, so I do not cover them here.

The number systems $\mathbb{N}$ and $\mathbb{Z}$ are not fields. Although both satisfy the last condition of Definition 10.1, $\mathbb{N}$ does not satisfy either of the other conditions and $\mathbb{Z}$ satisfies only the first. Because arithmetic cannot be done in $\mathbb{N}$ or $\mathbb{Z}$, the rest of this chapter works in $\mathbb{Q}$, as Key Point 10.1 suggests.

So in $\mathbb{Q}$ we can do arithmetic, and all that arithmetic is good for, like accounting, basic physics, and many other quantitative disciplines. From work in these areas, natural questions arise. How many haircuts must a barber give in a month to break even on the costs of his shop? What will my monthly payment be on a loan of this much money over this much time? What is the escape velocity for a rocket with given specifications? Answering these questions is the business of algebra, building on the foundation of arithmetic. In algebra, you express a question's requirements in an equation using the language of arithmetic and an "unknown" (such as x). Then you "solve" the equation for the unknown quantity. Though notation has changed over time, this kind of mathematical work has been going on for millenia. So surely mathematicians must know how to solve every equation of arithmetic by now. But this is not so, and it is the problem that Abel and Galois addressed.

The question is about the most basic algebraic equations we can write, with only the operations of arithmetic, rational numbers, and one variable for expressing an unknown quantity; I always use the variable x. I call this family of equations "equations from arithmetic." Here is a simple example.

$$2x + 1 = 0$$

To "solve" the equation means to find a number that can replace x to make the equation true. This equation is easy to solve, using the operations of arithmetic on both sides at once.

$$\text{subtract 1:} \quad 2x = -1$$
$$\text{divide by 2:} \quad x = -\tfrac{1}{2}$$

But what about more difficult equations from arithmetic? Even this simple family contains some pretty hairy equations, like this one. (And you can imagine worse!)

$$\frac{(12 - x)(13 + \tfrac{x}{2})}{19 - \tfrac{x+1}{x-1}} = 100 + \frac{50}{x - 9}$$

But all equations from arithmetic can be simplified dramatically. The above equation, for instance, simplifies to this one.

$$x^4 + 4x^3 + 3157x^2 - 31354x + 31192 = 0$$

Still tricky to solve, but with two advantages. First, all fractions, parentheses, and nested expressions have been removed and the equation looks much neater. Second, all equations of the above form have a centuries-old, proven technique for solving them.

The left side of the above equation is a ***polynomial.*** A polynomial is a list of ***terms*** added together; each term includes a ***coefficient,*** an x, and an exponent on the x. Here are some polynomials, with example terms explained.

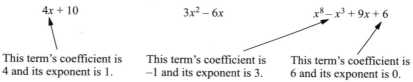

$4x + 10$	$3x^2 - 6x$	$x^8 - x^3 + 9x + 6$
This term's coefficient is 4 and its exponent is 1.	This term's coefficient is -1 and its exponent is 3.	This term's coefficient is 6 and its exponent is 0.

We use exponents when writing polynomials, but exponentiation is not really a new operation; because the exponents will always be nonnegative whole numbers, they are just shorthand for repeated multiplication.

Polynomials have the general form

$$a_n x^n + a_{n-1} x^{n-1} + \cdots + a_2 x^2 + a_1 x + a_0,$$

in which each a_i is a coefficient and n is the highest power of x, called the ***degree*** of the polynomial. The degrees of the three example polynomials above are 1, 2, and 8, from left to right. As those examples indicate, some of the coefficients other than a_n may be zero.

Exercise 10.4 asks you to prove that *any* equation from arithmetic simplifies to a polynomial on one side and zero on the other. To get better acquainted with that fact, you may wish to try that exercise now. Because of that simplification, a method for solving equations of a polynomial equal to zero could be used to solve any equation from arithmetic. This is so important, I stress it as my second key point.

Key Point 10.2. A method for solving an equation of a polynomial equal to zero would actually solve every equation from arithmetic, because any equation from arithmetic simplifies to that form.

I therefore focus our attention on equations that set a polynomial equal to zero, called ***polynomial equations***. The solutions to such equations are called ***roots*** of the polynomial. Because our equations come from arithmetic, their coefficients are all rational numbers. Polynomials of degree 1 (like the one I solved on page 223) are called linear polynomials. Those of degree 2 are called quadratic, and algebra students learn to solve them with the quadratic formula. It says that any equation $ax^2 + bx + c = 0$ has two solutions, expressed together in the formula

$$x = \frac{-b \pm \sqrt{b^2 - 4ac}}{2a}.$$

The quadratic formula has been known since at least as early as the twelfth century. Although elementary algebra classes usually end their study of polynomial equations with the quadratic formula, the curiosity of mathematicians does not stop there, especially given the importance of solving polynomial equations just stated in Key Point 10.2. In the sixteenth century, Girolamo Cardano published a formula for solving cubic equations, those made from third-degree polynomials, and therefore having the form

$$ax^3 + bx^2 + cx + d = 0.$$

Although the cubic formula is lengthier than the quadratic formula (filling up several lines), it doesn't involve any new concepts. It uses only the operations of arithmetic and radical signs, just like the quadratic formula. The quadratic formula involves a square root and the cubic formula involves both square and cube roots, but both types of roots are written using radical signs. The quadratic formula finds two solutions, while the cubic finds three.

And mathematics marches on. In the same century Lodovico Ferrari developed a way to solve any quartic polynomial equation (fourth degree), and that solution also involves just the operations of arithmetic and radicals (in his case, sometimes fourth roots). It, too, is lengthy, but it finds all four solutions to any quartic polynomial equation.

The war on polynomial equations seemed to be going well, but that's where progress stopped. Although some later mathematicians developed formulas for solving *some* types of quintic polynomial equations (fifth degree), no method was found to handle *every* quintic. In fact, there were some quintic polynomials, like $x^5 + 10x^4 - 2$, whose solutions no one could find. This brings us to the big question I've been building up to since the first page of this chapter.

Key Point 10.3. The work of Abel and Galois answered this big question: "Is there a formula for solving fifth degree polynomial equations?"

In 1824, at the age of 22, Neils Abel found a particular quintic polynomial that he proved could *never* be solved using only the operations of arithmetic and radicals. It's not that the equation had no solutions, but rather the operations of arithmetic together with radicals *could not express* those solutions. In other words, the language for writing the solutions for polynomial equations of degrees 1 through 4 is insufficient for writing the solutions to polynomial equations of degree 5.

It has since become common to call a polynomial "solvable" or "solvable by radicals" when its roots can be written using the operations of arithmetic and radicals, and "unsolvable" or "unsolvable by radicals" otherwise. Shortly after Abel's revelation, the budding 19-year-old prodigy Évariste Galois introduced the subject of group theory to classify exactly which polynomials were solvable by radicals. This chapter is about the role of group theory in that work. Because all that follows depends on the answer Abel and Galois gave to the big question, I state their answer again, more fully, as my next key point.

Key Point 10.4. The operations of arithmetic and radicals have been used to write formulas to solve polynomial equations of degrees 1 through 4, but they cannot be used to write a "quintic formula," for solving polynomial equations of degree 5.

10.2 More big questions

Figure 10.1 showed us that learning the operations of arithmetic broadens our number system, first to the integers and then to the rationals. Algebra also broadens our view, because the fundamental operation of algebra, solving equations, often introduces new kinds of numbers. Solving a simple linear equation does not result in anything but rational numbers, as in the example on page 223. But solving a quadratic equation often requires taking a square root.

Some square roots of rational numbers give other rational numbers, such as $\sqrt{\frac{9}{4}} = \frac{3}{2}$, but not all. The number $\sqrt{2}$ is in the field of real numbers $\mathbb{R}$, but is outside $\mathbb{Q}$. Furthermore, some quadratic equations have solutions that aren't even real numbers; consider

Decimal expansions of real numbers

Rational numbers		Irrational numbers	
Finite	Repeating	Non-repeating pattern	No clear pattern
1.612	$\frac{1}{3} = 0.3333\cdots$	$1.2345678910\cdots$	$\pi \approx 3.14159\cdots$
5.0	$\frac{100}{11} = 9.0909\cdots$	$0.101001000\cdots$	$e \approx 2.71828\cdots$
19.32651	$\frac{1}{7} = 0.142857\cdots$	$0.1223334444\cdots$	$\sqrt{2} \approx 1.41421\cdots$

Table 10.1. Real numbers can be written as decimal expansions, of which there are many different kinds. This table splits out those decimals that represent rational numbers from those that represent irrational numbers.

$x^2 - 2x + 2 = 0$. The quadratic formula gives the solutions

$$x = \frac{2 \pm \sqrt{4-8}}{2} = \frac{2 \pm 2i}{2} = 1 \pm i.$$

Solving this simple polynomial with integer coefficients takes us all the way outside $\mathbb{R}$, into the complex numbers $\mathbb{C}$. (Recall that i is shorthand for $\sqrt{-1}$, so $\sqrt{-4} = i\sqrt{4} = 2i$.) Let's review how $\mathbb{Q}$, $\mathbb{R}$, and $\mathbb{C}$ differ.

The rational numbers are those that can be written as a ratio $\frac{a}{b}$ of two whole numbers a and b (with $b \neq 0$). Another (equivalent) way to define $\mathbb{Q}$ is those numbers whose decimal expansions are either finite or repeating. Many real numbers are in neither of these two categories, and are therefore irrational. Table 10.1 shows example decimal expansions inside and outside $\mathbb{Q}$. To move beyond $\mathbb{R}$ requires the introduction of the so-called "imaginary" number i, which satisfies the equation $i^2 = -1$. Just as arithmetic expands a child's number system from $\mathbb{N}$ to $\mathbb{Z}$ to $\mathbb{Q}$, so algebra expands students' number systems from $\mathbb{Q}$ to $\mathbb{R}$ to $\mathbb{C}$, as illustrated in Figure 10.2. (Technically, solving polynomial equations cannot generate every element of $\mathbb{R}$, but I come to that shortly.)

The irrational numbers that matter most to the big question are square roots, cube roots, and other radicals. If we can take square roots then we can use the quadratic formula to solve any quadratic equation. If we can also take cube roots, we can use the cubic formula to solve any cubic equation. If we have fourth roots, the quartic formula

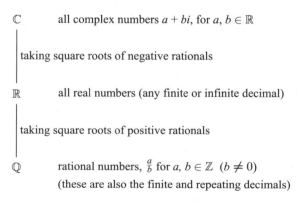

Figure 10.2. The Hasse diagram of the containment of the field $\mathbb{Q}$ in the field $\mathbb{R}$, and $\mathbb{R}$ in $\mathbb{C}$

10.2. More big questions

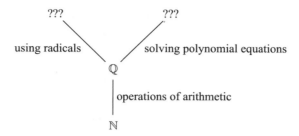

Figure 10.3. The big question boils down to a comparison of two operations that lead out from $\mathbb{Q}$, the operation of introducing radicals and the operation of solving polynomial equations. Where does each lead, and how do they relate?

will solve any quartic equation. You can see how it would be natural to think that fifth roots should make it possible to write a formula for solving quintic polynomial equations. But Abel and Galois's answer to the big question is the opposite. It is at this point—degree 5—that radicals stop being sufficient to solve polynomial equations.

This suggests we view the big question in a new light, as a comparison of the power of two operations, using radicals and solving polynomial equations. Figure 10.3 shows these two operations leading from $\mathbb{Q}$, and asks whether their destinations are different.

We can improve that figure in two ways. First, it should be clear that every radical is a solution to some polynomial equation; any $\sqrt[n]{a}$ is a solution to $x^n - a = 0$. So I should redraw Figure 10.3 to indicate that the unknown field on the top left is contained in the unknown field on the top right. Second, some real and complex numbers are not the solution to *any* algebraic equation; such numbers are called ***transcendental,*** and those that are solutions to polynomial equations are called ***algebraic.*** The numbers π and e are well-known transcendental numbers. So we know that neither of the unknown fields in Figure 10.3 is all of $\mathbb{R}$ or $\mathbb{C}$. Figure 10.4 is a more accurate version, in which A stands for the algebraic numbers.

Figure 10.4 helps us phrase the big question in terms of the relative power of two basic operations. *Does introducing radicals reach as far beyond $\mathbb{Q}$ as solving polynomials does? Or, can polynomials be solved using just arithmetic and radicals?* The big question is about the relationship between these two fundamental mathematical activities.

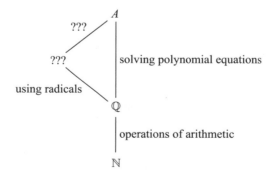

Figure 10.4. An updated version of Figure 10.3, using A to stand for the algebraic numbers, and recognizing that every $\sqrt[n]{a}$ is algebraic.

Phrasing the big question this way shows its similarity to other big mathematical questions, hence the title of this section. There are three famous questions that the ancient Greeks asked in the field of geometry. Can the basic operations of geometry, performed with a ruler and compass, be used to square a circle, double a cube, or trisect an angle? You can see their similarity to our big question; each is a question of how far certain operations can reach. These three questions can also be answered using Galois theory, due to the tight relationship between algebra and geometry. This is my final key point.

Key Point 10.5. Galois Theory is useful when analyzing operations that reach out from $\mathbb{Q}$ to new fields. Our big question can be answered using Galois theory, as can other historical questions in mathematics.

I mention these geometric questions to show that Galois theory has power beyond the one question I cover in this chapter. Readers interested in the application of Galois theory to geometric problems can refer to sources such as [11] and Chapter 13 of [1]. This completes the background material, and we can now dig into the work that answered the big question.

10.3 Visualizing field extensions

Two kinds of diagrams will play an important role in this chapter. I use diagrams like Figure 10.4 to illustrate and compare operations that reach outside $\mathbb{Q}$. They are actually a kind of Hasse diagram. In Section 6.6.3 we used Hasse diagrams to illustrate the relationships among a group's subgroups. Vertical lines connected smaller subgroups (lower in the diagram) up to larger ones (higher in the diagram) in which they were contained. Now I use Hasse diagrams to show the relationships among fields that extend $\mathbb{Q}$, and vertical or diagonal lines still indicate containment.

The second visualization technique important in this chapter is based on the fact that all polynomials whose coefficients are in $\mathbb{C}$ have all their roots in $\mathbb{C}$ as well. For this reason $\mathbb{C}$ is called an ***algebraically closed field***, meaning that the operation of solving equations never reaches outside it. Thus all the numbers and fields in this chapter are inside $\mathbb{C}$, and a way to visualize the field $\mathbb{C}$ would enable us to visualize any of those numbers and fields.

Exercise 10.8 guides you through proving that any complex number can be written as $a + bi$ for some real numbers a and b. Here are some complex numbers in this form.

$$1 + i \qquad 6 - 2i \qquad 3\tfrac{1}{2} + \tfrac{9}{11}i \qquad \sqrt{2} + \sqrt{3}i$$

Because complex numbers have this two-part structure, we use points in the familiar two-dimensional coordinate plane to represent complex numbers, plotting $a + bi$ at the point (a, b). Figure 10.5 shows where some example complex numbers lie on the complex plane. Take a moment to verify your understanding of this visualization technique, since many future figures depend on it. Notice that the x axis is the real number line.

We can use graphs like Figure 10.5 to visualize the solutions to polynomial equations, putting all of a polynomial's roots together on one graph. Figure 10.6 shows four examples, a graph for each of four different polynomials. See if you can spot any patterns in Figure 10.6. One interesting thing to notice is that whether a polynomial is solvable is not

10.3. Visualizing Field Extensions

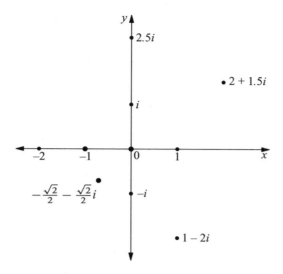

Figure 10.5. Example complex numbers plotted as points in the complex plane. Notice that all the real numbers are on the x axis and all real multiples of i are on the y axis.

related to how complicated it looks on paper. I stated earlier that the deceptively simple polynomial $x^5 + 10x^4 - 2$ is unsolvable, yet the lengthy quintic on the bottom right of Figure 10.6 is solvable, and all its solutions are listed there.

But a more important pattern you may have spotted is that the root sets in Figure 10.6 have some symmetry. Each is a mirror reflection of itself over the x axis, and a vertical flip would therefore respect the shape of the root set. Seeing symmetry in polynomial root sets is the first step in Galois theory. The following definition will help us prove that this mirror symmetry appears in the root set of any polynomial.

Definition 10.2 (complex conjugate). For any complex number $a + bi$, we call $a - bi$ its *complex conjugate* (or sometimes just its conjugate, for short). If c is any complex number, we write $\overline{c}$ to mean its complex conjugate.

For example, $1 + i$ and $1 - i$ are conjugates, so we could write $\overline{1+i} = 1 - i$. And if $c = \frac{\sqrt{2}}{2} + \frac{\sqrt{2}}{2}i$, one of the four roots to the bottom left polynomial in Figure 10.6, then $\overline{c} = \frac{\sqrt{2}}{2} - \frac{\sqrt{2}}{2}i$. Note that not all four roots of that polynomial are one another's complex conjugates.

$$\overline{c} \neq -\frac{\sqrt{2}}{2} + \frac{\sqrt{2}}{2}i \quad \text{and} \quad \overline{c} \neq -\frac{\sqrt{2}}{2} - \frac{\sqrt{2}}{2}i$$

Complex conjugation negates the imaginary part only, and is therefore the vertically flipped version of the original point. Real numbers are their own complex conjugates, for example $\overline{6} = 6$ and $\overline{0} = 0$.

Theorem 10.3 (Complex Conjugate Root Theorem). *If r is a root of a polynomial, then its conjugate $\overline{r}$ is a root of the same polynomial. Therefore every polynomial's root set has mirror symmetry over the x axis of the complex plane.*

Proof. Take any polynomial, written in the general form

$$a_n x^n + a_{n-1} x^{n-1} + \cdots + a_2 x^2 + a_1 x + a_0.$$

Polynomial: $x^2 - 2x + 2$
Roots: $1 \pm i$

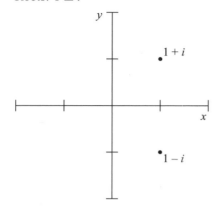

Polynomial: $12x^3 - 44x^2 + 35x + 17$
Roots: $-\frac{1}{3}, 2 \pm \frac{1}{2}i$

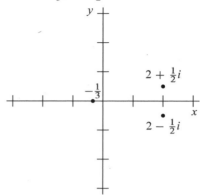

Polynomial: $x^4 + 1$
Roots: $\pm \frac{\sqrt{2}}{2} \pm \frac{\sqrt{2}}{2}i$

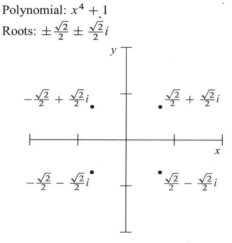

Polynomial: $8x^5 - 28x^4 - 6x^3 + 83x^2 - 117x + 90$
Roots: $-2, \frac{3}{2}, 3, \frac{1}{2} \pm i$

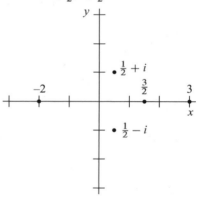

Figure 10.6. The root sets of four solvable polynomials (degrees 2 through 5) plotted in the complex plane

If one of its roots is r, then plugging r into the polynomial yields zero.

$$a_n r^n + a_{n-1} r^{n-1} + \cdots + a_2 r^2 + a_1 r + a_0 = 0$$

From this assumption I will prove that $\bar{r}$ is also a root. This involves several steps, but each is only basic algebra. Taking the complex conjugate of both sides of the above equation gives

$$\overline{a_n r^n + a_{n-1} r^{n-1} + \cdots + a_2 r^2 + a_1 r + a_0} = \bar{0},$$

because of course $\bar{0} = 0$. Now I want to simplify the left side, which is a conjugate of a lengthy sum. Therefore I'd like to prove that conjugation can be separated over each

10.4. Irreducible polynomials

term of a sum of complex numbers, as in the following equation.
$$\overline{(a+bi)+(c+di)} = \overline{a+bi} + \overline{c+di}$$

This is not very hard to prove; applying Definition 10.2 and some simple algebra shows that the two sides of this rule are equal.
$$\overline{(a+bi)+(c+di)} = \overline{(a+c)+(b+d)i} = (a+c)-(b+d)i$$
$$\overline{a+bi} + \overline{c+di} = a-bi+c-di = (a+c)-(b+d)i$$

Applying the rule simplifies the original polynomial equation a bit.
$$\overline{a_n r^n} + \overline{a_{n-1} r^{n-1}} + \cdots + \overline{a_2 r^2} + \overline{a_1 r} + \overline{a_0} = 0$$

Each term in this sum is a conjugate of two multiplied expressions, an a_i and an r^i. So to simplify further, I would like to split the conjugation over the two terms in each multiplication, using a rule like this one.
$$\overline{(a+bi)(c+di)} = \overline{(a+bi)}\,\overline{(c+di)}$$

Exercise 10.5 asks you to prove this rule, and I apply it here without proving it.
$$\overline{a_n}\,\overline{r^n} + \overline{a_{n-1}}\,\overline{r^{n-1}} + \cdots + \overline{a_2}\,\overline{r^2} + \overline{a_1}\,\overline{r} + \overline{a_0} = 0$$

Because each coefficient a_i is a real number, it is its own conjugate, so the equation simplifies even further.
$$a_n \overline{r^n} + a_{n-1} \overline{r^{n-1}} + \cdots + a_2 \overline{r^2} + a_1 \overline{r} + a_0 = 0$$

Finally, natural number exponents just mean repeated multiplication, so the multiplication rule used earlier allows me to split up powers as well.
$$a_n \overline{r}^n + a_{n-1} \overline{r}^{n-1} + \cdots + a_2 \overline{r}^2 + a_1 \overline{r} + a_0 = 0.$$

This last equation shows what I set out to prove, that $\overline{r}$ is a root of the same polynomial. □

This symmetry of root sets can also be expressed using group actions (as defined in Definition 9.1). The above theorem tells us that the group C_2 acts on any polynomial root set S via the homomorphism $\psi : C_2 \to Perm(S)$, with $\psi(1)$ being the permutation that sends each root r to $\overline{r}$. This is only the beginning of the symmetries we will find in root sets, and such symmetries are the foundation of Galois theory. To find other such symmetries, we will need to get acquainted with the concept of irreducible polynomials.

10.4 Irreducible polynomials

One of the first techniques algebra students learn for solving polynomial equations is factoring. For example, we do not need to resort to the quadratic formula to solve the following equation; we can factor instead.
$$x^2 - x - 6 = 0$$

$$(x-3)(x+2) = 0$$
$$x - 3 = 0 \quad \text{or} \quad x + 2 = 0$$
$$x = 3 \quad \text{or} \quad x = -2$$

Factoring is breaking a larger-degree polynomial down into smaller-degree polynomials multiplied together; they are the *factors,* and they are easier to solve because they're smaller. In this case the factors are linear polynomials, which are very easy to solve. If we had known the two roots 3 and -2 beforehand, they would have told us how to factor the polynomial, because a quadratic with roots a and b factors as $(x-a)(x-b)$.

In general, if we know how to find even one rational root r of any polynomial, then we can factor the linear polynomial $(x-r)$ out of the original. For example, take the cubic polynomial

$$12x^3 - 44x^2 + 35x + 17$$

from Figure 10.6; one of whose roots is $-\frac{1}{3}$. We can factor out the corresponding linear polynomial $(x + \frac{1}{3})$, which has $-\frac{1}{3}$ as a root. To stick with integer coefficients, we could use $(3x + 1)$, which has the same root. Factoring can be done by polynomial division, which you can learn from Exercise 10.6, and yields

$$(3x + 1)(4x^2 - 16x + 17).$$

The quadratic factor cannot be factored further (into two linear polynomials) because its roots are not rational and we are only considering polynomials with rational coefficients. We could factor it if we allowed complex coefficients; its roots are $2 \pm \frac{1}{2}i$, so it factors as

$$\left(x - \left(2 + \tfrac{1}{2}i\right)\right)\left(x - \left(2 - \tfrac{1}{2}i\right)\right).$$

Because $4x^2 - 16x + 17$ cannot be factored into polynomials with rational coefficients, we call it ***irreducible,*** or more specifically ***irreducible over*** $\mathbb{Q}$.

Irreducible polynomials are of interest to us precisely because they are the polynomials whose roots are outside of $\mathbb{Q}$. If a polynomial can be factored into linear factors over $\mathbb{Q}$, like the first example in this section, then its roots are part of that factoring and must therefore be in $\mathbb{Q}$. Solving such a polynomial does not require expanding our number system beyond $\mathbb{Q}$. If a polynomial cannot be factored at all, then none of its roots are in $\mathbb{Q}$, and solving it requires expanding to some number system that contains the roots. Then of course there are polynomials in between these two extremes, like the one we just considered; one of its roots was in $\mathbb{Q}$, but to find the other two we had to move outside $\mathbb{Q}$.

For polynomial roots outside $\mathbb{Q}$, we need new notation (besides just arithmetic and natural numbers) to write them down. Thus the radical sign was invented. There is no rational number solution to $x^2 - 2 = 0$, and so mathematicians created a new symbol, $\sqrt{2}$, and adopted the convention that that symbol stands for the number whose square is 2, the solution to the equation $x^2 - 2 = 0$.

Of course then whenever we encounter any irreducible polynomial $x^2 - a$ (for $a > 0$ in $\mathbb{Q}$) we express its root as $\sqrt{a}$. And the root of any irreducible $x^n - a$, for any integer $n > 1$, we write as $\sqrt[n]{a}$, which allows cube roots, fourth roots, and so on.[2] Radicals represent solutions to this particular family of irreducible polynomials, $x^n - a$, for $a > 0$ in $\mathbb{Q}$.

---

[2] Polynomials like $x^n - a$ have n roots, but only one of them is a positive real number. That is the one we express as $\sqrt[n]{a}$. For more details on this ambiguity (and what the other roots look like), try Exercise 10.17.

10.5. Galois groups

Combining radicals with the other field operations, we can express solutions to equations outside that family as well. We've seen that the quadratic formula combines one radical with several arithmetic operations to solve any second-degree polynomial equation. Although we haven't looked at the formulas for solving cubic and quartic equations, they, too, are only arithmetic operations and radical signs (though they are large expressions!). So the new notation $\sqrt[n]{a}$ has more power than just solving $x^n = a$; in fact, combined with the existing arithmetic operations, it has been used to create formulas for solutions to all polynomials up to degree 4!

But we begin to see the problem that causes the quintic to be unsolvable. This handy $\sqrt[n]{a}$ notation was invented only to solve *some* irreducible polynomials, and thus we should not be surprised if it is not enough to solve *all* irreducible polynomials. In fact, perhaps the surprise should be that it can be used to solve all polynomials of degrees 1 through 4! This is what Figure 10.4 represents; we know that what we can reach by adding radicals to $\mathbb{Q}$ is part of A, but it is not surprising that radicals are not enough to reach all of A.

Since irreducible polynomials play such an important role, we will want to be able to tell when a polynomial is irreducible. There is not an easy method that works all the time, but the following theorem helps. It is the first of several facts outside of group theory that I introduce without proof.[3]

Theorem 10.4 (Eisenstein Criterion). *A polynomial with integer coefficients is irreducible if we can find a prime number p that satisfies both of the following requirements.*

1. *The coefficient of the highest power of x is not a multiple of p, but all the other coefficients are.*

2. *The constant term (the term without an x) is not a multiple of p^2.*

For example, the theorem says that the following polynomial is irreducible, using the prime $p = 2$.

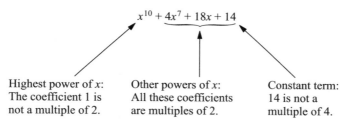

Highest power of x:
The coefficient 1 is
not a multiple of 2.

Other powers of x:
All these coefficients
are multiples of 2.

Constant term:
14 is not a
multiple of 4.

10.5 Galois groups

Now comes the real meat of the chapter, where I use both irreducible polynomials and group theory to analyze operations like radicals that reach outside $\mathbb{Q}$. Such operations are called ***field extensions,*** and we have already seen in Theorem 10.3 that they have symmetry, which group theory can describe.

---

[3]Mathematical texts do not usually state results without proving them. But my purpose with this theorem and those that follow is not to convince you that they are true, but to show you the beautiful theory they create. If you are curious about the proofs, as I hope you are, refer to the books I recommend at the end of the chapter.

We study polynomial roots (and their symmetries) by examining what it is like to extend $\mathbb{Q}$ just enough to contain one polynomial's roots. I begin with a simple example, x^2-2, whose two roots are $\pm\sqrt{2}$. Once we understand the operation of adding one radical to $\mathbb{Q}$ (in this case $\sqrt{2}$), I move on to adding more than one radical (Section 10.5.4), then adding radicals that are not square roots (Section 10.5.5). I then show how Galois proved powerful and beautiful facts about these symmetries.

10.5.1 A small field extension, $\mathbb{Q}(\sqrt{2})$

How far beyond $\mathbb{Q}$ must I expand my view to solve $x^2 - 2 = 0$? It is not enough to add the single number $\sqrt{2}$; I want the result to be a number system capable of arithmetic, a field. Furthermore, I want it to be the smallest field outside $\mathbb{Q}$ that contains $\sqrt{2}$, because I want to know just how far I am forced to reach to solve $x^2-2=0$. We call such a field $\mathbb{Q}(\sqrt{2})$; it includes $\mathbb{Q}$, $\sqrt{2}$, and all numbers obtainable from $\sqrt{2}$ using the operations of arithmetic. For example, all of the following numbers are in $\mathbb{Q}(\sqrt{2})$.

$$-\sqrt{2} \qquad 6+\sqrt{2} \qquad \left(\sqrt{2}+\frac{3}{2}\right)^3 \qquad \frac{\sqrt{2}}{16+\sqrt{2}}$$

Even so, every element of $\mathbb{Q}(\sqrt{2})$ can be written compactly. Although some of those shown above look not-so-compact (and we could imagine huge expressions involving lots of natural numbers, arithmetic, and $\sqrt{2}$), they all simplify to a rather small form. For example, expanding the cube $\left(\sqrt{2}+\frac{3}{2}\right)^3$ and combining like terms (remembering that $(\sqrt{2})^2 = 2$ and $(\sqrt{2})^3 = 2\sqrt{2}$) yields a two-term result.

$$\left(\sqrt{2}+\tfrac{3}{2}\right)^3 = \left(\sqrt{2}\right)^3 + \tfrac{9}{2}\left(\sqrt{2}\right)^2 + \tfrac{27}{4}\sqrt{2} + \tfrac{27}{8} = 12\tfrac{3}{8} + 8\tfrac{3}{4}\sqrt{2}$$

Exercise 10.9 covers tricks for simplifying other expressions from $\mathbb{Q}(\sqrt{2})$; curious readers may wish to try that exercise now. The result is that any element of $\mathbb{Q}(\sqrt{2})$ can be simplified to the two-term form $a + b\sqrt{2}$ for some $a, b \in \mathbb{Q}$. Thus $\mathbb{Q}(\sqrt{2})$ is rather like $\mathbb{C}$, whose elements have the two-term form $a + bi$, with $a, b \in \mathbb{R}$.

The number of terms in the expression for a general element of an extension field is called the ***degree*** of the extension. We write $\left[\mathbb{Q}(\sqrt{2}) : \mathbb{Q}\right]$ to mean the degree of $\mathbb{Q}(\sqrt{2})$ as an extension of $\mathbb{Q}$, so $\left[\mathbb{Q}(\sqrt{2}) : \mathbb{Q}\right] = 2$. Though I did not need it for this example, the following theorem will prove useful in determining the degree of other (larger) extensions. Exercise 10.12 asks you to do a portion of its proof.

Theorem 10.5. *The degree of an extension $\mathbb{Q}(r)$ always matches the degree of the irreducible polynomial to which r is a root.*

For example, $\mathbb{Q}(\sqrt{2})$ is a degree-2 extension and $\sqrt{2}$ is a root of the degree-2 irreducible polynomial $x^2 - 2$.

Extension fields are not easy to visualize directly. On a graph of the complex plane, both $\mathbb{Q}$ and $\mathbb{Q}(\sqrt{2})$ look just like $\mathbb{R}$, the x axis. So rather than visualize field extensions themselves in the complex plane, I return to Hasse diagrams and depict relationships among fields. Figure 10.7 shows $\mathbb{Q}(\sqrt{2})$ as larger than $\mathbb{Q}$, but smaller than $\mathbb{R}$.

10.5. Galois groups

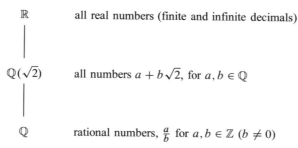

Figure 10.7. The field $\mathbb{Q}(\sqrt{2})$ is larger than $\mathbb{Q}$ and smaller than $\mathbb{R}$, but equal to neither of them.

The fact that $\mathbb{Q}(\sqrt{2}) \neq \mathbb{R}$ means that there are still irrational numbers outside of $\mathbb{Q}(\sqrt{2})$. Although adding $\sqrt{2}$ to $\mathbb{Q}$ required adding many irrational numbers besides just $\sqrt{2}$, it did not require adding every irrational number. For example, $\sqrt{3}$ is not in $\mathbb{Q}(\sqrt{2})$, as you already know if you have tried Exercise 10.9. Now that we know what $\mathbb{Q}(\sqrt{2})$ contains, let's see how Galois theory can show us its symmetry.

10.5.2 The symmetries of $\mathbb{Q}(\sqrt{2})$

In the polynomial root sets of Figure 10.6, the vertical mirror symmetry of complex conjugation is visually evident. But since we are not visualizing $\mathbb{Q}(\sqrt{2})$ in the complex plane, how can we measure its symmetries? More to the point, what do we mean by "symmetry" in this case?

In Chapter 3, I said that an object has symmetry when it looks the same from two different points of view. In the case of $\mathbb{Q}(\sqrt{2})$, we cannot speak of points of view, because we have no way to look at the field with our eyes. So in place of our "eyes" we use arithmetic, the field $\mathbb{Q}$. Any two roots that can't be distinguished *using equations of arithmetic* we will say "look the same" to us. More specifically, we're interested in those ways that the roots of a polynomial can be rearranged without making any true equations of arithmetic become false, or vice versa.

Let me show you what I mean. The polynomial $x^2 - 2$ has two roots, $\pm\sqrt{2}$, so we seek an action of some group G on the roots, $\psi : G \to Perm(\{\sqrt{2}, -\sqrt{2}\})$. Because the group $Perm(\{\sqrt{2}, -\sqrt{2}\})$ has only two elements, our task is simple; either the roots are interchangeable, in which case G is C_2 and ψ is an isomorphism, or they aren't, in which case G is the trivial group (no symmetry). The group G is therefore the symmetries of the polynomial's roots, and is called the *Galois group* of the polynomial. Here is a precise definition.

Definition 10.6 (Galois group). If r_1 through r_n are the roots of a polynomial and we specify an action of G on them by an embedding $\psi : G \to Perm(\{r_1, \ldots, r_n\})$ whose image is exactly those permutations undetectable by equations of arithmetic, then G is the **Galois group** of the polynomial.

The question is therefore this: Can we use equations of arithmetic to tell the roots $\pm\sqrt{2}$ apart? If we can, then they are not interchangeable and these roots have no symmetry. Otherwise, they are interchangeable, and the roots have a small amount of symmetry (but as much symmetry as you can expect from two things—they can be swapped). Since arithmetic does not include the square root symbol, I will need to call the roots r_1 and r_2 in order to be able to mention them in an equation of arithmetic.

We might try to force $r_1 = \sqrt{2}$ and $r_2 = -\sqrt{2}$ using the equation $r_1 = r_2 + 2\sqrt{2}$. This would certainly distinguish one root from the other, but unfortunately it cheats by including the symbol $\sqrt{2}$, so it's not an equation from $\mathbb{Q}$. Sticking to equations of arithmetic allows true statements like $r_1 r_2 = -2$, but this equation fails to distinguish the two roots. You could assign the values $\pm\sqrt{2}$ to r_1 and r_2 in either order and this equation would remain true. The equation $r_1 + r_2 = 0$ has the same problem; it is an equation from arithmetic, but doesn't tell the roots apart.

Try as I may, I will never find any way to distinguish the roots of $x^2 - 2$ using only equations of arithmetic. (A theorem supporting this appears in Section 10.5.4.) The two roots "look the same" from $\mathbb{Q}$'s point of view. The group G of symmetries of this polynomial's roots is isomorphic to C_2, because the two roots are interchangeable as far as $\mathbb{Q}$ is concerned. The action $\psi(1)$ interchanges the two roots.

10.5.3 Symmetries of field extensions

This symmetry extends beyond just the roots themselves, to all of $\mathbb{Q}(\sqrt{2})$, because $\mathbb{Q}(\sqrt{2})$ is generated by those roots. All elements of the field $\mathbb{Q}(\sqrt{2})$ are of the form $a + b\sqrt{2}$, so interchanging $\pm\sqrt{2}$ turns any element $a + b\sqrt{2}$ into $a - b\sqrt{2}$. Because we're only rearranging the roots, the elements and operations of $\mathbb{Q}$ are left unmoved. Mapping $a + b\sqrt{2}$ to $a - b\sqrt{2}$ interchanges $\pm\sqrt{2}$, but leaves a and b unchanged, and if we think of $a - b\sqrt{2}$ as $a + b(-\sqrt{2})$, we see that the addition was also unaltered. Any rearrangement of a polynomial's roots can be expanded to a symmetry of the whole extension field of $\mathbb{Q}$ generated by those roots, one that leaves elements and operations of $\mathbb{Q}$ unaltered. Galois theory is a study of the symmetries both of roots and of their field extensions.

In Chapter 8, I called ways to rearrange the elements of a group while respecting its operation automorphisms of the group. So we call ways to rearrange a field while respecting its operations *field automorphisms*. If $\phi : \mathbb{Q}(\sqrt{2}) \to \mathbb{Q}(\sqrt{2})$ is the above example, $\phi(a + b\sqrt{2}) = a - b\sqrt{2}$, then it does respect the operations of $\mathbb{Q}$, by satisfying the following two equations for any a and b in $\mathbb{Q}(\sqrt{2})$. (See Exercise 10.15.)

$$\phi(a + b) = \phi(a) + \phi(b) \qquad \phi(a \cdot b) = \phi(a) \cdot \phi(b)$$

Beyond being a field automorphism, ϕ also keeps every element of $\mathbb{Q}$ unmoved. Like all symmetries we'll study in polynomial roots, it only rearranges numbers outside $\mathbb{Q}$. We say that ϕ leaves $\mathbb{Q}$ *fixed,* or ϕ *fixes* $\mathbb{Q}$.

Complex conjugation is another field automorphism that fixes $\mathbb{Q}$. (Recall Theorem 10.3.) Note how similar complex conjugation is to the map ϕ; it maps $a + bi$ to $a - bi$. Figure 10.8 illustrates complex conjugation in two ways, on the left as a permutation of the roots of a polynomial, and on the right as a field automorphism. Both ways of thinking of the action have natural visualizations, as shown.

But not all field automorphisms are easy to draw in the complex plane. For example, the permutation interchanging the roots $\pm\sqrt{2}$ is easy to draw, as in Figure 10.9, but the corresponding automorphism ϕ of $\mathbb{Q}(\sqrt{2})$ cannot be drawn clearly. The reason is that ϕ does not keep numbers near one another. For example, it maps $\sqrt{2}$ to $-\sqrt{2}$, but

10.5. Galois groups

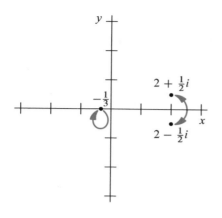

Complex conjugation acts on the roots of $12x^3 - 44x^2 + 35x + 17$ by this permutation.

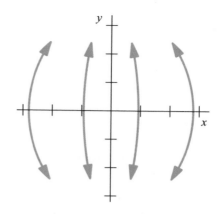

Complex conjugation is an automorphism of $\mathbb{C}$ that can be shown as a vertical flip over the x axis.

Figure 10.8. Complex conjugation depicted in two ways, on the left as a permutation of the roots of a specific polynomial ($12x^3 - 44x^2 + 35x + 17$ from Figure 10.6), and on the right as an automorphism of $\mathbb{C}$ that fixes $\mathbb{R}$ (and therefore $\mathbb{Q}$).

the rational number $\frac{7}{5}$, which is near to $\sqrt{2}$, is kept fixed. This same problem occurs all along the x axis, and so no simple geometric operation like the flip of Figure 10.8 suffices to represent ϕ. For this reason, I primarily speak of polynomial root symmetries as permutations of the roots, as in Definition 10.6, and I illustrate them as in Figure 10.9.

Galois theory is about computing and using these groups of symmetries. We computed C_2 as the Galois group of $x^2 - 2$, and the homomorphism $\psi : C_2 \to Perm(\{\sqrt{2}, -\sqrt{2}\})$ defining the action is an isomorphism. It is common to use the term "Galois group"

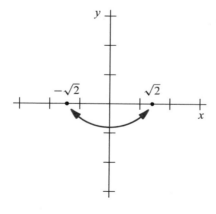

Figure 10.9. The permutation that interchanges the roots $\pm\sqrt{2}$ of $x^2 - 2$, whose extension to an automorphism of $\mathbb{Q}(\sqrt{2})$ has no easy visualization.

loosely, and sometimes refer to the image of ψ (a group of permutations) as the Galois group, or refer to the corresponding group of automorphisms as the Galois group. These three groups are isomorphic (since ψ is an embedding), so this bit of imprecision does no harm.

In Section 10.6, we'll use a polynomial's Galois group to determine that polynomial's solvability by radicals. But first, let's get a better feel for Galois groups by computing two more.

10.5.4 The symmetries of $\mathbb{Q}(\sqrt{2}, \sqrt{3})$

I stated earlier that $\sqrt{3}$ is not in $\mathbb{Q}(\sqrt{2})$, but adding $\sqrt{3}$ to $\mathbb{Q}(\sqrt{2})$ with *another* field extension provides a good example of a larger Galois group. This time, we are not extending the field $\mathbb{Q}$, but the field $\mathbb{Q}(\sqrt{2})$. We can express this as $\mathbb{Q}(\sqrt{2})(\sqrt{3})$ to show the two-step extension, or using the more common shorthand $\mathbb{Q}(\sqrt{2}, \sqrt{3})$.

Because $\sqrt{3}$ is a root of a degree-2 polynomial, we might expect that this is a degree-2 extension of $\mathbb{Q}(\sqrt{2})$. This means it should contain elements of the form $a + b\sqrt{3}$, with a and b coming from $\mathbb{Q}(\sqrt{2})$ rather than from $\mathbb{Q}$ (that is, each of a and b is a two-term $c + d\sqrt{2}$, for some $c, d \in \mathbb{Q}$). You can show that this is correct in Exercise 10.11. Thus elements of $\mathbb{Q}(\sqrt{2}, \sqrt{3})$ all look like

$$(a + b\sqrt{2}) + (c + d\sqrt{2})\sqrt{3},$$

with $a, b, c, d \in \mathbb{Q}$. Expanding that form obtains the four-term expression

$$a + b\sqrt{2} + c\sqrt{3} + d\sqrt{6},$$

for representing elements of $\mathbb{Q}(\sqrt{2}, \sqrt{3})$ with coefficients from $\mathbb{Q}$. So $\mathbb{Q}(\sqrt{2}, \sqrt{3})$ is a degree-2 extension of $\mathbb{Q}(\sqrt{2})$, but a degree-4 extension of $\mathbb{Q}$. That is,

$$\left[\mathbb{Q}(\sqrt{2}, \sqrt{3}) : \mathbb{Q}(\sqrt{2})\right] = 2$$

because we added $\sqrt{3}$ to $\mathbb{Q}(\sqrt{2})$ as a solution to the degree-2 equation $x^2 - 3 = 0$, but

$$\left[\mathbb{Q}(\sqrt{2}, \sqrt{3}) : \mathbb{Q}\right] = 4.$$

This suggests the pattern expressed in the following theorem.

Theorem 10.7. *Successive extensions multiply degrees.*

$$[\mathbb{Q}(a, b) : \mathbb{Q}] = [\mathbb{Q}(a, b) : \mathbb{Q}(a)] [\mathbb{Q}(a) : \mathbb{Q}]$$

This theorem is best shown in a diagram like Figure 10.10. On the bottom is the field $\mathbb{Q}$, to which we add $\sqrt{2}$ to move up to the degree-2 extension $\mathbb{Q}(\sqrt{2})$, and then add $\sqrt{3}$ to move up to the degree-4 extension $\mathbb{Q}(\sqrt{2}, \sqrt{3})$. The full extension $\mathbb{Q}(\sqrt{2}, \sqrt{3})$ is a degree-4 extension because it is two successive degree-2 extensions, and $2 \cdot 2 = 4$.

The field $\mathbb{Q}(\sqrt{2}, \sqrt{3})$ can be obtained in other ways that make sense based on the general form $a + b\sqrt{2} + c\sqrt{3} + d\sqrt{6}$ for elements of the field. In addition to the two-step extension $\mathbb{Q}(\sqrt{2}, \sqrt{3})$ in Figure 10.10 any of the following two-step extensions also reaches the same field, as you can prove in Exercise 10.11.

$\mathbb{Q}(\sqrt{3}, \sqrt{2}) \qquad \mathbb{Q}(\sqrt{2}, \sqrt{6}) \qquad \mathbb{Q}(\sqrt{6}, \sqrt{2}) \qquad \mathbb{Q}(\sqrt{3}, \sqrt{6}) \qquad \mathbb{Q}(\sqrt{6}, \sqrt{3})$

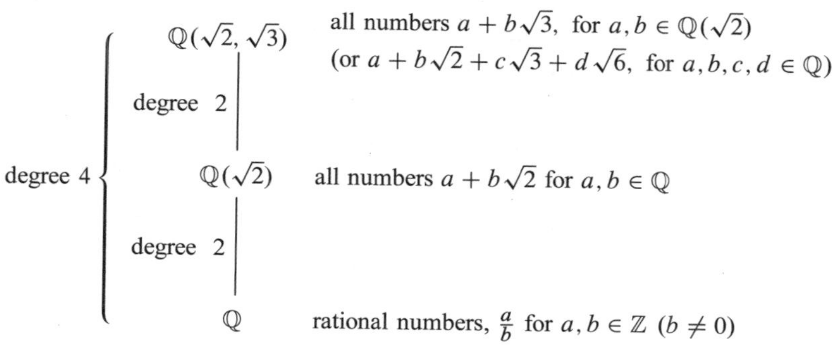

$\mathbb{Q}(\sqrt{2}, \sqrt{3})$
all numbers $a + b\sqrt{3}$, for $a, b \in \mathbb{Q}(\sqrt{2})$
(or $a + b\sqrt{2} + c\sqrt{3} + d\sqrt{6}$, for $a, b, c, d \in \mathbb{Q}$)
degree 2
degree 4
$\mathbb{Q}(\sqrt{2})$
all numbers $a + b\sqrt{2}$ for $a, b \in \mathbb{Q}$
degree 2
$\mathbb{Q}$
rational numbers, $\frac{a}{b}$ for $a, b \in \mathbb{Z}$ $(b \neq 0)$

Figure 10.10. Extending $\mathbb{Q}$ with the degree-2 extension $\sqrt{2}$ and then the degree-2 extension $\sqrt{3}$ to yield the degree-4 extension $\mathbb{Q}(\sqrt{2}, \sqrt{3})$

I illustrate this in Figure 10.11, which shows all the intermediate fields between $\mathbb{Q}$ and $\mathbb{Q}(\sqrt{2}, \sqrt{3})$.

The Galois group of the extension $\mathbb{Q}(\sqrt{2}, \sqrt{3})$ can be seen as all automorphisms of $\mathbb{Q}(\sqrt{2}, \sqrt{3})$ that fix $\mathbb{Q}$. The following theorem helps us find them.

Theorem 10.8. *Take any polynomial irreducible over $\mathbb{Q}$ and any two of its roots, r_1 and r_2.*

1. *There is an isomorphism $\phi : \mathbb{Q}(r_1) \to \mathbb{Q}(r_2)$ that replaces every r_1 with r_2, but fixes $\mathbb{Q}$. For example, in a degree-2 extension, ϕ would look like this.*

$$\phi(a + br_1) = a + br_2$$

2. *This remains true if $\mathbb{Q}$ is replaced with any field between $\mathbb{Q}$ and $\mathbb{C}$.*

This theorem is best understood in action. Let's take stock of the irreducible polynomials in our current situation. The irreducible polynomial for $\sqrt{2}$ is $x^2 - 2$, and the irreducible polynomial for $\sqrt{3}$ is $x^2 - 3$. These two polynomials together have the four roots $\pm\sqrt{2}$ and $\pm\sqrt{3}$. A polynomial with all four roots is $(x^2-2)(x^2-3)$, or x^4-5x^2+6, but it is not irreducible; Theorem 10.8 cares about the irreducible ones.

$\mathbb{Q}(\sqrt{2}, \sqrt{3})$
all numbers $a + b\sqrt{2} + c\sqrt{3} + d\sqrt{6}$
2
2
2
all numbers
$a + b\sqrt{2}$
$\mathbb{Q}(\sqrt{2})$
$\mathbb{Q}(\sqrt{3})$
all numbers
$a + b\sqrt{3}$
$\mathbb{Q}(\sqrt{6})$
all numbers
$a + b\sqrt{6}$
2
2
2
$\mathbb{Q}$

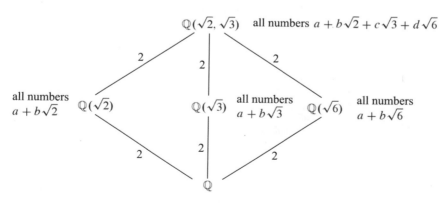

Figure 10.11. Several routes from $\mathbb{Q}$ to $\mathbb{Q}(\sqrt{2}, \sqrt{3})$. Each line is labeled with the degree of its extension.

Theorem 10.8 is about isomorphisms, and we seek automorphisms; part (2) of the theorem resolves that distinction. I use part (2) of the theorem to replace $\mathbb{Q}$ with $\mathbb{Q}(\sqrt{3})$, and the irreducible polynomial I consider first is $x^2 - 2$, with roots $\pm\sqrt{2}$. So there is an isomorphism between $\mathbb{Q}(\sqrt{3})(\sqrt{2})$ and $\mathbb{Q}(\sqrt{3})(-\sqrt{2})$ that maps each $a + b\sqrt{2}$ to $a - b\sqrt{2}$ (with $a, b \in \mathbb{Q}(\sqrt{3})$). Of course, $\mathbb{Q}(\sqrt{3})(\sqrt{2})$ and $\mathbb{Q}(\sqrt{3})(-\sqrt{2})$ are both just $\mathbb{Q}(\sqrt{2}, \sqrt{3})$, so this is one of the automorphisms we seek. Let's call it ϕ_2, because it swaps $\pm\sqrt{2}$ and leaves $\pm\sqrt{3}$ fixed.

Applying the same theorem to $\mathbb{Q}(\sqrt{2})$ and the irreducible polynomial $x^2 - 3$ gives an automorphism of $\mathbb{Q}(\sqrt{2}, \sqrt{3})$ that switches $\pm\sqrt{3}$ and leaves $\pm\sqrt{2}$ fixed; call it ϕ_3. We know enough about ϕ_2 and ϕ_3 to deduce how they treat $\sqrt{6}$.

$$\begin{aligned}
\phi_2(\sqrt{6}) &= \phi_2(\sqrt{2} \cdot \sqrt{3}) & & \text{because } \sqrt{6} = \sqrt{2} \cdot \sqrt{3} \\
&= \phi_2(\sqrt{2})\phi_2(\sqrt{3}) & & \text{because } \phi_2 \text{ respects multiplication} \\
&= (-\sqrt{2})(\sqrt{3}) & & \text{because } \phi_2 \text{ swaps } \pm\sqrt{2} \text{ and fixes } \pm\sqrt{3} \\
&= -\sqrt{6}
\end{aligned}$$

A similar set of steps can show the following two facts, about how ϕ_2 and ϕ_3 operate on elements of $\mathbb{Q}(\sqrt{2}, \sqrt{3})$.

$$\phi_2(a + b\sqrt{2} + c\sqrt{3} + d\sqrt{6}) = a - b\sqrt{2} + c\sqrt{3} - d\sqrt{6}$$

$$\phi_3(a + b\sqrt{2} + c\sqrt{3} + d\sqrt{6}) = a + b\sqrt{2} - c\sqrt{3} - d\sqrt{6}$$

Combinations of ϕ_2 and ϕ_3 generate the Galois group of $\mathbb{Q}(\sqrt{2}, \sqrt{3})$. Obviously ϕ_2 is an order-two element, since it is just a swap of two field elements, and ϕ_3 is order-two as well. But what is their product?

$$\begin{aligned}
\phi_2(\phi_3(a + b\sqrt{2} + c\sqrt{3} + d\sqrt{6})) &= \phi_2(a + b\sqrt{2} - c\sqrt{3} - d\sqrt{6}) \\
&= a - b\sqrt{2} - c\sqrt{3} + d\sqrt{6}
\end{aligned}$$

Continued algebra along these lines would reveal that the four automorphisms $\{e, \phi_2, \phi_3, \phi_2\phi_3\}$ are a group; its multiplication table and Cayley diagram appear in Figure 10.12. It is isomorphic to V_4, and its action on the four roots is shown in Figure 10.13.

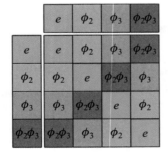
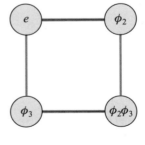

Figure 10.12. Multiplication table and Cayley diagram for the Galois group of the extension field $\mathbb{Q}(\sqrt{2}, \sqrt{3})$ over $\mathbb{Q}$

10.5. Galois groups

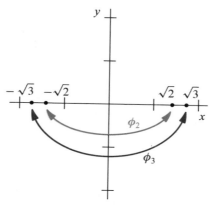

Figure 10.13. The action of V_4 on the roots of $x^4 - 5x^2 + 6$. The homomorphism $\psi : V_4 \to \mathrm{Perm}(\{\sqrt{2}, -\sqrt{2}, \sqrt{3}, -\sqrt{3}\})$ specifies the action, assuming $V_4 = \langle h, v \rangle$.

Theorem 10.8, which helped us find this Galois group, raises the importance of irreducible polynomials. By saying that there is an automorphism sending any root of an irreducible polynomial to any other, it says that $\mathbb{Q}$ can't tell apart the roots of an irreducible polynomial. In other words, setting an irreducible polynomial equal to zero is the most specific statement we can make in the language of $\mathbb{Q}$ about that polynomial's roots.

10.5.5 The symmetries of $\mathbb{Q}(\sqrt[3]{2})$

The final example Galois group I compute is for $x^3 - 2$. It obviously has the root $\sqrt[3]{2}$, but if you have tried Exercise 10.17, then you know that there are two other roots as well,

$$\frac{\sqrt[3]{2}}{2}\left(-1 + \sqrt{3}i\right) \quad \text{and} \quad \frac{\sqrt[3]{2}}{2}\left(-1 - \sqrt{3}i\right).$$

I graph all three roots in Figure 10.14, and name them r_1, r_2, and r_3 for easier writing.

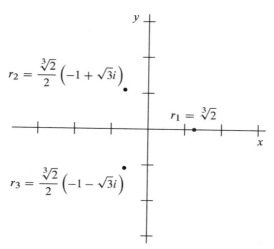

Figure 10.14. The three complex roots of $x^3 - 2$

Consider $\mathbb{Q}(r_1)$, which is $\mathbb{Q}(\sqrt[3]{2})$. We do not expect to represent every element of this field as some $a + b\sqrt[3]{2}$ for $a, b \in \mathbb{Q}$, because the degree of the extension is 3. Indeed, multiplying two such elements together does not yield another.

$$(a + b\sqrt[3]{2})(c + d\sqrt[3]{2}) = ac + (bc + ad)\sqrt[3]{2} + bd(\sqrt[3]{2})^2$$

We also need a term for $(\sqrt[3]{2})^2$, making the general form for elements of $\mathbb{Q}(r_1)$

$$a + b\sqrt[3]{2} + c(\sqrt[3]{2})^2.$$

Therefore $\mathbb{Q}(r_1)$ does not contain the roots r_2 or r_3, because $\mathbb{Q}(\sqrt[3]{2})$ is inside the real numbers, and neither r_1 nor r_2 is real. This situation is therefore quite different than the degree-2 cases we have seen, in which adding one root of an irreducible polynomial automatically added the other. To include all the roots of $x^3 - 2$ requires another field extension. I leave some of its details for Exercise 10.18, in which you are asked to verify that

$$\mathbb{Q}(r_1, r_2) = \mathbb{Q}(r_1, r_3) = \mathbb{Q}(r_2, r_3) = \mathbb{Q}(r_1, r_2, r_3)$$

and that elements from this field have the six-term form

$$a + b\sqrt[3]{2} + c(\sqrt[3]{2})^2 + d\sqrt{3}i + e\sqrt[3]{2}\sqrt{3}i + f(\sqrt[3]{2})^2\sqrt{3}i.$$

This six-term expression indicates that $\mathbb{Q}(r_1, r_2, r_3)$ is a degree-6 extension of $\mathbb{Q}$, and thus a degree-2 extension of $\mathbb{Q}(r_1)$. The full relationship among these fields (based on work you can do in Exercise 10.18) is shown in Figure 10.15. Even though $\sqrt{3}i$ is not a root of $x^3 - 2$, the Hasse diagram includes the field $\mathbb{Q}(\sqrt{3}i)$ because $\sqrt{3}i$ can be obtained from r_1, r_2, and r_3 using the operations of arithmetic. This is clear from the six-term form above, which can yield $\sqrt{3}i$ by letting all coefficients but d equal zero. Therefore $\mathbb{Q}(\sqrt{3}i)$ is one of the intermediate fields between $\mathbb{Q}$ and $\mathbb{Q}(r_1, r_2, r_3)$.

When an extension includes all the roots of an irreducible polynomial, we call it a **normal extension**. Thus $\mathbb{Q}(r_1)$ is not normal, but $\mathbb{Q}(r_1, r_2, r_3)$ is. As we compute the Galois group of this normal extension, we have the following useful theorem.

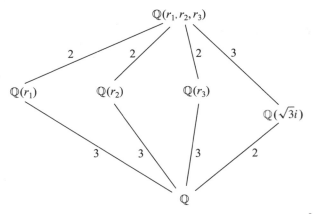

Figure 10.15. The relationship among extension fields of $\mathbb{Q}$ with roots of $x^3 - 2$. As in Figure 10.11, each line is labeled with the degree of its extension.

10.6. The heart of Galois theory

Theorem 10.9. *The degree of a normal extension equals the order of its Galois group.*

This theorem makes it easier to compute the Galois group of $\mathbb{Q}(r_1, r_2, r_3)$. Because there are three roots, the Galois group (which permutes those roots) must be isomorphic to some subgroup of S_3. Theorem 10.9 says that the order of the Galois group must match the degree of the extension, 6, and so we conclude that it must be all of S_3. The homomorphism $\psi : S_3 \to \textit{Perm}(\{r_1, r_2, r_3\})$ specifying how this Galois group acts on the roots is an isomorphism. I illustrate its action in Figure 10.16.

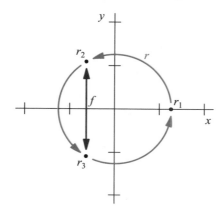

Figure 10.16. The action of S_3 on the roots of $x^3 - 2$

This completes our tour of Galois groups. Now it is time to see the use to which Galois put these tools, determining which polynomials are solvable. The following section introduces the two central theorems of Galois theory that help us prove that there is no formula for solving quintic polynomials by radicals.

10.6 The heart of Galois theory

Figure 10.15 is a Hasse diagram, showing relationships among field extensions of $\mathbb{Q}$. We first encountered Hasse diagrams in Section 6.6.3, when showing relationships among subgroups. Take a moment to compare the Hasse diagram in Figure 10.15 with the Hasse diagram of the subgroups of its Galois group, S_3, in Exercise 6.22 on page 112. What relationships do you see?

The two Hasse diagrams (one of subgroups and one of field extensions) are identical in shape, except that one of them is upside down. Applying a vertical flip to either the field extension diagram or the subgroup diagram makes the two look identical. Will the symmetry never stop! This same symmetry holds for the Galois group V_4 that we computed in Section 10.5.4, whose field extension diagram is in Figure 10.11. The Hasse diagram for subgroups of V_4 is in the solution to Exercise 6.23, on page 270. The vertical mirror symmetry still holds in this case, but is a bit redundant because each Hasse diagram itself has vertical mirror symmetry. This vertical mirror symmetry brings us to a key theorem from Galois's work.

Theorem 10.10 (mirror theorem). *The following two facts are true for any polynomial with roots $r_1, r_2, \ldots, r_n$ and Galois group G.*

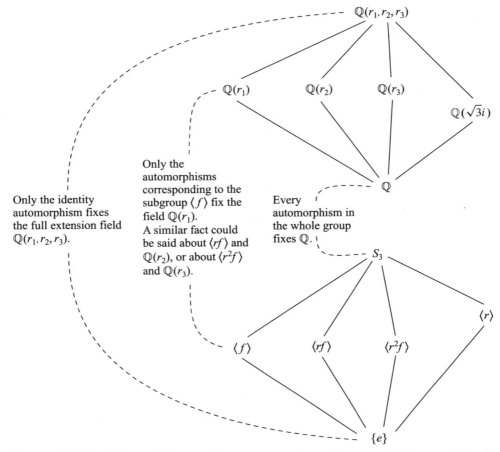

Figure 10.17. The vertical mirror image symmetry from Theorem 10.10 illustrated. Each subgroup of S_3 is the reflection of an extension field of $\mathbb{Q}$, but not just any extension field—the field fixed by that subgroup of automorphisms.

1. The Hasse diagram for the subgroups of G becomes identical to the Hasse diagram of the extension fields between $\mathbb{Q}$ and $\mathbb{Q}(r_1, \ldots, r_n)$ if you flip it vertically.

2. Each subgroup $H < G$ is not only the reflection of an extension field of $\mathbb{Q}$, but it is the reflection of the exact field that the automorphisms in H hold fixed.

The second part of Theorem 10.10 requires some explanation; I exemplify it in Figure 10.17. It tells us that the symmetry of the two Hasse diagrams is not superficial, but meaningful. We don't use groups to measure the symmetry in field extensions just because mathematicians like patterns, but because those patterns say something about the field extensions. In fact, the Galois group can tell us everything we need to know about the field extension. Galois groups aren't just pretty patterns; they answer the big question. One final theorem shows us how.

Theorem 10.11. *A field extension contains only elements expressible by radicals if and only if the following criterion holds about its Galois group G.*

10.6. The Heart of Galois Theory

There is a chain of subgroups from $\{e\}$ up to G, each one normal in the next,

$$\{e\} \triangleleft N_1 \triangleleft N_2 \triangleleft \cdots \triangleleft N_n \triangleleft G,$$

and each quotient along the chain is abelian.

$$\frac{N_1}{\{e\}} \quad \frac{N_2}{N_1} \quad \frac{N_3}{N_2} \quad \cdots \quad \frac{G}{N_n}$$

Although this may seem like a strange condition, a few remarks may make it more intuitive. First, it talks about a chain of subgroups between $\{e\}$ and G, which forms a path through a Hasse diagram of the subgroups of G, one that corresponds (by Theorem 10.10) to a chain of field extensions between $\mathbb{Q}$ and the field in question. Theorem 10.11 also requires the steps along the chain to be simple ones, from a subgroup to another in which it is normal, and the quotient abelian. The theorem says that putting such a requirement on the Galois group translates to an analogous requirement on the field extensions; they cannot be too complex either. Specifically, the theorem guarantees that the corresponding field extensions only add numbers expressible by radicals.

Because of the connection to polynomials, we call groups that fit the criteria of Theorem 10.11 **solvable groups.** Any abelian group G is solvable, because the tiny chain $\{e\} \triangleleft G$ satisfies Theorem 10.11; $\frac{G}{\{e\}}$ is abelian. Also, we've computed the Galois groups of two solvable polynomials already. So we expect both those groups (V_4 and S_3) to be solvable. The group V_4 is solvable because it is abelian. The group S_3 has a normal subgroup $\langle r \rangle$ that is abelian, so the following chain of normal subgroups shows its solvability.

$$\{e\} \triangleleft \langle r \rangle \triangleleft S_3$$

Both quotients $\frac{\langle r \rangle}{\{e\}} \cong C_3$ and $\frac{S_3}{\langle r \rangle} \cong C_2$ are abelian groups. In Figure 10.18 I highlight this chain of subgroups as a path through the Hasse diagram of subgroups of S_3, and label the corresponding quotients. The same evidence can be visualized in greater detail using homomorphisms, as in Figure 10.19, which shows the embedding of $\{e\}$ into $\langle r \rangle$ and the corresponding abelian quotient $\frac{\langle r \rangle}{\{e\}}$, as well as the embedding of $\langle r \rangle$ into S_3 and the corresponding abelian quotient $\frac{S_3}{\langle r \rangle}$.

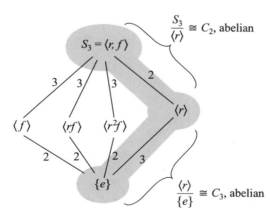

Figure 10.18. The path through the subgroup diagram of S_3 demonstrating its solvability. Each step goes from a subgroup up to another in which it is normal with abelian quotient.

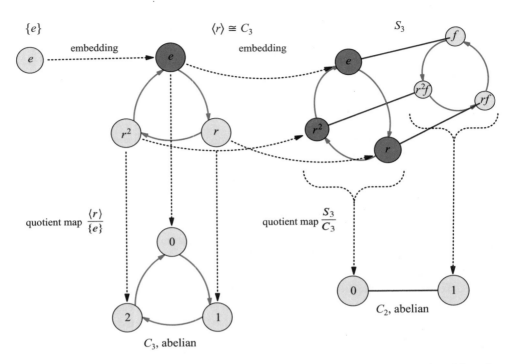

Figure 10.19. Solvability of S_3 illustrated, using $\{e\} \triangleleft \langle r \rangle \triangleleft S_3$. The chain $\{e\} \triangleleft \langle r \rangle \triangleleft S_3$ appears across the top of the diagram, and the corresponding quotients are below.

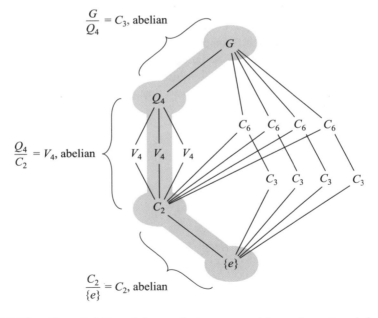

Figure 10.20. The solvability of the smallest group requiring a three-step chain of normal subgroups between $\{e\}$ and the full group of order 24. I name the subgroups by groups to which they are isomorphic, rather than define G in terms of generators. The chain $\{e\} \triangleleft C_2 \triangleleft Q_4 \triangleleft G$ does not include a subgroup isomorphic to V_4, but passes directly from C_2 to Q_4.

10.7. Unsolvability

The solvability of larger groups sometimes requires a longer chain of subgroups, but the smallest such group is of order 24. The corresponding diagram of homomorphisms is too tangled to be enlightening, but the path through the Hasse diagram appears in Figure 10.20. In fact, solvable groups are actually very common; it's the *unsolvable* ones that are hard to find. And that's what's left to do.

Armed with Galois's classification of which polynomials are solvable, we need to do what Abel did, find one specific unsolvable quintic polynomial. We need a quintic polynomial with roots r_1 through r_5 such that the Galois group of $\mathbb{Q}(r_1,\ldots,r_5)$ is not solvable. This is a job that requires mostly group theory, so let's leverage our knowledge of group theory to see how this story ends.

10.7 Unsolvability

We need an unsolvable group that describes the symmetries among the roots of some quintic polynomial. Those symmetries will form a subgroup of S_5, because we can view them as permutations of the five roots. A systematic search through S_5 yields many subgroups, and Exercise 10.26 asks you to show that all of them are solvable, except one. The smallest non-solvable group is one with no normal subgroups at all, A_5.

10.7.1 An unsolvable group

In Section 7.5, analyzing the conjugacy classes of A_4 revealed its only normal subgroup. I now analyze the conjugacy classes of A_5 to show that it has *no* normal subgroups.

We learned in Chapter 5 that A_5 has order 60, so we seek a class equation with 60 on the left and a sum of sizes of conjugacy classes on the right. Knowing that the identity element is in a conjugacy class by itself, we begin with this equation.

$$1 + [\text{some partition of the other 59 elements}] = 60$$

Figures 5.26 and 5.29 showed that A_5 is the group of symmetries of the icosahedron and the dodecahedron. I use the symmetries of the icosahedron to provide the words and images for describing conjugacy classes in A_5 (though the dodecahedron would work just as well).

The icosahedron permits a one-fifth clockwise turn around any one vertex, as shown in Figure 10.21. (To be specific, the axis of rotation is the line through the vertex and the icosahedron's center.) The icosahedron has 12 vertices, so there are 12 such rotations.

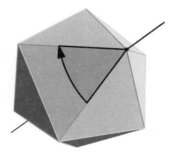

Figure 10.21. The icosahedron permits a one-fifth turn about any vertex.

They are conjugate to one another just as are the rotations of the tetrahedron, illustrated in Figure 7.33. I do not make a similar illustration here to prove it, because it would involve labeling all the icosahedron's vertices. It should be clear that the simple idea in Figure 7.33 can be transferred to the icosahedron. This updates our class equation with a conjugacy class of order 12.

$$1 + 12 + [\text{some partition of the other 47 elements}] = 60$$

Another 12 elements can be found by considering the squares of the same one-fifth turns, which are therefore two-fifths turns. They are all conjugate for the same reason that the one-fifth turns are, bringing the class equation to

$$1 + 12 + 12 + [\text{some partition of the other 35 elements}] = 60.$$

But considering the cubes of the one-fifth turns (the three-fifths turns) produces no new elements. A three-fifths clockwise turn is a two-fifths *counter*clockwise turn, and so a three-fifths turn about a vertex is the same as a two-fifths turn about the vertex on the opposite side of the icosahedron. Similarly, a four-fifths turn is just a one-fifth turn about the opposite vertex. So all the actions that involve rotating around a vertex have been counted in the two 12-element conjugacy classes.

The icosahedron also permits rotations about the center of each face, as shown in Figure 10.22. The axis of rotation is the line through the face's center and the icosahedron's center. Each of these one-third clockwise turns is another element of A_5, and they are all conjugates for the same reason that the one-fifth turns were. The icosahedron has 20 faces, giving a new conjugacy class of 20 elements, and updating our class equation to

$$1 + 12 + 12 + 20 + [\text{some partition of the other 15 elements}] = 60.$$

Since we have only 15 elements left to classify, clearly the two-thirds turns of each face do not form another 20-element conjugacy class. This is because a two-thirds turn is the same as a one-third turn of the opposite face, so we have already counted them.

The icosahedron allows a one-half turn about the axis through the center of any edge and the icosahedron's center, as shown in Figure 10.23. However, a one-half turn about any edge is the same as a one-half turn about the opposite edge, so the 30 edges yield only 15 distinct group elements. These elements are conjugates, making the final class equation

$$1 + 12 + 12 + 20 + 15 = 60.$$

I use this class equation to show that A_5 has no normal subgroups.

Figure 10.22. The icosahedron permits a one-third turn of any face.

Figure 10.23. The icosahedron permits a one-half turn that flips any edge.

10.7. Unsolvability

A normal subgroup's order must be a sum of sizes of conjugacy classes. Furthermore, we must ensure that the identity element is present if we hope to form a subgroup. Thus the only sizes achievable are these.

All classes, the whole group	Four conjugacy classes
$1 + 12 + 12 + 15 + 20 = 60$	$1 + 12 + 15 + 20 = 48$
	$1 + 12 + 12 + 20 = 45$
	$1 + 12 + 12 + 15 = 40$

Three conjugacy classes	Two conjugacy classes	One conjugacy class
$1 + 15 + 20 = 36$	$1 + 20 = 21$	$1 = 1$
$1 + 12 + 20 = 33$	$1 + 15 = 16$	
$1 + 12 + 15 = 28$	$1 + 12 = 13$	
$1 + 12 + 12 = 25$		

But most of these totals do not describe subgroups, because they do not divide the order of the group, which is 60. The only totals from this list dividing 60 are 1 and 60, the normal subgroups $\{e\}$ and A_5, which we knew were there before we began. Thus no other normal subgroups exist.

The only chain of normal subgroups between $\{e\}$ and A_5 is therefore the short one $\{e\} \triangleleft A_5$, which does not meet the requirement that $\frac{A_5}{\{e\}}$ be abelian. The group A_5 is not solvable. Furthermore, no group containing A_5 can be solvable, because the smallest first step in any chain of normal subgroups in such a group would be the invalid step $\{e\} \triangleleft A_5$. Thus S_5, S_6, S_7, and so on are all unsolvable groups.

10.7.2 An unsolvable polynomial

Our question narrows to this: Does any quintic polynomial have roots whose Galois group contains A_5? I have already claimed that this is so for the polynomial $x^5 + 10x^4 - 2$. From my analysis of this polynomial, you will see that it is easy to create others like it, and Exercise 10.28 gives you the opportunity to do so. Let's see why $x^5 + 10x^4 - 2$ is unsolvable.

It is irreducible by the Eisenstein Criterion (Theorem 10.4), using $p = 2$. Call its five roots r_1 through r_5. Then $\mathbb{Q}(r_1)$ is a degree-5 extension because the polynomial has degree 5. Perhaps it is a normal extension, and equal to the field $\mathbb{Q}(r_1, r_2, r_3, r_4, r_5)$, or perhaps it is not, and $\mathbb{Q}(r_1, r_2, r_3, r_4, r_5)$ is much larger. But the degree of the extension $\mathbb{Q}(r_1, r_2, r_3, r_4, r_5)$ is either 5 or some multiple of 5, because Theorem 10.7 says

$$[\mathbb{Q}(r_1, r_2, r_3, r_4, r_5) : \mathbb{Q}] = [\mathbb{Q}(r_1, r_2, r_3, r_4, r_5) : \mathbb{Q}(r_1)] \cdot [\mathbb{Q}(r_1) : \mathbb{Q}]$$
$$= [\mathbb{Q}(r_1, r_2, r_3, r_4, r_5) : \mathbb{Q}(r_1)] \cdot 5.$$

Because it is a normal extension, the order of the Galois group of $\mathbb{Q}(r_1, r_2, r_3, r_4, r_5)$ matches the extension's degree, some multiple of 5. Now group theory begins to help us.

Cauchy's Theorem guarantees that the Galois group contains an element of order 5. That element permutes the five roots, and the only permutations of order 5 in S_5 are cycles of the five roots, like this example.

The Galois group must also contain the complex conjugacy automorphism, as all Galois groups do. How this automorphism permutes the five roots depends on how many of those roots are real numbers. We can find the number of real roots by graphing the polynomial on a calculator or computer, giving a plot like the one on the left of Figure 10.24. It shows three points on the real number line where the polynomial equals zero, meaning three roots are real and two are complex. If you prefer to reason out this fact without using technology, see Exercise 10.27.

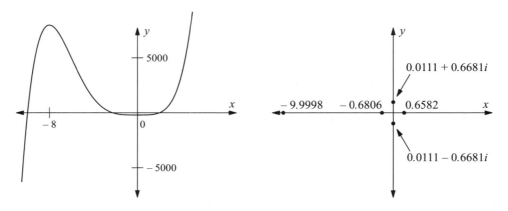

Figure 10.24. On the left, the graph of $x^5 + 10x^4 - 2$ shows that it has exactly three real roots, and therefore two complex ones. On the right, those five roots are plotted, and decimal approximations of their values given.

By Theorem 10.3, these two non-real roots must be conjugates, so the complex conjugacy automorphism swaps them. It corresponds to a permutation like this one.

Although we only know two of the permutations in the Galois group, it is enough. The work of Exercise 7.32 proves that a subgroup of S_5 containing two permutations like these (a five-cycle and a transposition) must be all of S_5. So this Galois group does indeed contain A_5. We have found an unsolvable Galois group! By Theorem 10.11, the polynomial is unsolvable as well. On the right of Figure 10.24 I plot the five roots of $x^5 + 10x^4 - 2$, but their values are given only approximately. I cannot write them precisely with radical notation because the polynomial is not solvable by radicals.

10.7.3 Conclusion

You now have a good idea why $x^5 + 10x^4 - 2$ is unsolvable, and therefore have a deeper appreciation of both the big question and its answer. Of course, I gave several facts from field theory without proof, so I cannot say that you know every detail of why the polynomial is unsolvable. But we have seen the beautiful and famous relationship between

10.7. Unsolvability

groups and fields from Theorems 10.10 and 10.11, and how group theory marshals the ideas of field theory to this historical result.

The group A_5 turned out to be central to the answer to the big question. Figure 10.25 is a larger version of a Cayley diagram of A_5 from Figure 5.29. A close look shows hints of the group's unsolvability all over the diagram; all visible subgroups look quite non-normal. For instance, consider the cyclic subgroup of order 5 generated by a red arrow. If that subgroup were normal, then all blue arrows leaving one of its cosets would go to another coset. But the complete opposite is true—each blue arrow goes to a different coset! This same fact is true about the order-3 subset obvious in the other Cayley diagram of A_5 in Figure 5.29, also generated by a red arrow. All the visible subgroups of A_5 are as far from normal as they can be.

This makes the Cayley diagram very interconnected, and gives it a certain appearance of stability. Each cycle is linked to as many others as possible, a feature that makes physical structures strong. This is not just appearance; carbon atoms in nature can bond in a structure just like the Cayley diagram in Figure 10.25, forming the molecule Carbon-60. This structure is called a fullerene, named after architect Richard Buckminster Fuller, who designed geodesic domes in a similar style.

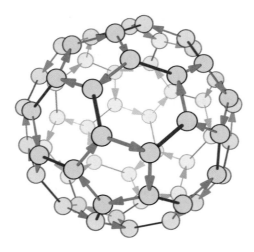

Figure 10.25. A Cayley diagram for A_5, showing its thorough interconnectedness

So in place of the details from field theory, I show you the structure on which the unsolvability of the quintic rests, the group A_5. And you've seen it in a way that Abel and Galois probably never did!

The unsolvability of polynomials shows us that groups are essential for understanding the relationships among the most basic operations of algebra. We have seen both the beauty and the utility of studying abstract patterns like groups. If you are interested in learning more about structures like fields, I encourage you to try textbooks such as [1], [8], or [9]. Two other books that do an excellent job of keeping the historical roots of field theory clear of modern debris are [6] and [11]. The latter is aimed at a broader audience than the former. Other books I find useful appear in the Bibliography.

10.8 Exercises

10.8.1 Basics

Exercise 10.1. For each of the following criteria, give an example number that satisfies it.

(a) a number in $\mathbb{N}$

(b) a number in $\mathbb{Z}$ but not $\mathbb{N}$

(c) a number in $\mathbb{Q}$ but not $\mathbb{Z}$

(d) a number in $\mathbb{R}$ but not $\mathbb{Q}$

(e) an algebraic number not in $\mathbb{R}$

(f) a number in $\mathbb{C}$ that's not algebraic

(g) a number in $\mathbb{C}$ that's not algebraic nor in $\mathbb{R}$

Exercise 10.2. Simplify each of these complex-number expressions.

(a) $(12 + i)(3 - i)$

(b) $2i(1 + i + i^3) - 9$

Exercise 10.3. Identify which of the following statements are true and which are false, and explain why.

(a) Saying a polynomial is "solvable" means that it has solutions.

(b) Every quartic polynomial is solvable.

(c) Every quintic polynomial is solvable.

(d) No quintic polynomial is solvable.

(e) Every sextic (degree 6) polynomial is solvable.

(f) No sextic polynomial is solvable.

(g) Because every Galois group contains the complex conjugacy operation, every Galois group has even order.

(h) Every polynomial $x^2 - a$ (for $a > 0$ in $\mathbb{Q}$) is irreducible.

(i) The Eisenstein Criterion can never be used to conclude that a polynomial is *factorable*. It only tells us when a polynomial is *not* factorable.

(j) If a quintic polynomial factors over $\mathbb{Q}$, then it is solvable.

(k) Every Galois group is finite.

(l) Every algebraic number is a real number.

(m) Every real number is an algebraic number.

Exercise 10.4. I say in the text that any equation made from the operations of arithmetic can be simplified to a polynomial equal to zero. The following simplification procedure supports my claim.

1. If needed, subtract to move the right side over to the left, leaving just zero on the right.

10.8. Exercises

2. Combine fractions until the left-hand side contains at most one fraction. (More on this step below.)
3. Multiply both sides of the equation by the fraction's denominator, leaving a polynomial equal to zero.

Answer the following questions to show that the above procedure works.

(a) Consider the equation $\frac{x-1}{x+\frac{1}{x}} = 2$. Applying step 1 yields $\frac{x-1}{x+\frac{1}{x}} - 2 = 0$, and step 2 then yields $-\frac{x^2+3x}{x^2+1} = 0$. Explain how step 2 was applied (that is, show the steps).

(b) In general, if step 1 results in two fractions subtracted, how should they be combined? What other algebra steps may be needed to get a large, complicated expression down to containing at most one fraction?

(c) What might equations from algebra that contain at most one fraction look like? Why does step 3 always turn them into polynomials?

(d) What equation results if we apply step 3 to the expression from part (a)?

(e) Explain how any polynomial equation whose coefficients are rational numbers can be turned into one with only integer coefficients. Test your method on $\frac{2}{9}x^7 - \frac{11}{8}x + \frac{1}{2} = 0$.

Exercise 10.5. Prove the missing step in Theorem 10.3, that

$$\overline{(a+bi)(c+di)} = \overline{(a+bi)}\,\overline{(c+di)}.$$

Exercise 10.6. Dividing polynomials is like dividing whole numbers. For example, to divide $2x^3 - 6x + 5$ by $x - 4$, we set up a division problem like this.

$$x - 4 \,\overline{\smash{\big)}\, 2x^3 + 0x^2 - 6x + 5}$$

Notice that I've included all powers of x, even those whose coefficients are zero. Proceeding as with division of whole numbers, we ask what polynomial can be multiplied by $x - 4$ to give the leading term $2x^3$. The answer is $2x^2$, so I put that at the beginning of the answer (on top), multiply it by $x - 4$, and subtract, as in division of whole numbers.

$$\begin{array}{r} 2x^2 \\ x - 4 \,\overline{\smash{\big)}\, 2x^3 + 0x^2 - 6x + 5} \\ \underline{2x^3 - 8x^2 } \\ 8x^2 \end{array}$$

Note that subtracting $-8x^2$ is the same as adding $8x^2$. Then bring down $-6x$ and continue the procedure. The final result is shown here.

$$\begin{array}{r} 2x^2 + 8x + 24 \\ x - 4 \,\overline{\smash{\big)}\, 2x^3 + 0x^2 - 6x + 5} \\ \underline{2x^3 - 8x^2 } \\ 8x^2 - 6x \\ \underline{8x^2 - 32x } \\ 24x + 5 \\ \underline{24x - 96} \\ 101 \end{array}$$

The remainder 101 is expressed in the answer as follows.

$$(2x^3 - 6x + 5) \div (x - 4) = 2x^2 + 8x + 24 + \frac{101}{x - 4}$$

Solve the following polynomial division problems. The solutions for the first two are available in the back of the book.

(a) $\frac{x^3-1}{x-1}$

(b) $(76x^7 + 20x^6 - 38x^4 - 10x^3 + 2) \div (4x^6 - 2x^3)$

(c) $(x^5 - 1) \div (x - 1)$

(d) $\frac{9x^3-7x^2+16x-10}{x^2+2}$

(e) $\frac{x^n-1}{x-1}$

(f) Factor the lower right polynomial in Figure 10.6 into irreducible polynomials.

(g) If a division (such as part (d)) does not come out even, but has a remainder, does that make the polynomial irreducible?

Exercise 10.7. Discern whether the following polynomials are irreducible. For those that are reducible, factor them.

(a) $x^2 - 14x + 51$

(b) $60x^2 + 50x - 10$

(c) $9x^3 - 36x^2 + x - 4$

10.8.2 Fields and extensions

Exercise 10.8. This exercise shows that the complex numbers $\mathbb{C}$, containing all numbers $a + bi$, for $a, b \in \mathbb{R}$, form a field, and explores arithmetic in that field.

(a) What is the sum $(a + bi) + (c + di)$ of two complex numbers, written in $a + bi$ form?

(b) What is the product $(a + bi)(c + di)$ of two complex numbers, written in $a + bi$ form?

(c) The additive inverse of any $a + bi$ is clearly $-a - bi$, which is obviously in the set $\mathbb{C}$. The multiplicative inverse of $a + bi$, however, is $\frac{1}{a+bi}$, and it is not obvious that this number is in $\mathbb{C}$. Prove that it is.

(d) Write all the following powers of i in $a + bi$ form.

$$i \qquad i^2 \qquad i^3 \qquad i^4 \qquad i^5$$

(e) The number i^n is equal to

_____	when $n \equiv 0 \bmod 4$,
_____	when $n \equiv 1 \bmod 4$,
_____	when $n \equiv 2 \bmod 4$,
and _____	when $n \equiv 3 \bmod 4$.

10.8. Exercises

(f) Show that $\sqrt{i}$ is the element $\frac{\sqrt{2}}{2} + \frac{\sqrt{2}}{2}i$ from $\mathbb{C}$, by showing that the result of squaring that element is i.

Exercise 10.9. This exercise proves several claims from the text about $\mathbb{Q}(\sqrt{2})$. I claimed that the set of elements $a + b\sqrt{2}$ for $a, b \in \mathbb{Q}$ is the smallest field containing $\mathbb{Q}$ and $\sqrt{2}$.

(a) Show that addition, $(a + b\sqrt{2}) + (c + d\sqrt{2})$, produces a new element of the same form.

(b) Show that multiplication, $(a + b\sqrt{2})(c + d\sqrt{2})$, produces a new element of the same form.

(c) What is the additive inverse of $a + b\sqrt{2}$?

(d) What is the multiplicative inverse of $a + b\sqrt{2}$?

(e) Show that no $a + b\sqrt{2}$ can equal $\sqrt{3}$, by simplifying $(a + b\sqrt{2})^2 = 3$ and finding no solutions. Recall that $\sqrt{2}$ and $\sqrt{3}$ are irrational.

(f) Show that no $a + b\sqrt{2}$ can equal $\sqrt[3]{2}$, by simplifying $(a + b\sqrt{2})^3 = 2$ and finding no solutions. Note that $\sqrt[3]{2}$ is irrational.

Exercise 10.10. In Sections 10.5.1 and 10.5.2, I analyzed $\mathbb{Q}(\sqrt{2})$, giving the formula $a + b\sqrt{2}$ for a typical element and computing its Galois group.

(a) Do a similar analysis for $\mathbb{Q}(\sqrt{3})$.

(b) Is $\mathbb{Q}(\sqrt{3})$ inside $\mathbb{Q}(\sqrt{2})$? Is $\mathbb{Q}(\sqrt{2})$ inside $\mathbb{Q}(\sqrt{3})$?

(c) Is $\mathbb{Q}(\sqrt{3})$ inside $\mathbb{Q}(\sqrt{2}, \sqrt{3})$?

Exercise 10.11. I say in the text that $\mathbb{Q}(\sqrt{2})(\sqrt{3}) = \mathbb{Q}(\sqrt{3})(\sqrt{2})$, and that it is a field whose elements have the form $a + b\sqrt{2} + c\sqrt{3} + d\sqrt{6}$. This exercise asks you to give the evidence for this claim.

(a) Show that all four terms are necessary to express elements of $\mathbb{Q}(\sqrt{2})(\sqrt{3})$.

(b) Show that the set of all such terms is closed under addition and contains the additive inverse of each of its elements.

(c) Show that it is also closed under multiplication and contains each element's multiplicative inverse.

(d) Show that $\mathbb{Q}(\sqrt{6})(\sqrt{2}) = \mathbb{Q}(\sqrt{6})(\sqrt{3}) = \mathbb{Q}(\sqrt{2}, \sqrt{3})$.

Exercise 10.12. Theorem 10.5 can be proven in two steps, the first of which is much harder than the second. I give both steps here, and ask you to prove the second. If you like a challenge, try proving the first as well.

First, show that every element of $\mathbb{Q}(r)$ can be written as some

$$c_0 + c_1 r + c_2 r^2 + \cdots + c_n r^n,$$

with each $c_i \in \mathbb{Q}$.

Second, show that n is always less than the degree of the polynomial for which r is a root.

Exercise 10.13.

(a) Make a diagram like Figure 10.17 for the field extension $\mathbb{Q}(\sqrt{2})$ and its Galois group, which is isomorphic to C_2.

(b) Make a diagram like Figure 10.17 for the field extension $\mathbb{Q}(\sqrt{2}, \sqrt{3})$ and its Galois group, which is isomorphic to V_4.

Exercise 10.14. Compute the Galois group of $\mathbb{Q}(i)$ as an extension of $\mathbb{Q}$.

Exercise 10.15. Recall the function $\phi : \mathbb{Q}(\sqrt{2}) \to \mathbb{Q}(\sqrt{2})$ from Section 10.5.3, defined by $\phi(a + b\sqrt{2}) = a - b\sqrt{2}$.

(a) Show that it is a field automorphism, meaning that it satisfies the following two equations.

$$\phi(a + b) = \phi(a) + \phi(b) \qquad \phi(a \cdot b) = \phi(a) \cdot \phi(b)$$

(b) Explain why it fixes $\mathbb{Q}$.

Exercise 10.16. Explain why the Galois group of any irreducible quadratic polynomial is isomorphic to C_2.

Exercise 10.17. Every polynomial $x^n - a$ (for $a > 0$ in $\mathbb{Q}$) has the solution $\sqrt[n]{a}$ in the positive real numbers. But it actually has $n - 1$ other solutions as well, most of which are not real numbers.

I define the function $c : \mathbb{R} \to \mathbb{C}$ by $c(\theta) = \cos(\theta) + i \sin(\theta)$. This function gives the point in the complex plane that is one unit distant from $(0, 0)$ in the direction of the angle θ, as shown here.

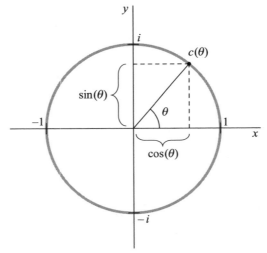

(a) Use the trigonometric identities

$$\cos(\alpha + \beta) = \cos\alpha \cos\beta - \sin\alpha \sin\beta$$
$$\text{and } \sin(\alpha + \beta) = \sin\alpha \cos\beta + \cos\alpha \sin\beta$$

to prove that $c(\alpha) \cdot c(\beta) = c(\alpha + \beta)$.

10.8. Exercises

(b) Explain why c is a homomorphism from the group $\mathbb{R}$ under addition to the group $\mathbb{C}$ under multiplication. Is it an embedding or a quotient map?

(c) What is $c(\alpha)^n$?

(d) Let $r_1 = \sqrt[3]{2}$, $r_2 = \sqrt[3]{2}c(\frac{2}{3}\pi)$, and $r_3 = \sqrt[3]{2}c(\frac{4}{3}\pi)$. Compare these values to those in Section 10.5.5.

(e) Prove that if $r = \sqrt[n]{a}$, then every $\sqrt[n]{a}\,c(\frac{k}{n}\cdot 2\pi)$ is also a root of $x^n - a$, for any natural number k. (Assume $a > 0$ in $\mathbb{Q}$.)

(f) Consider the group of roots of $x^6 - 1$ under the operation of multiplication. To what group is it isomorphic? What elements generate it?

(g) Consider the group of roots of $x^n - 1$ under the operation of multiplication. To what group is it isomorphic? What elements generate it?

Exercise 10.18. Section 10.5.5 analyzed $\mathbb{Q}(r_1)$, finding the expression for a typical element and computing the Galois group.

(a) Do the same for $\mathbb{Q}(r_2)$. What are the similarities and differences?

(b) Do the same for $\mathbb{Q}(r_3)$. To which of $\mathbb{Q}(r_1)$ or $\mathbb{Q}(r_2)$ is it more similar, and how?

(c) Show that the expression for elements of $\mathbb{Q}(r_1, r_2, r_3)$ is the six-term one given in Section 10.5.5, by analyzing $\mathbb{Q}(r_1)(r_2)$, $\mathbb{Q}(r_1)(r_3)$, or a similar field.

(d) For each of the irrational numbers that appears in that six-term form, find its irreducible polynomial over $\mathbb{Q}$, and the smallest extension field of $\mathbb{Q}$ containing it.

(e) Arrange all the fields you found into a Hasse diagram.

Exercise 10.19. Use the irreducible polynomials you found in Exercise 10.18 to create the Galois group of $\mathbb{Q}(r_1, r_2, r_3)$ as a group of automorphisms (using Theorem 10.8). Write each as a function operating on a typical element of $\mathbb{Q}(r_1, r_2, r_3)$, as in

$$\phi(a + b\sqrt[3]{2} + c(\sqrt[3]{2})^2 + d\sqrt{3}i + e\sqrt[3]{2}\sqrt{3}i + f(\sqrt[3]{2})^2\sqrt{3}i)$$
$$= a + b\sqrt[3]{2} + c(\sqrt[3]{2})^2 - d\sqrt{3}i - e\sqrt[3]{2}\sqrt{3}i - f(\sqrt[3]{2})^2\sqrt{3}i.$$

Specify the isomorphism ψ mentioned in Section 10.5.5, between S_3 and this Galois group.

Exercise 10.20. Which path of field extensions between $\mathbb{Q}$ and $\mathbb{Q}(r_1, r_2, r_3)$ corresponds to the solvable sequence for S_3? Why did I not create $\mathbb{Q}(r_1, r_2, r_3)$ along this route in the chapter?

Exercise 10.21. Prove that if ϕ is an automorphism of any field extending $\mathbb{Q}$ (and thus ϕ respects addition and multiplication), then ϕ fixes $\mathbb{Q}$.

10.8.3 Polynomials and solvability

Exercise 10.22. Find the irreducible polynomial for each of the following algebraic numbers.

(a) $\sqrt{15}$

(b) $\sqrt[5]{1+\sqrt{2}}$

(c) $\sqrt{10}+\sqrt{11}$

(d) $\sqrt[3]{2}+10$

Exercise 10.23.

(a) Create a polynomial with these four roots.

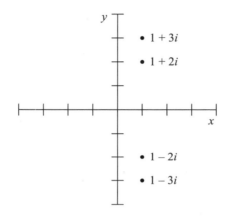

(b) Is it irreducible?

(c) What is its Galois group?

(d) Write an equation of arithmetic that distinguishes the inner two roots from the outer two.

Exercise 10.24. Consider the cubic polynomial on the top right of Figure 10.6.

(a) Write an equation of arithmetic that distinguishes one of its roots from the other two.

(b) Can an equation of arithmetic distinguish one of the roots of $x^3 - 5$ from the other two?

(c) Compute the Galois group of the polynomial in part (a).

Exercise 10.25. Given the roots r_1, r_2, and r_3 of $x^3 - 2$ defined in Section 10.5.5, recall the non-normal extension $\mathbb{Q}(r_1)$ of $\mathbb{Q}$, and the normal extension $\mathbb{Q}(r_1, r_2, r_3)$. In the text we found that $[\mathbb{Q}(r_1, r_2, r_3) : \mathbb{Q}] = 6$ and $[\mathbb{Q}(r_1) : \mathbb{Q}] = 3$, so it must be the case that $[\mathbb{Q}(r_1, r_2, r_3) : \mathbb{Q}(r_1)] = 2$. Find the corresponding degree-2 polynomial with coefficients from $\mathbb{Q}(r_1)$ whose two roots are r_2 and r_3.

10.8. Exercises

Exercise 10.26. Every subgroup of S_5 (except A_5) is isomorphic to one of the following groups. Show that each is solvable.

Cyclic groups: $C_1, C_2, C_3, C_4, C_5, C_6$

Dihedral groups: D_2 (which is V_4), D_3 (which is S_3), D_4, D_6

Other groups: A_4 (Cayley diagrams for which appear in Exercise 4.6 and Figure 5.27), a 20-element group whose Cayley diagram appears on the left of Exercise 5.15, and the 24-element group whose Cayley diagram appears below.

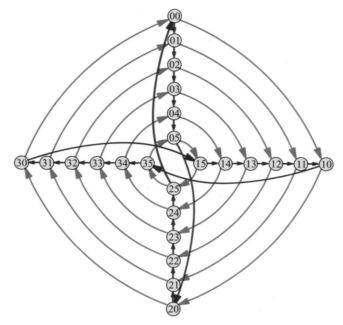

Exercise 10.27. How can calculus be used to show that there are three real and two imaginary roots of $x^5 + 10x^4 - 2$? Hint: Set the derivative equal to zero to find the peaks and valleys in the curve.

Exercise 10.28. Create irreducible polynomials of degree 5, as assigned here.

(a) Make one with only one real root.

(b) Make one with exactly three real roots.

(c) Make one with five real roots.

(d) Are any of the polynomials you created unsolvable?

10.8.4 Finite fields

Exercise 10.29. Here are diagrams of the finite fields of orders 5 and 8. Each is shown as two Cayley diagrams overlaid on the same set of nodes; the solid arrows are the Cayley diagram for addition, and the dashed ones are the Cayley diagram for multiplication. The diagrams for multiplication do not include zero.

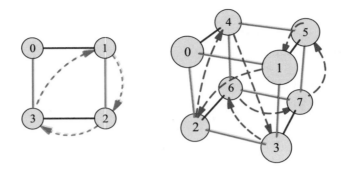

(a) Are the addition and multiplication operations in finite fields addition and multiplication mod some number?

(b) There is never more than one finite field of a given order. Create the Cayley diagrams for the finite fields of orders 3, 4, 7, and 11.

Exercise 10.30. Why can the group C_4 under addition not be made into a finite field by overlaying a multiplicative structure on $\{1, 2, 3\}$? Why do C_6 and C_{15} have the same problem?

A
Answers to selected Exercises

Chapter 1. What is a group?

Exercise 1.1. It is a group, as follows.

Rule 1.5 is satisfied because the problem specified a predefined list of actions that contained just one action.

Rule 1.6 is satisfied because swapping two things *back* is the same as swapping them in the first place.

Rule 1.7 is satisfied because swapping two things has a very predictable outcome.

Rule 1.8 is satisfied because any sequence of swaps is possible; swapping the two coins 100 times doesn't prevent you from doing it again, or 100 more times.

Exercise 1.3. It is not a group. Rule 1.8 is not satisfied; you cannot chain together any sequence of moves you like. If you try to move marbles from your left pocket to your right pocket twenty times in a row, you will find that it is not possible.

Exercise 1.5. No, it does not, as we can clearly see by the fact that we have encountered several finite groups already! But more to the point, although Rule 1.8 requires that any combination of actions be an action, it does not require that it be a *new* action. For the group in Exercise 1.1, for instance, there are infinitely many action sequences we could list, "swap the coins once," "swap the coins twice," "swap the coins three times," etc., but that long, infinite list really only contains two *different* things. Swapping the coins once produces the same result as swapping them three times, so we do not consider them different actions in the group. Swapping the coins twice or four times has the same result as not swapping them at all.

Exercise 1.7.

(a) Swapping the coins twice in a row is equivalent to the action "doing nothing."

(b) Every group will have such an action, because every action is reversible, and thus any action can be combined with its reversal to produce an action that does nothing.

Exercise 1.10. Many answers are valid here; this one is an example. Consider the list of actions containing only one thing, this action: Light a match and let it burn out. Assuming we're doing this with reliable matches in a controlled environment, the result is predictable. Assuming matches will continue to be manufactured indefinitely, you can do as many of this action in a row as you like. But it's definitely not reversible.

Exercise 1.12. Many answers are valid here; one example is the situation described in Exercise 1.3.

Chapter 2. What do groups look like?

Exercise 2.1. The two generators are the horizontal flip and vertical flip. The other actions in the puzzle are the non-action (do nothing) and the combined action "horizontal flip then vertical flip."

Exercise 2.3. Arrows that point from a node to itself represent the non-action. Typically such arrows are not included in the diagram, since they add clutter, but not any useful information.

Exercise 2.4.

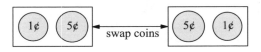

Exercise 2.8.

(a) This group is called D_4, and there are several ways to represent it with a Cayley diagram. Here is one.

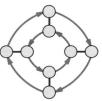

(b) The moves in the rectangle puzzle are restricted by the shape of the rectangle itself. Each move returns the rectangle to the same area it originally occupied, with just the label numbers different. Rotating a rectangle (that isn't a square) one quarter-turn clockwise does not keep it in the same area, but rotating a square a quarter-turn does.

I was asking you to think ahead a bit with that question; I don't introduce those reasons in the text until Chapter 3.

Exercise 2.9.

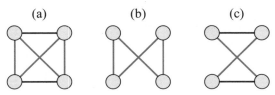

In both parts (b) and (c), the diagram shows that the generators drawn are sufficient because the diagram is connected. (From any one location, you can get to any other.)

Answers to Chapter 3

Exercise 2.12. The following is one example answer; it represents the cycle of activities for a typical newborn baby.

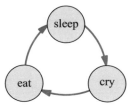

Exercise 2.13. The generators have the special status of being represented by the diagram's arrows. (And like all other elements of the group, the generators are also represented as nodes.)

Exercise 2.15. If following arrows in a diagram corresponds to doing actions, then following arrows backwards corresponds to undoing the actions. So in order for us to be able to follow arrows backwards in a diagram, we cannot have two arrows of the same type pointing to the same destination from two different sources. For instance, in the following diagram, it is not clear how to undo the action of the solid black arrow when in configuration C. Should you go to A or to B?

Exercise 2.17. It guarantees that from every point on the map, there will always be an arrow of each type leaving that point. This can be verified by considering what it would be like if it were not true; let's say we had some point P on our map that was lacking an outbound arrow corresponding to some action A. Then the sequence of moves leading from the start to P followed by action A would be invalid, which Rule 1.8 forbids.

Chapter 3. Why study groups?

Exercise 3.6. In the Cayley diagram in Figure A.1, the arrows labeled "$\frac{1}{4}$ turn" mean a quarter turn as if you had twisted the white molecule on top clockwise, spinning the atom around the vertical axis.

Exercise 3.8. A solution to this problem is not provided, but here is a hint. You may find it easier to explore the molecule's symmetry group if you know that it has the same symmetries as a regular pentagon. (Think of laying a pentagon flat with its center at the center of the molecule, each of its five corners half way between two of the purple atoms, vertically. Just as the molecule can be spun or flipped, so can the pentagon, and vice versa.)

Exercise 3.11. The solution to part (d) is shown in Figure A.2. It is a pattern that repeats vertically in blocks of four; each block of four is a little Klein 4-group, like the group of the rectangle puzzle from Chapter 2. The curved arrows represent the action of sliding the pattern to the right by the width of one leaf.

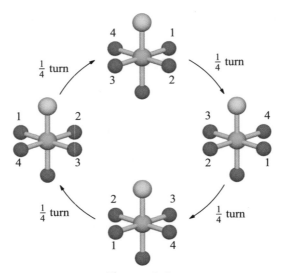

Figure A.1.

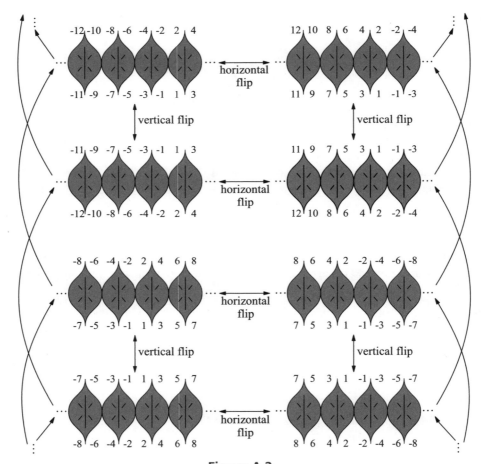

Figure A.2.

Answers to Chapter 3

The particular numbering scheme you used is unimportant; it need not be exactly like mine. But of course, you must label each of the four corners of each leaf, since they are all alike.

Exercise 3.13.

(a) The following Cayley diagram shows the group generated by the contra dance figure called "Circle right." Note that the dancers are moving to their own right sides, which from our point of view means they move counterclockwise.

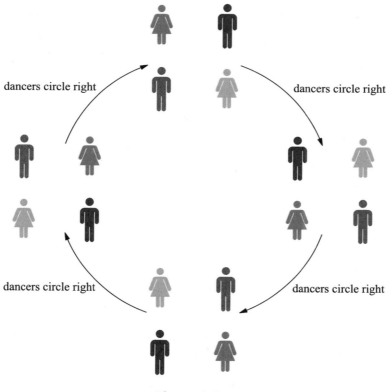

Figure A.3.

Rather than moving the dancers around the square, as in the diagram above, I could have drawn each square with the dancers in their original position, but with arrows to show where they would end up. This makes it easier to see how repetitions of the "Circle right" figure can yield other figures, and therefore makes it easier to answer the second part of this question. Figure A.4 does so.

(b) It contains the "Right and left through," as well as two other figures not shown in Figure 3.15. One of those figures doesn't look very interesting, because the dancers don't change positions at all. But keep in mind that these illustrations only show the results of the figures, not the steps taken during them; a figure may be interesting to watch or to dance, even though it does not result in a shuffling of the dancers at its end.

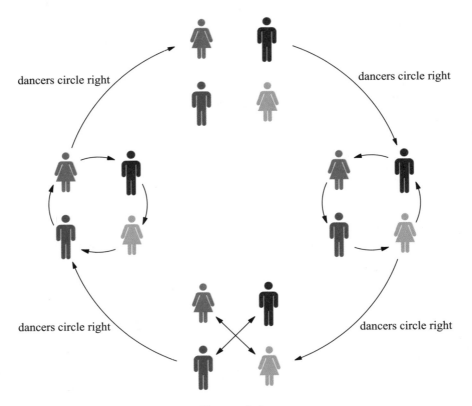

Figure A.4.

Chapter 4. Algebra at last

Exercise 4.5.

(a) Beginning at the element x in the diagram, follow a red arrow twice. Each red arrow represents one multiplication on the right by a, so this computes $x \cdot a \cdot a$, or $x \cdot a^2$.

(b) From Exercise 4.4 we learned that $k = j \cdot i$, which is performed by following the j arrow (blue) and then the i arrow (red). So to multiply $x \cdot k$, begin at the node for x and follow a blue arrow and then a red one.

Exercise 4.10(a). Hint: Check associativity of $A \cdot A \cdot B$.

Exercise 4.13(a). The operator is not associative, as the following computation demonstrates.
$$4 \cdot (3 \cdot 2) = 4 \cdot 1 = 2 \neq 4 = 2 \cdot 2 = (4 \cdot 3) \cdot 2$$

Exercise 4.14. The element s appears in the table, but is not in the row or column headings. Is it in the group or isn't it?

Exercise 4.17. Hint: It may help to consider a specific example. Why can the following two-piece diagram not be viewed as one Cayley diagram for an eight-element group? Which part of Definition 4.2 would be violated?

Answers to Chapter 5

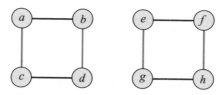

(Note that you could *add* arrows to make it a Cayley diagram, but as is, it is not one.)

Exercise 4.18. It leads you to conclude that a group can have only one identity element.

Exercise 4.19. Hint: For this exercise and those that follow it, use the fact from Exercise 4.15 liberally. The answer to part (a) is shown here.

	0	1
0	0	1
1	1	0

Exercise 4.22. Order matters in S_3, and the two groups shown in the bottom row, but order does not matter in the other three groups shown in that figure.

Exercise 4.23. S_3

Exercise 4.24. Read on to Chapter 5 to find out!

Exercise 4.26(a). $e^{-1} = e$, $a^{-1} = a^4$, $(a^2)^{-1} = a^3$, $(a^3)^{-1} = a^2$, and $(a^4)^{-1} = a$.

Exercise 4.27(a).
$$a^3 x = a^2$$
$$(a^3)^{-1} a^3 x = (a^3)^{-1} a^2$$
$$x = a^4$$

Exercise 4.32(b). No, none of the elements except 1 and -1 have inverses.

Exercise 4.33.

(b) No, one element is missing an inverse. Which is it?

(g) Hint: The answer is not just that they are infinite. After all, we've seen Cayley diagrams and multiplication tables for infinite groups illustrated using representative finite portions.

Chapter 5. Five families

Exercise 5.6(c).

	e	r	f	fr
e	e	r	f	fr
r	r	e	fr	f
f	f	fr	e	r
fr	fr	f	r	e

Exercise 5.9. If we view A_n as all the squares of elements from S_n, then squaring the one permutation in S_1, the identity, gives back that same permutation. Therefore $A_1 = S_1$ and so the order of both is 1.

Exercise 5.20. The group of order 1 is in several of these families; it is C_1, S_1, and A_1.
The group of order 2 is in several of these families; it is C_2, S_2, and D_1.
The group of order 3 is in two of these families; it is both C_3 and A_3.
The non-abelian group of order 6 is in two of these families; it is both S_3 and D_3.

Exercise 5.23. Answers will vary; for instance, the blade on a blender, some windmills.

Exercise 5.24. Answers will vary; for instance snowflakes and stars, like those shown on the left below, and truncated polygons, like those shown on the right below.

It is interesting to note that if we take D_n to be the group of symmetries of the truncated n-gon, then it is natural to make D_2 a long, tall rectangle, which is appropriate.

Exercise 5.28.

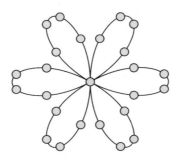

Exercise 5.30(a).

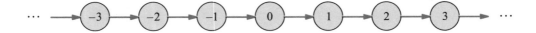

Exercise 5.31. See the Frieze group in Figure 3.13.

Exercise 5.35(a). If the equation $a \cdot b = c$ is true in a group, then in a Cayley diagram for that group, starting at the node for the element a and following the arrow(s) for the element b will end up at the node for the element c. In short, from a doing b arrives at c.

Exercise 5.38. Hint: Show that $ab = ba$ based on the facts $a^2 = e$, $b^2 = e$, and $(ab)^2 = e$.

Answers to Chapter 6

Exercise 5.40(a).

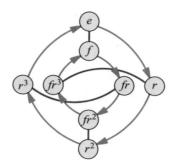

Exercise 5.43(c). Hint: Square every element of S_4.

Exercise 5.44(a). Hint: Check your results against the first 12 values of n from the table below. The table covers values of n from 1 to 20, and m is the smallest value such that a copy of C_n exists in S_m. But finding the pattern is difficult from the numbers alone; computing the embeddings themselves will probably be more fruitful.

n	m	n	m	n	m	n	m
1	1	6	5	11	11	16	16
2	2	7	7	12	7	17	17
3	3	8	8	13	13	18	11
4	4	9	9	14	9	19	19
5	5	10	7	15	8	20	9

Chapter 6. Subgroups

Exercise 6.1. The rightmost one is a Cayley diagram of the group C_6, reorganized to use generators 2 and 3. The leftmost one satisfies all requirements except for regularity; the blue arrows do not behave the same everywhere. The remaining diagram lacks blue arrows exiting either of the nodes in its bottom row.

Exercise 6.3(a).

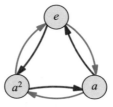

Exercise 6.12(a). Two. (The explanation is up to you.)

Exercise 6.14. Yes. If n is the order of the desired subgroup, try the generator $\frac{|G|}{n}$.

Exercise 6.16(b). The coset Hb is the set of destinations of the b-arrows that lead from the elements of H. There are never two different b-arrows leading into the same node because the actions represented by the generators are reversible (Exercise 2.15). There are never two b-arrows leaving the same node because the actions represented by the arrows are deterministic (Exercise 2.16). Thus each b-arrow leads to a different node from each of the elements of H, so there are exactly as many nodes in Hb as in H.

Exercise 6.17(d). Yes, always. For example, $\langle 2n \rangle < \langle n \rangle$ for any n.

Exercise 6.23(a). The diagram for V_4 looks like this.

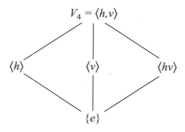

Exercise 6.26.

(a) Every multiplication table for V_4 is organized by an order-2 subgroup, because every element in V_4 has order 2. So whichever one appears next to the identity in the row/column ordering will generate a subgroup of order 2 in the upper-left corner of the table.

(b) Your column headings should read as follows.

0	2	4	8	1	3	5	7

Exercise 6.28. The answers to both parts are "yes," but I leave you to find the examples.

Exercise 6.30. Hint: Look at subgroups of the groups S_3, A_4, and D_{10}.

Chapter 7. Products and quotients

Exercise 7.3(b). False. For instance, the element $(3, 0)$ is in $C_4 \times C_3$, but it is not in $C_3 \times C_4$, because 3 is not in C_3.

Exercise 7.4(c). Hint: Start from Figure 7.6, which shows C_2^3. Your final answer may require some careful drawing of arrows.

Exercise 7.8.
  (c) Hint: Use the information in Section 7.1.4.
  (d) Hint: Can you tell whether it is abelian? How can you tell?

Exercise 7.9. Hint: Recall Exercise 6.31.

Exercise 7.14(a). The following diagram is for the rewiring group of C_5, with blue arrows standing for the operation of doubling each red arrow. Another way to say it is that the action in this rewiring group moves the red arrow that had stood for a to stand for a^2.

Answers to Chapter 7

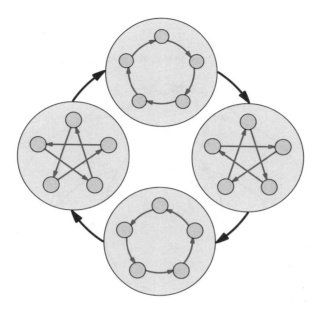

Exercise 7.14(d). Interestingly enough, the rewiring group of S_3 is isomorphic to S_3 itself. It is shown below; green arrows in the rewiring group represent reversing the red arrows in S_3, and orange arrows in the rewiring group represent advancing the blue arrow in S_3 to where the path red-then-blue had been.

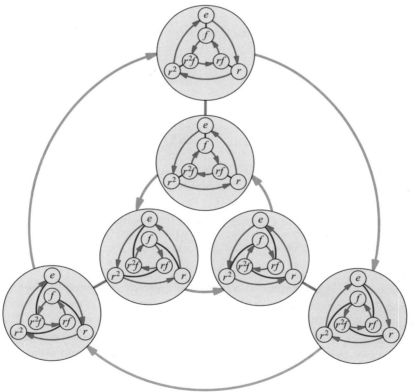

Exercise 7.15.
(a) D_4
(d) It is produced from the diagram above, the solution to Exercise 7.14(a).

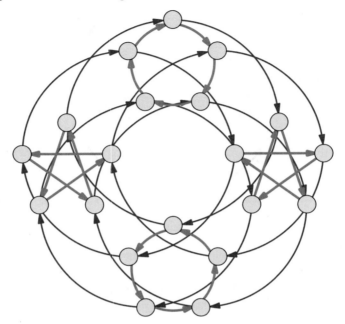

Exercise 7.16. It is the infinite dihedral group from Figure 3.13.

Exercise 7.18(b).

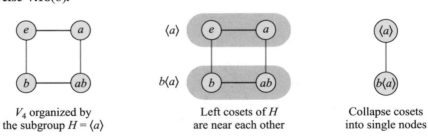

V_4 organized by
the subgroup $H = \langle a \rangle$

Left cosets of H
are near each other

Collapse cosets
into single nodes

Exercise 7.21. Your answer to Exercise 6.29 part (e) should prove helpful here.

Exercise 7.25(d). Hint: Refer to Exercise 7.23.

Exercise 7.33.
(a) Because D_3 is isomorphic to S_3, you have already computed its class equation in Exercise 7.32. The class equation for D_1 (isomorphic to C_2) is easy to compute, and then doing one more (D_5) should make the pattern clear.

(b) Because D_2 is isomorphic to V_4, you have already computed the class equations for the first two odd D_n in Exercise 7.31. Doing one more (D_6) should make the pattern clear.

(d) Hint: See Exercise 7.29.

Chapter 8. The power of homomorphisms

Exercise 8.4(c). There are four. Here is one way to describe them.

$$\theta_1(x) = e \qquad \theta_2(x) = x \qquad \theta_3(x) = 4 - x \qquad \theta_4(x) = 2x \bmod 4$$

Exercise 8.6. Yes, all four parts are true, but I leave the explanations to you. Be sure to base your explanations on Definition 8.1.

Exercise 8.9(c). This is true as long as the group is finite. When the group is infinite, consider the function in Exercise 8.5.

Exercise 8.11(c). Here is one way to show the infinite quotient using a multiplication table. The ellipses indicate infinite portions of the table that have been left out of this finite representation.

		−6	−3	0	3	6		−5	−2	1	4	7		−4	−1	2	5	8	
	⋱	⋮	⋮	⋮	⋮	⋮	⋰	⋱	⋮	⋮	⋮	⋮	⋮	⋰	⋱	⋮	⋮	⋮	⋮	⋮	⋰
−6		−12	−9	−6	−3	0		−11	−8	−5	−2	1		−10	−7	−4	−1	2	
−3		−9	−6	−3	0	3		−8	−5	−2	1	4		−7	−4	−1	2	5	
0		−6	−3	0	3	6		−5	−2	1	4	7		−4	−1	2	5	8	
3		−3	0	3	6	9		−2	1	4	7	10		−1	2	5	8	11	
6		0	3	6	9	12		1	4	7	10	13		2	5	8	11	14	
	⋰	⋮	⋮	⋮	⋮	⋮	⋱	⋰	⋮	⋮	⋮	⋮	⋮	⋱	⋰	⋮	⋮	⋮	⋮	⋮	⋱
	⋱	⋮	⋮	⋮	⋮	⋮	⋰	⋱	⋮	⋮	⋮	⋮	⋮	⋰	⋱	⋮	⋮	⋮	⋮	⋮	⋰
−5		−11	−8	−5	−2	1		−10	−7	−4	−1	2		−9	−6	−3	0	3	
−2		−8	−5	−2	1	4		−7	−4	−1	2	5		−6	−3	0	3	6	
1		−5	−2	1	4	7		−4	−1	2	5	8		−3	0	3	6	9	
4		−2	1	4	7	10		−1	2	5	8	11		0	3	6	9	12	
7		1	4	7	10	13		2	5	8	11	14		3	6	9	12	15	
	⋰	⋮	⋮	⋮	⋮	⋮	⋱	⋰	⋮	⋮	⋮	⋮	⋮	⋱	⋰	⋮	⋮	⋮	⋮	⋮	⋱
	⋱	⋮	⋮	⋮	⋮	⋮	⋰	⋱	⋮	⋮	⋮	⋮	⋮	⋰	⋱	⋮	⋮	⋮	⋮	⋮	⋰
−4		−10	−7	−4	−1	2		−9	−6	−3	0	3		−8	−5	−2	1	4	
−1		−7	−4	−1	2	5		−6	−3	0	3	6		−5	−2	1	4	7	
2		−4	−1	2	5	8		−3	0	3	6	9		−2	1	4	7	10	
5		−1	2	5	8	11		0	3	6	9	12		1	4	7	10	13	
8		2	5	8	11	14		3	6	9	12	15		4	7	10	13	16	
	⋰	⋮	⋮	⋮	⋮	⋮	⋱	⋰	⋮	⋮	⋮	⋮	⋮	⋱	⋰	⋮	⋮	⋮	⋮	⋮	⋱

Exercise 8.12(f). Here is an illustration for the extension of part (a) with two new maps satisfying the requirements.

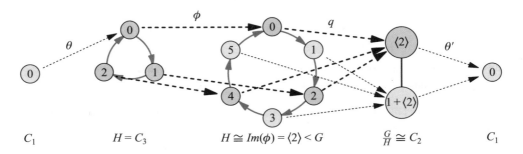

Exercise 8.13.

(b) Technically, no. We get a group of cosets that is *isomorphic* to a subgroup of G.

(c) No. For instance, $\frac{Q_4}{\langle -1 \rangle} \cong V_4$, and yet there is no embedding from V_4 into Q_4.

Exercise 8.16. Hint: If H is the commutator subgroup, consider an element in the conjugate subgroup gHg^{-1}, say $gaba^{-1}b^{-1}g^{-1}$. Multiply this element by a few commutators to yield an element of H. What can you then conclude?

Exercise 8.17. In each case, the commutator subgroup is highlighted in red; it maps to $\{e\}$ in the abelianization.

(a)

(b)

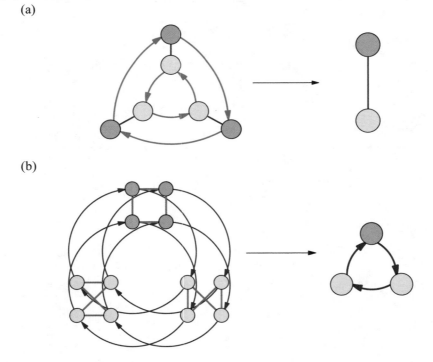

Exercise 8.23(a). Hint: Inspect a few elements from $C_3 \times C_3$ or $C_5 \times C_5$, using a Cayley diagram. Infer the answer, then justify it.

Answers to Chapter 8

Exercise 8.24. Hint: Does the orbit of $(1, 1)$ in $C_n \times C_m$ pass through the element $(0, 1)$?

Exercise 8.28. You'll find the answer shown in Exercise 7.7 part (c).

Exercise 8.29.

(b) $D_n \cong Z_n \rtimes_\theta C_2$ for a certain θ. (What is θ?)

(d) We have seen such groups before in Exercise 7.7(c). There is another in Exercise 8.30.

Exercise 8.32.

(b) Here are two rows from the table so that you can check some of your work.

r^2	r^2	r^2	r^2	r	r	r
f	f	r^2f	rf	f	r^2f	rf

(c) All of an element's conjugates are in a row, so the row lists all the elements in a conjugacy class (sometimes more than once). Therefore the rows show all the conjugacy classes of the group.

(d) A column shows all the conjugates by the specific element at the head of the column. It therefore shows the whole inner automorphism created by that element. For instance, reading down the column that is headed by r will tell you where the inner automorphism $\theta(x) = rxr^{-1}$ maps e, then r, then r^2, and so on. This makes answering part (e) easy.

Exercise 8.35. Hint: Use the direct product process for multiplication tables in Definition 7.3 as a starting point. Then come up with a notion of rewiring for multiplication tables.

Exercise 8.38 (b) and (c). Hint: Divide C_8 by C_2.

Exercise 8.42.

The formula for multiplication of two-by-two matrices will help with parts (a) and (b); it is given here.

$$\begin{bmatrix} a_1 & b_1 \\ c_1 & d_1 \end{bmatrix} \cdot \begin{bmatrix} a_2 & b_2 \\ c_2 & d_2 \end{bmatrix} = \begin{bmatrix} a_1a_2 + b_1c_2 & a_1b_2 + b_1d_2 \\ c_1a_2 + d_1c_2 & c_1b_2 + d_1d_2 \end{bmatrix}$$

Exercise 8.43(a). Hint: Show that it is an embedding by showing that the only element to which θ maps (e, e) is e.

Exercise 8.44.

(a) By Theorem 8.8, any abelian group of order 4 must be a direct product of cyclic groups whose orders multiply to 4. The only possibilities of whole numbers multiplying to 4 are $1 \cdot 4$ and $2 \cdot 2$. The first corresponds to $C_1 \times C_4$, or just C_4, and the second corresponds to $C_2 \times C_2$, which is isomorphic to V_4. Thus all abelian groups of order 4 are isomorphic either to C_4 or V_4, and no other.

(b) The final answer for this part is given here, but justifying it is up to you. There are three groups to which any abelian group of order 8 must be isomorphic, $C_2 \times C_2 \times C_2$, $C_2 \times C_4$, and C_8.

(d) By Theorem 8.8, any abelian group of order 30 must be a direct product of cyclic groups whose orders multiply to 30. The possibilities are therefore $C_1 \times C_{30}$, $C_2 \times C_{15}$, $C_3 \times C_{10}$, $C_5 \times C_6$, and $C_2 \times C_3 \times C_5$. The first of these ($C_1 \times C_{30}$) is obviously just C_{30}. But the rest are as well, based on Theorem 8.7. Thus every abelian group of order 30 is isomorphic to just one, C_{30}.

Exercise 8.45(c). Hint: Rewrite the equation as follows.

$$g^{ap^k} = g^{bm} g$$
$$(g^{p^k})^a = (g^m)^b g$$
$$g = (g^m)^{-b} (g^{p^k})^a$$

What order does the element g^m have? What order does g^{p^k} have?

(d) Hint: Apply Exercise 8.43.

Exercise 8.46(a). Hint: The following figure illustrates the homomorphism ϕ together with its associated quotient map and isomorphism (as in Figure 8.13).

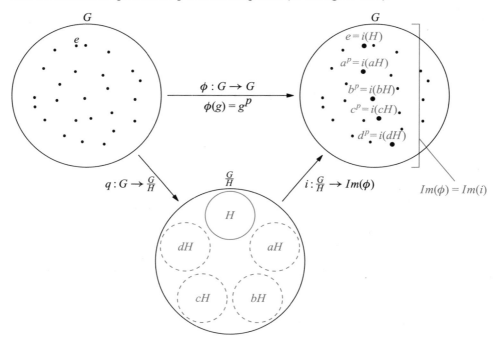

If we organize the right copy of G by $Im(\phi)$ and emphasize the fact that $Im(\phi)$ is a cycle, it also allows us to emphasize the fact that $\frac{G}{H}$ is a cycle, as shown in the figure below. Consider the element a, which in the following figure satisfies the equation $\langle aH \rangle = \frac{G}{H}$. Try to show that $\langle a \rangle = G$. (Is H a subgroup of $\langle a \rangle$?)

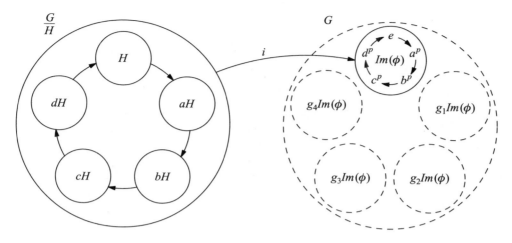

(c) Just one. You provide the reason.

(d) The chain is illustrated as follows. It ends in a cycle, because ϕ_{n+1} would send all elements to the identity.

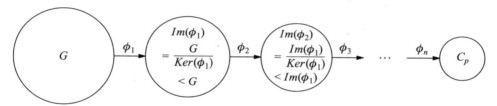

Exercise 8.47(b). Hint: Recall Exercise 6.8.

(c) Their only common element is the identity. Prove this by considering what any other element of K being in C would imply.

(e) Hint: Apply Exercise 8.37 part (a).

(g) If a copy of C always remains, then the quotients must be eliminating the other parts of the group. Continuing the process will therefore eventually lead to a G_n that is just a copy of C, as shown in the following illustration.

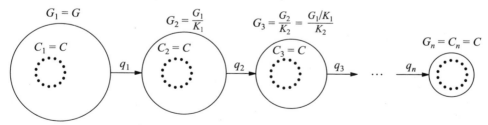

A specific example may be more helpful. In the following example, $p = 2$, so the abelian group is a product of cyclic groups whose orders are all powers of 2. Notice that not every quotient undoes a direct product (e.g., q_2).

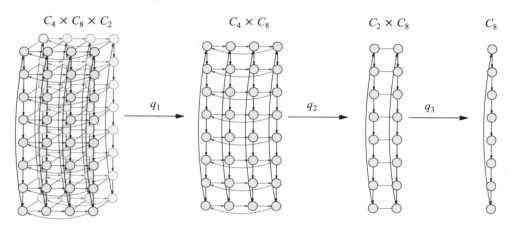

(h) The following illustration shows the quotient operation, assuming G_{i+1} is isomorphic to the direct product $C_{i+1} \times H_{i+1}$ for some subgroup $H_{i+1} < G_{i+1}$ that we have not yet determined. The quotient map q_i as stated in the problem divides by K_i, so each coset of K_i in the left diagram maps to one element in the right diagram.

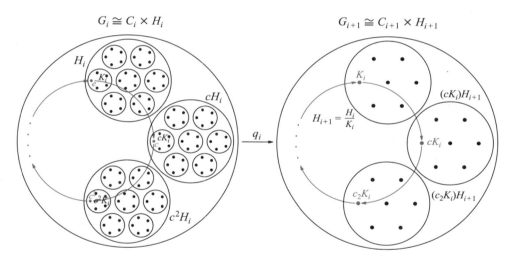

Consider the subgroup $H_i < G_i$, which maps to the subgroup $H_{i+1} < G_{i+1}$. Use the above illustration to explain why $G_i = C_i H_i$, and C_i and H_i overlap only at the identity element. Then apply Exercise 8.43.

Exercise 8.49. Each subscript is a power of a prime.

Exercise 8.50. Hint: Assume for simplicity's sake that the generators are ordered so that $g_1, \ldots, g_k$ are of finite order and $g_{k+1}, \ldots, g_n$ are of infinite order.

Chapter 9. Sylow Theory

Exercise 9.3(b). This is the question that Theorem 9.4 answers.

Exercise 9.12. Such a counterexample was given in the text, in Section 6.5.

Exercise 9.15(a). Hint: Let $n = |g|$. If n is larger than 1 and smaller than p, then it is not a factor of p. Dividing p by n gives some remainder $r < n$, so that $p = mn + r$. What is g^{mn+r}?

Exercise 9.16(a). Hint: Consider the case when $p = 5$. If $a \cdot b \cdot c \cdot d \cdot (abcd)^{-1} = e$ is an equation whose left side is in S, then what does the permutation $\phi(1)$ do to it? What does this say about any $\phi(n)$?

Exercise 9.17(a). Hint: Argue as in Theorem 7.7 and Observation 8.3, showing the closure is the original set.

Exercise 9.21. Hint: Apply the Fundamental Theorem of Abelian Groups.

Exercise 9.22(c). Hint: You cannot prove that the Sylow 3-subgroup must be normal, because in A_4 it is not. You cannot prove that the Sylow 2-subgroup must be normal, because in the group $C_4 \rtimes_\theta C_3$ of Exercise 8.30, it is not. But you can prove that either one or the other will be normal in any group of order 12.

Exercise 9.23.

(a) They are the four subgroups generated by the following four cyclic permutations.

(b) There are ten. Did you find them all?

Exercise 9.27. There are two, the abelian group C_{21} (which is isomorphic to $C_3 \times C_7$) and the non-abelian semidirect product group $C_3 \rtimes_\theta C_7$ based on the homomorphism $\theta : C_3 \to Aut(C_7)$ by

$$\theta(n) = \text{the automorphism that maps } a \text{ to } a^{(2^n)}.$$

A Cayley diagram for this non-abelian group is shown here.

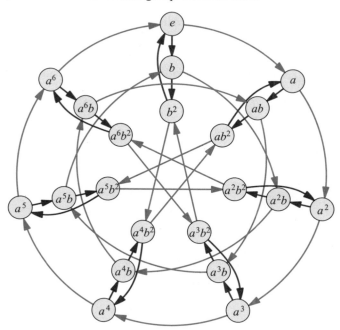

Exercise 9.28. As part of analyzing whether your work is correct, compare your results with the following correct list.

Abelian groups: C_{12} and $C_2 \times C_6$
Non-abelian groups: D_6, A_4, and the $C_4 \rtimes_\theta C_3$ from Exercise 8.30

Chapter 10. Galois Theory

Exercise 10.3.

(g) False, because complex conjugation might be equal to the identity, if all the polynomial's roots are real numbers.

(i) True, a polynomial that doesn't fit the theorem's requirements may still be irreducible.

(j) True. All factorable quintic polynomials are solvable because the existing methods for polynomials of degrees 1 through 4 can be applied to the factors. For example, because the quintic equation

$$3x^5 + 6x^4 + 13x^3 + 21x^2 + 28x + 14 = 0,$$

factors into

$$(x^2 + 2x + 2)(3x^3 + 7x + 7) = 0,$$

it can be solved. One of the factors must equal zero, so we have two simpler equations.

$$x^2 + 2x + 2 = 0 \qquad 3x^3 + 7x + 7 = 0$$

For the first, the quadratic formula gives $-1 \pm i$. For the second, the cubic formula gives solutions whose exact representations are a bit lengthy, but which are approximately equal to $0.394 \pm 1.67i$ and -0.789.

Exercise 10.5. Hint: The left side simplifies like this:

$$\overline{(a+bi)}\,\overline{(c+di)} = (a-bi)(c-di) = ac - bci - adi + bdi^2 = (ac-bd) - (bc+ad)i$$

(Recall that $i^2 = -1$.) Show that the right side simplifies to the same expression.

Exercise 10.6.

(a)
$$\begin{array}{r}
x^2 + x + 1 \\
x-1 \overline{\smash{\big)}\, x^3 + 0x^2 + 0x - 1} \\
\underline{x^3 - x^2} \\
x^2 \\
\underline{x^2 - x} \\
x - 1 \\
\underline{x - 1} \\
0
\end{array}$$

(b)
$$\begin{array}{r}
19x + 5 \\
4x^6 + 0x^5 + 0x^4 - 2x^3 \overline{\smash{\big)}\, 76x^7 + 20x^6 + 0x^5 - 38x^4 - 10x^3 + 0x^2 + 0x + 2} \\
\underline{76x^7 + 0x^6 + 0x^5 - 38x^4} \\
20x^6 + 0x^5 + 0x^4 - 10x^3 \\
\underline{20x^6 + 0x^5 + 0x^4 - 10x^3} \\
0 + 2
\end{array}$$

Answers to Chapter 10

(e) What pattern do you find in parts (a) and (c)?

(g) No, one uneven division cannot conclude irreducibility. Dividing by a different polynomial may come out evenly.

Exercise 10.8.

(b) Hint: Use the fact that $i^2 = -1$ to eliminate i^2 from the product.

(c) Hint: Multiply $\frac{1}{a+bi}$ by $\frac{a-bi}{a-bi}$ and simplify.

Exercise 10.9.

(a) $(a + b\sqrt{2}) + (c + d\sqrt{2}) = a + b\sqrt{2} + c + d\sqrt{2} = (a + c) + (b + d)\sqrt{2}$

(d) Hint: Proceed as in Exercise 10.8 part (c).

(e)
$$(a + b\sqrt{2})^2 = 3$$
$$a^2 + 2ab\sqrt{2} + 2b^2 = 3$$
$$2ab\sqrt{2} = 0$$
$$a = 0 \text{ or } b = 0$$

The first of these options leads to $2b^2 = 3$, or $b = \frac{\sqrt{6}}{2}$, which is not an option because $b \in \mathbb{Q}$ and $\sqrt{6} \notin \mathbb{Q}$. The second of these options leads to $a^2 = 3$, which gives $a = \sqrt{3} \notin \mathbb{Q}$, and yet of course $a \in \mathbb{Q}$. Thus neither can happen.

Exercise 10.11(c). Computing an element's multiplicative inverse is messy, but doable.

$$\frac{1}{a + b\sqrt{2} + c\sqrt{3} + d\sqrt{6}}$$
$$= \frac{1}{(a + b\sqrt{2}) + (c + d\sqrt{2})\sqrt{3}} \cdot \frac{(a + b\sqrt{2}) - (c + d\sqrt{2})\sqrt{3}}{(a + b\sqrt{2}) - (c + d\sqrt{2})\sqrt{3}}$$
$$= \frac{(a + b\sqrt{2}) - (c + d\sqrt{2})\sqrt{3}}{(a + b\sqrt{2})^2 - (c + d\sqrt{2})^2(\sqrt{3})^2}$$
$$= \frac{(a + b\sqrt{2}) - (c + d\sqrt{2})\sqrt{3}}{a^2 + 2ab\sqrt{2} + 2b^2 - 3c^2 - 6cd\sqrt{2} - 6d^2}$$
$$= \underbrace{\frac{a + b\sqrt{2}}{\text{same denominator}}}_{} - \underbrace{\frac{c + d\sqrt{2}}{\text{same denominator}}}_{}\sqrt{3}$$

The two fractions are members of $\mathbb{Q}(\sqrt{2})$ and so by Exercise 10.9 they simplify to some $w + x\sqrt{2}$ and $y + z\sqrt{2}$, making this last expression of the appropriate form,

$$w + x\sqrt{2} - y\sqrt{3} - z\sqrt{6}.$$

Exercise 10.12. If r is a root of a degree-m polynomial from $\mathbb{Q}$, write

$$a_m r^m + a_{m-1} r^{m-1} + \cdots + a_0 = 0.$$
$$a_m r^m = -a_{m-1} r^{m-1} - a_{m-2} r^{m-2} - \cdots - a_0$$
$$r^m = \frac{-a_{m-1}}{a_m} r^{m-1} - \frac{a_{m-2}}{a_m} r^{m-2} - \cdots - \frac{a_0}{a_m}$$

Thus all powers r^m or above can be reduced, so only powers less than m are needed.

Exercise 10.17.
(c) $c(\alpha)^n = \underbrace{c(\alpha)c(\alpha)\cdots c(\alpha)}_{n \text{ times}} = c(\underbrace{\alpha + \alpha + \cdots + \alpha}_{n \text{ times}}) = c(n\alpha)$

(f) The group of roots is isomorphic to C_6, which is generated by the elements 1 and 5. The group of roots itself is generated by either $c(\frac{\pi}{3})$ or $c(\frac{5\pi}{3})$.

Exercise 10.20. It requires starting with $\sqrt{3}i$, which is not a root of $x^3 - 2$. Also, I was not focused on solvability at that time; it was clear that $\mathbb{Q}(r_1, r_2, r_3)$ is an extension by radicals because I wrote r_1, r_2, and r_3 using radicals.

Exercise 10.21. Hint: Show that ϕ fixes 0 and 1, then proceed to all of $\mathbb{Q}$ from there.

Exercise 10.22(b). If r is the given root, $r = \sqrt[5]{1 + \sqrt{2}}$, then use algebra to eliminate the radical signs.

$$r^5 = 1 + \sqrt{2}$$
$$r^5 - 1 = \sqrt{2}$$
$$(r^5 - 1)^2 = 2$$
$$r^{10} - 2r^5 + 1 = 2$$
$$r^{10} - 2r^5 - 1 = 0$$

So the polynomial is $x^{10} - 2x^5 - 1$.

Exercise 10.24(a). One of the roots is $-\frac{1}{3}$; the other two are not. Write $x = -\frac{1}{3}$.

Exercise 10.25. Hint: Factor $x - \sqrt[3]{2}$ out of $x^3 - 2$. Recall that you are permitted to use any coefficient in $\mathbb{Q}(r_1)$.

Exercise 10.27. The derivative is $5x^4 + 40x^3$.

$$5x^4 + 40x^3 = 0$$
$$5x^3(x + 8) = 0$$
$$x = 0 \text{ or } x = -8$$

Thus the polynomial has only two places where it changes direction (from uphill to downhill or vice versa). Thus there are at most three places it might cross the x axis, since between any two such crossings it must change direction. We can tell that the polynomial does indeed cross the x axis exactly three times by plugging into the polynomial four values, 0, -8, a large number (say 100), and a small number (say -100). The results are in the following table.

x	polynomial
-100	< -9 billion
-8	8190
0	-2
100	> 10 billion

Answers to Chapter 10

Thus the function goes from negative to positive, back to negative, and then positive one final time, for a total of three crossings of the x axis, and thus three real roots.

Exercise 10.28.

(a) See Exercise 10.17.

(c) I describe the general technique and one possible answer; you must find your own polynomial. Let the derivative of the polynomial we seek have four different roots a, b, c, d, so there may be five crossings of the x axis. That derivative is then $(x-a)(x-b)(x-c)(x-d)$.

Multiply out the derivative and integrate to get the original polynomial (introducing a new constant e). Multiply by a common denominator to have only integer coefficients. Find values of a, b, c, d, e that satisfy the Eisenstein criterion. (You have a lot of freedom, so guess-and-check is a somewhat effective method.)

Now we have to verify that the peaks and valleys of your polynomial are on opposite sides of the x axis. So plug a, b, c, d into your polynomial and see what y values they yield. Modify e to ensure they straddle the x axis, while preserving the Eisenstein criterion.

When following this procedure, I obtained the polynomial

$$6x^5 - 15x^4 - 10x^3 + 30x^2 - 1,$$

but yours may be different.

(d) Your answer to part (b) is, for the reasons in Section 10.7.

Bibliography

[1] Michael Artin. *Algebra*. Prentice Hall, Englewood Cliffs, NJ, 1991.

[2] David J. Benson. *Music: a Mathematical Offering*. Cambridge University Press, Cambridge CB2 2RU, UK, 2007.

[3] Hans Ulrich Besche, Bettina Eick, and Eamonn O'Brien. The SmallGroups library — a GAP package. 2002.

[4] S. Bhagavantam and T. Venkatarayudu. *Theory of Groups and its Application to Physical Problems*. Academic Press, New York, 1969.

[5] Larry Copes. Representations of contra dance moves. On the World Wide Web at www.edmath.org/copes/contra/representations.html, January 2003.

[6] Edgar Dehn. *Algebraic Equations, An Introduction to the Theories of Lagrange and Galois*. Columbia University Press, New York, 1930.

[7] Persi Diaconis. *Group representations in probability and statistics*. Institute of Mathematical Statistics Lecture Notes—Monograph Series, 11. Institute of Mathematical Statistics, Hayward, CA, 1988.

[8] John B. Fraleigh. *A First Course in Abstract Algebra*. Addison-Wesley, Reading, MA, seventh edition, 2002.

[9] Joseph Gallian. *Contemporary Abstract Algebra*. Houghton Mifflin Company, 2004.

[10] Jonathan Goss. Point group symmetry. On the World Wide Web at www.phys.ncl.ac.uk/staff/njpg/symmetry/, September 2005.

[11] Charles Robert Hadlock. *Field Theory and its Classic Problems*. Number 19 in Carus Mathematical Monographs. Mathematical Association of America, 1978.

[12] B.A. Kennedy, D.A. McQuarrie, and C.H. Brubaker, Jr. Group theory and isomerism. *Inorganic Chemistry*, 3(2):265–268, February 1964.

[13] Wilhelm Magnus and Israel Grossman. ***Groups and their Graphs***. Anneli Lax New Mathematical Library. Mathematical Association of America, 1964.

[14] Gabriel Navarro. On the fundamental theorem of finite abelian groups. ***American Mathematical Monthly***, 110(2):153–154, February 2003.

[15] Ivars Peterson. Contra dances, matrices, and groups. On the World Wide Web at `www.sciencenews.org/articles/20030308/mathtrek.asp`, March 2003.

[16] Gert Sabidussi. Vertex-transitive graphs. ***Monatshefte für Mathematik***, 68(5):426–438, October 1964.

[17] David S. Schonland. ***Molecular Symmetry***. D. Van Nostrand, London, 1965.

[18] Daniel Shanks. ***Solved and Unsolved Problems in Number Theory***. American Mathematical Society Chelsea Publishing, Providence, RI, fourth edition, 2001.

[19] Uri Shmueli, editor. ***International Tables for Crystallography***, volume A. Springer, 5th edition, April 2005.

[20] John M. Sullivan. Classification of finite abelian groups. Course notes available on the World Wide Web, Spring 2003.

[21] Ed Turner and Karen Gold. Rubik's groups. ***American Mathematical Monthly***, 92(9):617–629, November 1985.

Index of Symbols Used

In most cases, refer to the Index for more information on a symbol.

Groups

C_n	cyclic group	Q_4	quaternion group
D_n	dihedral group	U_n	U-groups, from page 188
V_n	Klein 4-group	$\mathbb{Z}_n$	alternate notation for C_n
S_n	symmetric group		(see page 65)
A_n	alternating group	$G_{4,4}$	see page 151

Number Systems

$\mathbb{N}$	natural numbers	$\mathbb{Q}$	rational numbers
$\mathbb{Z}$	integers	$\mathbb{Q}^*$	non-zero rational numbers
$\mathbb{R}$	real numbers	$\mathbb{Q}^+$	positive rational numbers
$\mathbb{C}$	complex numbers		

Functions

$Perm$	permutation group	$N_G(H)$	normalizer of H in G
Ker	kernel of a homomorphism	sin, cos	trigonometric functions
Im	image of a homomorphism	$\phi: G \to H$	ϕ is a homomorphism from G to H
Aut	automorphism group		
Orb	orbit of an element	$\phi(x) = y$	defining a homomorphism
$Stab$	stabilizer of an element	ϕ, θ, τ, ψ	various homomorphisms

Molecules

$B(OH)_3$	Boric acid	SF_5Cl	sulfur chloride pentafluoride
C_2H_4	ethylene	$Fe(C_5H_5)_2$	eclipsed ferrocene
C_6H_6	benzene		

Other symbols

$+$	addition, sometimes mod n	$\in$	element of a set
$-$	subtraction	$\{\ \}$	set brackets
$\cdot$	multiplication, general group operator (see page 48)	e	group identity
		$\langle\ \rangle$	generator notation
$*$	example operator (see page 48)	$|\ |$	group/element order
$\times$	direct product	aH, Ha	left and right cosets
G^n	repeated direct product	HK	multiplying subgroups, from page 189
(a, b)	element in a product group, point in complex plane	$\mapsto$	notation for defining functions/homomorphisms
$\rtimes$	semidirect product		
$\div$	division	$\cong$	isomorphic to
$\frac{G}{H}$	group quotient	$\equiv_n$	congruent mod n
$<$	less than, subgroup	i	imaginary number $\sqrt{-1}$
$\triangleleft$	normal subgroup	$a + bi$	complex number
$\sqrt{\ }\ \sqrt[n]{\ }$	square root, n^{th} root	$\bar{c}$	complex conjugate of c
$!$	factorial	$\mathbb{Q}(\ )$	field extension of $\mathbb{Q}$

Index

Abel, Neils, 36, 69, 221, 223, 225, 227, 247, 251
abelian, 59, 68–71, 88–90, 92, 94, 114, 122, 148, 152, 175–177, 188–193, 211–213, 219, 220, 222, 245, 249, 275, 277
    quotient, 245, 249
abelianization, 183, 274
action
    diagrams, 196–197, 200, 202, 218
    of a group on a set, *see* group actions
algebra, 41, 44–45, 48, 50, 59–62, 69, 94, 97–98, 102, 120, 126–129, 205, 220, 221, 223–225, 230, 231, 253
algebraic numbers, 227, 252, 258
algebraically closed field, 228
alternating group A_n, 22, 47, introduced on 80, 88, 90
ambiguous, *see* Cayley diagram, ambiguous arrows
applications of group theory, 25–40, 81, 194, 221–260
arrows, *see* Cayley diagram arrows
art
    group theory in, 31
    on bedroom walls, 7
associativity, 49–52, 55, 56, 266
automorphism
    field, 236–241, 244, 250, 256, 257
    group Aut, footnote on 98, 178, 180, 181, 185
    inner, 186, 213, 214, 275
axioms, 6
axis, 32
    x, 282, 283
    in a multiplication table, 48

    in the complex plane, 228, 229, 234, 237
    of rotation, 14, 247, 248, 263

benzene C_6H_6, 38
Big Book, The, 11–13
big question of Galois Theory, 221, 225, 227, 228
bilateral symmetry, *see* symmetry, bilateral
body-centered cubic, 30
Boric acid $B(OH)_3$, 28, 64, 74

calculator, *see* Cayley diagram as a calculator
canceling, 61, 142, 205
carbon, 38, 251
Cardano, Girolamo, 224
Cauchy's Theorem, 199–206, 208–211, 216, 220, 249
Cayley diagram, introduced on 18
    ambiguous arrows, 137–138, 142, 166
    arrows, introduced on 19–20
    as a calculator, 44–46
    connected, 34, 57, 172, 266
    infinite, *see* infinite Cayley diagram
    layout, 94, 101, 106, 113, 125, 132–135, 138, 149, 163, 171, 175
    parallel arrows, 95, 175
Cayley's Theorem, 83–86, 95–96
Cayley, Arthur, 18
cell, *see* multiplication table cells
chain
    of field extensions, 245
    of normal subgroups, 245, 249
class equation, 146, 153, 154, 205, 247–248
classifying groups, 220
    abelian, 275

order 6, 203–205
order 8, 211–213
order 15, 216–217
clock, 65, 169
    arithmetic, *see* modular arithmetic
clockwise, 4, 12, 13, 23, 28, 37, 50, 64, 65, 74, 75, 95, 129, 141, 144–146, 153, 247, 248, 262, 263
closure, 56, 255, 279
codes, *see* error-correcting codes
codomain, *see* function, codomain of a
coefficient, 224, 228, 231, 232, 238, *see* integer coefficients
coins, 7, 261
collapsing cosets, *see* group quotient and quotient maps
color spectrum, 66
column, *see* multiplication table rows and columns
common multiples, 173, 175
commutative, 52, 59, 69–71, 120, 128,
    see abelian
commutator, 183, 274
compass, *see* ruler and compass
complex
    conjugate $\bar{c}$, *see* conjugacy, complex
    numbers $\mathbb{C}$, 163, 188, 241, 223, 226–231, 250, 254
    Complex Conjugate Root Theorem, 229, 236, 250, 253
component of a pair, 120, 123, 127
configurations
    of light switches, 19, 20
    of Rubik's Cube, 4, 5, 11–12
    of the rectangle puzzle, 15–19, 42, 195–197
    of the triangle, 75
    starting, 14, 19, 42–44, 75
congruence $\equiv_n$, *see* mod and modular arithmetic
conjugacy
    classes, 145–147, 152–155, 247–249
    complex, 229–231, 235–237, 250, 252, 280
    operation, 142–147, 152–155, 186, 197, footnote on 197, 198, 208, 210, 213–217
conjugation table, 186, 275
connected, *see* Cayley diagram, connected
connecting corresponding nodes, *see* technique for constructing (semi)direct products using Cayley diagrams
constant term, 233
constructing products, *see* technique for constructing
contra dance, *see* dancing

converse, *see* Lagrange's Theorem
copies of a subgroup, *see* coset
coset, introduced on 102
    left, introduced on 103
    representative, 103, 140
    right, introduced on 103
cosine function, *see* trigonometry
counterclockwise, *see* clockwise
crystals, 29–31, 34
cube, 30, 80, 82, 153, *see* Rubik's Cube
    doubling a, 228
    root $\sqrt[3]{}$, 225, 226, 232
cubic
    formula, 224, 226
    polynomial, 224
cycle graph, 68, 71, 77–78, 88–92, 114, 154, 162, 184, 274
cyclic group C_n, 22, 29, 47, introduced on 64

dancing, 34–35, 40, 195, 265
decimals, 226
degree
    multiplying, 238
    of a field extension, *see* field, degree of an extension
    of a polynomial, *see* polynomial degree
denominator, 253, 281, 283
derivative, 259, 282, 283
deterministic, 5, 6, 15, 24, 270
diagonal, 30, 66, 67, 70, 71, 93, 130
dihedral group D_n, 24, 55, 74–78, 90–94, 154, 183
dimension of an abelian group, 175, 176
direct product, *see* product, direct
disguise a (sub)group, 101, 134, 172, 175, 184
divides (evenly into), introduced on 107, 110, 173, 190, 200, 203, 204, 206, 208, 209, 211, 213, 215, 216, 219, 249
dodecahedron, 81–83, 247
domain, *see* function, domain of a
doubling a cube, *see* cube, doubling a

eclipsed ferrocene $Fe(C_5H_5)_2$, 38
Eisenstein Criterion, 233, 249, 252, 283
element (of a set or group) $\in$, introduced on 48
embeddings, introduced on 163, 157–167, 180–182, 185, 196, 235, 245, 257, 269, 274, 275
equations
    class, *see* class equation
    of arithmetic, 223–227, 252–253, *see* polynomial equations

Index

solving, *see* equations of arithmetic and group, solving equations in
equivalence relation, 154
error-correcting codes, 37
ethylene C_2H_4, 38
Euler number, 50
existence of subgroups, *see* First Sylow Theorem
expanding number systems, 222, 232
explore (when mapmaking), 15–17, 19, 26, 102, 263
extensions, *see* field extensions

factorable, 252, 280, *see* irreducible
factorial !, 80
factors
    in a product group, 120, 121, 123, 127–128, 148, 149, 189
    in a quotient group, 132, 134, 150
    independence, 123–125
    of a polynomial, 232, 280
families of groups, 63, *see* cyclic, abelian, dihedral, symmetric, alternating, and U-groups
field, 221–223, 226–228, 234, 235, 238, 242, 251, 254, 255
    degree of an extension, 234, 238–239, 242–243, 249
    extension, 228, 233–234, 236, 238–239, 242–245, 249, 256, 257, *see* expanding number systems
    finite, 223, 259
    fixed, 236, 237, 239, 240, 244, 256, 257, 282
    intermediate, 239, 242
    operations, 233, 236
    successive extensions, 238
fifth root $\sqrt[5]{\ }$, 227
figures, *see* dancing
finite
    decimal expansion, 226
    field, *see* field, finite
    group, *see* group, finite
First Sylow Theorem, introduced on 208
fixed field, *see* field, fixed
flip
    180-degree in space, 146, 248
    and non-flip, 76
    horizontal/vertical, 14–20, 23, 28, 29, 32, 41, 42, 65, 74–75, 195, 196, 229, 237, 243, 244
folding

    a multiplication table, *see* multiplication table, folding
    a tetrahedron, 91
fourth root $\sqrt[4]{\ }$, 225, 226, 232
frieze pattern, 31–34, 39–40, 263, 268
Fuller, Richard Buckminster, 251
function, 158–159, 161, 163, 169, 171, 177, 180, 187, 195, 256
    codomain of a, 159–167, 178, 180–182
    domain of a, 159–167, 180–182
    image Im, 159, 163–165, 167, 178, 183
    image Im, 195, 196, 235
    that is not a homomorphism, 161
Fundamental Homomorphism Theorem, 167
Fundamental Theorem of Abelian Groups, 70, 175, 177, 189, 193, 279

Galois groups, 233–245, 249–250, 252, 255–258
Galois, Évariste, 36, 217, 221, 223, 225, 227, 234, 243, 247, 251
generators, *see* group generators and homomorphism, generating
geodesic domes, 251
geometry, 228
glide reflection, 32, 37
graph, footnote on 98, 228, 230, 234, 241, 250
Greeks, 228
grid (abelian Cayley diagrams), 118, 121, 124, 175
group
    actions, 194–197, 200, 202, 206–208, 215, 216, 218–219, 231, 235–237, 241, 243
    classification, *see* classifying groups
    crystallographic, *see* crystals
    families, *see* families of groups
    finite, 22, 47, 94, 107, 175, 188, 189, 252, 261
    frieze, *see* frieze pattern
    generators, 6–8, 15, 19, 21, 23–24, 34, 40–41, 59–60, 89, 90, 92, 93, 100–102, 111, 141–142, 153, 174–178, 188, 196, 222, 263
    infinite, *see* infinite groups
    of even order, 252
    of roots, 256–257, 282
    order, 64, 80, 88–90, 94, 107–108, 110, 114–115, 140, 146, 148, 175, 189–194, 198–213, 215–220, *see* classifying groups
    p-group, *see* p-group
    presentations, 92
    quotient, 77, 117, 132–139, 150–152, 171, 172, 183, 187, 209, 245, 273, 277

simple, 219
solvable, 245–247, 249–250, 257, 259
solving equations in, 60–61
structure, 19–20, 35, 80, 84, 90, 99, 117, 120, 125, 129, 131–135, 150, 157–160, 163–166, 168, 175, 177, 194, 203, 214, 251
wallpaper, *see* wallpaper groups
Group Explorer, ix, 21, 24, 38, 46, 55, 58, 59, 90, 94, 95, 101, 154, 217

Hasse diagram, 112–113, 226, 228, 234, 242–247, 257, 270
  path through a, 245–247
hidden subgroups, *see* subgroups, hidden
highlighting a subgroup, *see* subgroup highlighting
homomorphism, introduced on 159
  generating, 160–161, 181
  image Im, *see* function, image of a
  interpretation, 195, 199, 202, 206, 218, 219
  kernel Ker, 166–169, 181–183
  kernel Ker, 219
  sequence of, 168, 169, 171, 182, 190, 191, 245, 247, 276, 277

icosahedron, 81–83, 247–248
identity element e, introduced on 50, 51, 63, 68, 78, 85, 86, 89, 94, 95, 98, 100, 102, 110, 128, 146, 151, 166–167, 175, 180, 181, 189, 196, 200, 222, 249
image Im, *see* function, image
independent factors, *see* factors, independence
index, *see* subgroup index
infinite
  Cayley diagram, 23, 32, 33, 149
  groups, 8, 23, 31, 33, 34, 92, 170, 192
  multiplication tables, 54, 273
  symmetrical objects, 31, 32
inflating
  cells, *see* technique for constructing direct products using multiplication tables
  nodes, *see* technique for constructing direct products using Cayley diagrams
integer coefficients, 226, 232, 233, 253, 283
integers $\mathbb{Z}$, 62, 65, 78, 170, 171, 172, 222
interpretation homomorphism, *see* homomorphism, interpretation
intersect, *see* informal word "overlap"
inverse
  elements, 50–51, 55–56, 60–61, 63, 80, 94, 141, 143, 201, 222, 254, 255, 267, 281
  multiplicative, 222, 254, 255

irrational, *see* rational
irreducible polynomial, *see* polynomial, irreducible
isomorphic $\cong$, footnote on 33, 35, 84, 85, 90, 120, 133, 134, 135, 148, 157, introduced on 163, 167–168, 171, 174, 175, 187, 188, 196, 204, 213, 217, 238, 271, 274, 275

kernel Ker, *see* homomorphism kernel
Klein 4-group V_4, 20, 21, 43, 263
Klein, Felix Christian, 20
knit product, *see* product, knit

Lagrange's Theorem, 105–110, 114, 146, 194, 199, 205, 218, 219, 278
  counterexample to converse, 108, 218, 278
Lagrange, Joseph Louis, 107
layout, *see* Cayley diagram layout and multiplication table layout
least common multiple, *see* common multiples
left cosets, *see* cosets
left multiplication, *see* multiplication
linear polynomial, 224

manipulations of an object, *see* technique for measuring symmetry
map, *see* function
  roadside, 123
  through the wilderness, 11–12
matrix, 188, 275
mirror symmetry, *see* symmetry, mirror
mod, 101, 188, 200, 206, 208, 210, 215–217, 254, 260, 273
modular arithmetic, 65–66, 169–171, 184
molecules, 27–31, 38–39, 64, 74, 251, 263
moves (in a game), 4–6, 11–19, 23, 24, 28, 31–32, 91, 261, 262
multiples, *see* common multiples
multiplication
  of variables, 44
  on the left, 104, 206, 207, 215, 219
  on the right, 84, 103, 104, 142, 143, 266
multiplication table, introduced on 45
  cells, 46, 66, 70, 76, 84–86, 105, 126
  folding, 71
  infinite, *see* infinite multiplication table
  layout, 76, 101, 113, 151
  rows and columns, 45–46, 55, 57, 66, 70, 84, 85, 105, 113, 126, 127, 266, 270
music, 34, 37

n-gon, regular, 24, 55, 74, 92, 268

Index

natural numbers $\mathbb{N}$, 221–223
navigating a Cayley diagram, 98, 123–125
no action, 23, 42, 44, 46, 50, 81, 196, 261
node, introduced on 20
non-commutative, *see* commutative
non-embedding, *see* embedding and quotient map
non-normal extension, *see* normal extension
non-proper subgroup, *see* proper subgroup
non-real roots, *see* roots, real
non-repeating decimals, 226
non-stable element, *see* stable element
non-trivial subgroup, *see* trivial subgroup
normal
    extension, 242, 249, 258
    subgroup, *see* subgroup, normal
normalizer $N_G(H)$, 139–142, 152, 206–210, 216
n^{th} power homomorphism, 182
number of subgroups, *see* Third Sylow Theorem
number theory, 37, 173
numbering an object's parts, *see* technique for measuring symmetry

Observations, 4, 5, 102, 103, 164–166
octahedron, 82, 93
one-dimensional abelian group, *see* dimension of an abelian group
operation
    binary, introduced on 48, 48–51, 55–57, 62, 126, 163, 169
    comparing, 227
    of arithmetic (the field $\mathbb{Q}$), 221–225, 234, 236, 252
    product, *see* product
    quotient, *see* group quotient
orbit
    in a group action Orb, 197–200, 206, 216–218
    of a group element, 66–68, 71, 74, 77, 89, 94, 100, 110, 158, 160, 173–175, 184, 204, 213, 275
Orbit-Stabilizer Theorem, 198–200, 218, 278
order
    of a (sub)group, *see* group order
    of actions, 8, 59, 69
    of an element, 94, 110, 115, 152, 154, 173, 175, 219, *see* group order
organization, *see* Cayley diagram layout and multiplication table layout
orientation, 15, 91, 145
overlap, 105, 106, 154, 184, 189, 278

p-group, 205–209, 213, 219
Pólya's technique, 36
page (in The Big Book), 12–13
paradigm (for approaching group theory), 1, 52, 97, 99, 126
parentheses, 49–50, 121, 223
particles, 36
partition, 105–107, 114, 145, 154, 165, 197, 200, 247–248
path
    in a Cayley diagram, 18, 43, 57, 67, 98, 103, 104, 123, 141, 142, 158, 160, 165, 205, 212
    through a Hasse diagram, *see* Hasse diagram, path through a
pentagon, regular, 24, 55, 74, 263
pentaradial symmetry, *see* symmetry, pentaradial
permutation, 78–86, 90, 93, 95–96, 110, 130, 153, 202, 219, 236–238, 247, 249–250, 268, 279
    group *Perm*, 194–197, 206, 231, 235
perpendicular
    arrows, *see* arrows, parallel
    in a direct product, 120, 121
physics, 36, 223
pinwheel, 64, 74, 91
plane
    complex, 228–229, 234–236, 256
    points in the, 120
Platonic solids, 63, 80–82
point in the plane, *see* plane, point in the
polygon
    regular, 24, 74, 78, 80, 81
    truncated, 268
polynomial, 36, 223–225
    degree, 224–225, 232–234, 252, 255, 259, 280, 281, *see* field, degree of an extension
    division, 232, 253–254, 280–281
    equations, 224–227, 252
    general form, 224
    irreducible, 231–234, 239, 241, 242, 254, 258–259
    quintic, 225, 247–250, 252
    roots, 224, 228–231, 235–238, 243
    solvable (by radicals), 225, 245, 247–250, 252
    unsolvable, *see* polynomial, solvable
predefined list of actions, 4, 6, 15, 24, 79
prime
    factorization, 185, 189

number, 92, 110, 149, 188, 190, 191, 203, 220, 233, *see* relatively prime numbers
order of a (sub)group, 199–200, 202, 204, 205, 210, 211, 219
powers of a, 189–191, 206, 208, 213, 215, 219, 278
product
   direct ×, 22, 47, introduced on 117–128, 128–135, 147–149, 171–175, 189, 191, 275, 277
   knit, 128
   semidirect ⋊, 77, 128–132, 135, 136, 137, 149–150, 177–180, 185–187, 279
propeller, 29, 64, 65, 74, 91
proper subgroup, *see* subgroup, proper

quadratic
   formula, 226, 231
   polynomial, 224
quartic
   formula, 226
   polynomial, 225
quasihedral group, 22, 47
quaternion group Q_4, 53, 54, 60
question marks, 194, 203, 216
quintic polynomial, *see* polynomial, quintic
quotient, *see* group quotient
   maps, 163–169, 181–182, 185, 191, 209, 276, 278

radicals $\sqrt{\ }$, 225–227, 232–234, 238, 243–245, 250, 282
rational numbers $\mathbb{Q}$, 62, 112, 152, 187, 223, 225–228, 232–245, 252, 255–258, 281–282
real
   numbers $\mathbb{R}$, 223, 225, 226, 227
   roots, *see* roots, real
rearranging, 7, 26, 35, 36, 78–79, 130, 135, 177, 194, 195, 235, 236
rectangle puzzle, 14–21, 23, 26, 27, 41–43, 195–197, 262
reflexive relation, 154
regular, *see* n-gon, pentagon, polygon, and regularity
regularity (of Cayley diagrams), 97–99, 102, 103, 108, 137, 139, 197, 205, 212, 213
relation, 154
relationship among subgroups, *see* Second Sylow Theorem
relatively prime numbers, 171–175, 184–185, 188, 190
remainder, 107, 170–171, 254

renaming isomorphism, 168, 169, 172, 204
repeating decimals, 226
representative, *see* coset representative
revealing subgroups, *see* subgroups, revealing
reversible, 4, 6, 15, 24, 50, 79, 261, 262, 270
rewiring, 129–132, 135, 137, 149–150, 177–179, 185–186, 271–272, 275
right cosets, *see* cosets
right multiplication, *see* multiplication
rings in a dihedral group, 67, 75–76, 95, 102, 178
roadside map, *see* map, roadside
roots
   of a polynomial, *see* polynomial roots
   real, 250, 259, 283
rotation, 74, *see* clockwise and turn
   90-degree, 6, 153
   120-degree, 91, 144, 146
   180-degree, 6, 91, 146, 153
   360-degree, 25
   axis, *see* axis of rotation
rotational symmetry, 64, 65, 74, 91
row, *see* multiplication table rows and columns
Rubik's Cube, 3–6, 11–15, 50
   solving, 3, 4, 11–13
Rubik, Ernö, 3, 4
ruler and compass, 228
rules
   breaking, 8
   of a group, 6

Second Sylow Theorem, introduced on 213
semidirect product, *see* product, semidirect
sequence (of actions), 5, 6, 15, 24, 44–45, 49–50, 144, 161, 261, 263
set, introduced on 48
sextic polynomial, 252
short exact sequence, 182
simplifying, 44–45, 49, 223, 224, 230, 234, 252
sine function, *see* trigonometry
solvable
   by radicals, *see* polynomial, solvable
   group, *see* group, solvable
   polynomial, *see* polynomial, solvable
solving equations, *see* equations of arithmetic and group, solving equations in
spectrum, *see* color spectrum
spin, 64, 74, 263
square
   dance, *see* dancing
   puzzle, 23, 24, 55
   root $\sqrt{\ }$, 225, 226, 232, 234, 236
   shape, 74, 78, 96, 262

squaring
- a circle, 228
- an element, 80, 81, 113, 248, 268, 269

stabilizer *Stab*, 217

stabilizer $Stab$, *see* stable element

stable element, 197–200, 202, 206–208, 215–216, 218, 219

stars
- sea, 25
- shape, 268

starting configuration or node, *see* configurations, starting

structure, *see* group structure

subgroup <, 97, introduced on 99
- for many terms related to subgroups, *see* group
- conjugate, *see* conjugacy operation
- hidden, 100
- highlighting, 73, 100, 103, 157, 158
- index, introduced on 107, 109, 110, 112, 114, 152, 199, 206, 208, 210, 215
- normal ◁, 105, 111, introduced on 121, 137–139, 142–144, 146–147, 149, 151–152, 166, 168, 170, 181–183, 187, 215, 216, 217, 219, 245–249, 251, 279, *see* normalizer
- *p*-subgroup, *see* *p*-group
- proper, 100, 101, 108, 114
- revealing, 100–101, *see* Cayley diagram layout and multiplication table layout
- trivial, 100, 112, 146, 167, 235

subtraction, example of non-associativity, 50

sulfur chloride pentafluoride SF_5Cl, 38

Sylow
- *p*-subgroup, 213–217, 219–279
- Theorems, 108, introduced on 208–216

symmetric
- group, 22, 47, introduced on 78–79, 88, 90, 96
- relation, 154

symmetry, 4–7, 25–34, 37–40, 64–65, 74, 81–82, 91–92, 98–99, 195–196, 229–231, 235–244, 247, 263–264, 268
- bilateral, 25, 74, 91
- mirror, 25, 229, 235, 243–244
- pentaradial, 25

technique
- for constructing direct products using Cayley diagrams, 118
- for constructing direct products using multiplication tables, 126
- for constructing semidirect products using Cayley diagrams, 179
- for constructing semidirect products using multiplication tables, 187
- for measuring symmetry, 26–34, 37–40, 81, 91, 195–196
- Pólya's, *see* Pólya's technique

term (in a polynomial), 224, 231, 233, 234, 238, 242, 255, 257

tetrahedron, 81–82, 91–92, 144–145, 155

Third Sylow Theorem, introduced on 215

three-dimensional abelian group, *see* dimension of an abelian group

topology, 37

transcendental numbers, *see* algebraic numbers

transitive relation, 154

triangle puzzle, 24, 55, 74–76

trigonometry, 256

trisecting an angle, 228

trivial subgroup, *see* subgroup, trivial

truncated, *see* cube, dodecahedron, icosahedron, *n*-gon, polygon, tetrahedron

turn (quarter, half, etc.), 23, 28, 247, 248, 262, 263

two-dimensional abelian group, *see* dimension of an abelian group

U-groups U_n, 188

unambiguous arrows, *see* arrows, ambiguous

uniform symmetry, *see* regularity

unit circle, 256

unknown, *see* variable

unsolvable
- by radicals, *see* polynomial, solvable
- group, *see* group, solvable
- polynomial, *see* polynomial, solvable

variable, 223
- (in)dependent, 158

vertex, 91, 144–146, 247–248

viewpoint
- diheral example, 75
- for approaching group theory, *see* paradigm

wallpaper groups, 34

wilderness, 11–12

x-axis, *see* axis in the complex plane and axis, x

y-axis, *see* axis in the complex plane

About the Author

Nathan Carter grew up in Northeastern Pennsylvania, earning a bachelor's in mathematics and computer science from the University of Scranton in 1999. He earned master's degrees in mathematics and computer science and a PhD in mathematics from Indiana University, Bloomington, Indiana. Nathan received the University of Scranton Excellence in Mathematics award in 1999, an Indiana University Rothrock Teaching Award in 2003, and a Bentley University Innovation in Teaching award in 2007. *Visual Group Theory* is his first book, based on lessons learned while writing the software *Group Explorer*. Like several of his research projects, it puts computers to work to improve mathematical understanding and education. He is a member of the American Mathematical Society and the Mathematical Association of America.